INTEGRATED COASTAL ZONE MANAGEMENT

Edited by

Erlend Moksness
Einar Dahl
and
Josianne Støttrup

A John Wiley & Sons, Ltd., Publication

This edition first published 2009
© 2009, Blackwell Publishing Ltd.

Blackwell Publishing was acquired by John Wiley & Sons in February 2007. Blackwell's publishing programme has been merged with Wiley's global Scientific, Technical, and Medical business to form Wiley-Blackwell.

Registered office
John Wiley & Sons Ltd., The Atrium, Southern Gate, Chichester, West Sussex, PO19 8SQ, United Kingdom

Editorial offices
9600 Garsington Road, Oxford, OX4 2DQ, United Kingdom
2121 State Avenue, Ames, Iowa 50014-8300, USA

For details of our global editorial offices, for customer services and for information about how to apply for permission to reuse the copyright material in this book please see our website at www.wiley.com/wiley-blackwell.

The right of the author to be identified as the author of this work has been asserted in accordance with the Copyright, Designs and Patents Act 1988.

All rights reserved. No part of this publication may be reproduced, stored in a retrieval system, or transmitted, in any form or by any means, electronic, mechanical, photocopying, recording or otherwise, except as permitted by the UK Copyright, Designs and Patents Act 1988, without the prior permission of the publisher.

Wiley also publishes its books in a variety of electronic formats. Some content that appears in print may not be available in electronic books.

Designations used by companies to distinguish their products are often claimed as trademarks. All brand names and product names used in this book are trade names, service marks, trademarks or registered trademarks of their respective owners. The publisher is not associated with any product or vendor mentioned in this book. This publication is designed to provide accurate and authoritative information in regard to the subject matter covered. It is sold on the understanding that the publisher is not engaged in rendering professional services. If professional advice or other expert assistance is required, the services of a competent professional should be sought.

Library of Congress Cataloging-in-Publication Data

Integrated coastal zone management /edited by Erlend Moksness, Einar Dahl, and Josianne Støttrup.
    p. cm.
    Proceedings of a conference held in Ardenal, Norway, in June 2007.
    Includes bibliographical references and index.
    ISBN 978-1-4051-3950-2 (hardback : alk. paper)   1. Coastal zone management–Congresses.
I. Moksness, Erlend.   II. Dahl, Einar.   III. Støttrup, Josianne.
    HT391.I4646 2009
    333.91′7–dc22

2008047434

A catalogue record for this book is available from the British Library.

Set in 10/12 pt Times by Aptara® Inc., New Delhi, India
Printed and bound in Singapore by Fabulous Printers Pte Ltd

1 2009

Contents

About the Editors   vi
Contributors   vii
Referees   xii
Preface   xv

Chapter 1   State of Fear or State of Oblivion? What Coastal Zones Are Telling Us about Global Change and Why We Need Integrated Coastal and Ocean Management on a Global Scale   1
Peter J. Ricketts

Section 1   Coastal Habitats

Chapter 2   The Challenge towards Sustainable Utilisation of Coastal Fish Resources   27
Josianne G. Støttrup

Chapter 3   Evaluating the Geomorphologic Stability of an Estuarine Sandy Beach   35
Filipa S.B.F. Oliveira and Catarina I.C. Vargas

Chapter 4   Macro-Algae Habitat as Fish Nursery: Evaluation of Function as Predation Refuge   50
Jun Shoji, Yasuhiro Kamimura and Chiaki Fujiki

Chapter 5   Predictive Probability Modelling of Marine Habitats – A Case Study from the West Coast of Norway   57
Ellen Soldal, Trine Bekkby, Eli Rinde, Vegar Bakkestuen, Lars Erikstad, Oddvar Longva and Martin Isæus

Section 2   Impacts on Coastal Systems

Chapter 6   Further Testing of the Approved EU Indicator to Measure the Progress in the Implementation of Integrated Coastal Zone Management in Europe   69
Alan H. Pickaver

Chapter 7	The Response of Hyperbenthos and Infauna to Hypoxia in Fjords along the Skagerrak: Estimating Loss of Biodiversity Due to Eutrophication *Lene Buhl-Mortensen, Eivind Oug and Jan Aure*	79
Chapter 8	The Effects of Human Impact on Benthic Foraminifera in the Augusta Harbour (Sicily, Italy) *Elena Romano, Luisa Bergamin, Maria Grazia Finoia, Maria Celia Magno, Isabel Mercatali, Antonella Ausili and Massimo Gabellini*	97
Chapter 9	Eco-Friendly Sustainable Shrimp Aquaculture in Bangladesh: Minimizing Coastal Degradation *Mohammad A.L. Siddique and John P. Volpe*	116
Chapter 10	Bioshields and Ecological Restoration in Tsunami-Affected Areas in India *Nibedita Mukherjee, Muthuraman Balakrishnan and Kartik Shanker*	131
Chapter 11	The Impact of Population Density and Urbanization in the Coastal Northeast Regions of the Baltic Sea: A Time-Spatial Description *Merle Looring*	145

Section 3 Integrated Coastal Zone Management (ICZM)

Chapter 12	Future Challenges in Environmental Policy Relative to Integrated Coastal Zone Management *Svein Jentoft*	157
Chapter 13	Integrated Coastal Zone Management and Sustainable Development of Coastal Area: A Short Overview of International Legal Framework *Md. Saiful Karim and Ridwanul Hoque*	170
Chapter 14	Lobster Reserves in Coastal Skagerrak – An Integrated Analysis of the Implementation Process *Alf Ring Pettersen, Even Moland, Esben Moland Olsen and Jan Atle Knutsen*	178
Chapter 15	Efficiency of Fishing Vessels Affected by a Marine Protected Area – The Case of Small-Scale Trawlers and the Marine Protected Area in Nha Trang Bay, Vietnam *Quach Thi Khanh Ngoc, Ola Flaaten and Nguyen Thi Kim Anh*	189

Chapter 16	Exploring the Future of Seas and Coasts: Scenarios within the Joint Research Project 'Zukunft Küste – Coastal Futures' *Andreas Kannen, Kai Ahrendt, Antje Bruns, Benjamin Burkhard, Doris Diembeck, Kira Gee, Bernhard Glaeser, Katharina Licht-Eggert, Tanja Michler, Ophelia Meyer-Engelhard, Corinna Nunneri, Sebastian Stragies and Wilhelm Windhorst*	207
Chapter 17	Integrated Coastal Zone Management (ICZM) in Guyana: Development Barriers, Opportunities and Recommendations *Robin McCall and Talia Choy*	219
Chapter 18	Strategies for the Beneficial Use of Dredged Material in Japan *Ryoji Naito and Yoshiyuki Nakamura*	237

Section 4   Coastal Governance

Chapter 19	A Practitioner's Perspective on Coastal Ecosystem Governance *Stephen Bloye Olsen*	253
Chapter 20	Problem Perceptions and Use of Technical Knowledge in Decision making for the Extension of Mainport Rotterdam *Saskia Hommes, Suzanne J.M.H. Hulscher, Jan P.M. Mulder and Henriëtte S. Otter*	266
Chapter 21	Local Coastal Zone Planning and Stakeholder Participation in Norway *Knut Bjørn Stokke and Sissel Hovik*	285
Chapter 22	The Evolution of Governance Mechanisms for the Eastern Scotian Shelf Integrated Management Initiative *Joseph Arbour*	295
Chapter 23	Climate Change, Coastal Communities and Governance: Developing Solutions for Change in Australia *Melissa Nursey-Bray*	306
Chapter 24	Map of Coastal Zone Vulnerabilities to Wave Actions Application to Aveiro District (Portugal) *Carlos Coelho, Maria-José Granjo and Cristina Segurado-Silva*	318
Chapter 25	Managing Coastal Vulnerability: New Solutions for Local Government *Timothy F. Smith, Benjamin Preston, Cassandra Brooke, Russell Gorddard, Deborah Abbs, Kathleen McInnes, Geoff Withycombe, Craig Morrison, Beth Beveridge and Tom G. Measham*	331
Index		341

Colour plates appear between pages 206 and 207

About the Editors

Erlend Moksness is a research director at the Institute of Marine Research (IMR) in Norway. He has been responsible in establishing a research and management advice programme on the coastal zone ecosystem at the same institute. His background is in recruitment in marine fish, fish ageing, stock enhancement of marine fishes and aquaculture of marine fishes. He has published 87 scientific articles and has been co-editor of 11 proceedings and scientific books.

Einar Dahl is a senior scientist at Institute of Marine Research (IMR) in Norway since 1973. He has been working not only with phytoplankton monitoring and ecology, including harmful blooms, but also with other subjects related to coastal ecology, such as nutrients, eutrophication and oxygen conditions. Since the last decade he has been head of a research group and later a programme at IMR, with focus on monitoring and ecology of the coastal zone.

Josianne G. Støttrup, PhD, is a senior scientist at the Technical University of Denmark, National Institute of Aquatic Resources (DTU Aqua, previously Danish Institute for Fisheries Research) since 1984. Her research focus has been on marine larval nutrition initially, but evolved to coastal juvenile fish habitats, flatfish ecology and more recently to integrated coastal management. She has led the marine coastal fisheries management programme since 1994 and now heads the Section of Coastal Ecology.

Contributors

Deborah Abbs, CSIRO Marine and Atmospheric Research, Private Bag 1, 107–121 Station Street, Aspendale VIC 3195, Australia.

Kai Ahrendt, Büro für Umwelt und Küste (BfUK), Kiel, Germany.

Nguyen Thi Kim Anh, Faculty of Economics, Nha Trang University, 02 Nguyen Dinh Chieu, Nha Trang, Vietnam.

Joseph Arbour, Defence Research and Development Canada (DRDC), 9 Grove Str., Dartmouth, Nova Scotia, Canada.

Jan Aure, Institute of Marine Research, 5817 Bergen, Norway.

Antonella Ausili, ISPRA, Institute for Environmental Protection and Research, Via di Casalotti, 300, 00166 Roma, Italy.

Vegar Bakkestuen, Department of Botany, NHM, University of Oslo, P.O. Box 1172 Blindern, N-0318 Oslo, Norway.

Muthuraman Balakrishnan, New Port and Drydock Development Project, Consolidated Contractors Company, P.B. 614, 100 Ruwi, Sultanate of Oman.

Trine Bekkby, Norwegian Institute for Water Research, Gaustadalléen 21, N-0349 Oslo, Norway.

Luisa Bergamin, ISPRA, Institute for Environmental Protection and Research, Via di Casalotti, 300, 00166 Roma, Italy.

Beth Beveridge, Sydney Coastal Councils Group, Inc., Level 12, 456 Kent Street, GPO Box 1591 Sydney, NSW 2001, Australia.

Cassandra Brooke, WWF Australia.

Antje Bruns, GKSS Research Centre, Max-Planck Str. 1, 21502 Geesthacht, Germany.

Lene Buhl-Mortensen, Institute of Marine Research, 5817 Bergen, Norway.

Benjamin Burkhard, Ecology Centre of Christian-Albrechts-University Kiel (ÖZK), Germany.

Talia Choy, University of Guyana, Department of Goverment and International Affairs, Faculty of Social Science, Lot 2, Sixth Street, Cummingslodge, Greater Georgetown, Guyana.

Carlos Coelho, Civil Engineering Department, University of Aveiro, Campus Universitário de Santiago, 3810-193 Aveiro, Portugal.

Doris Diembeck, Ecology Centre of Christian-Albrechts-University Kiel (ÖZK), Germany.

Lars Erikstad, Norwegian Institute for Nature Research, Gaustadalléen 21, N-0349 Oslo, Norway.

Maria Grazia Finoia, ISPRA, Institute for Environmental Protection and Research, Via di Casalotti, 300, 00166 Roma, Italy.

Ola Flaaten, Department of Economics and Management, Norwegian College of Fishery Science, University of Tromsø, 9037 Breivika, Tromsø, Norway.

Chiaki Fujiki, Takehara Fisheries Research Laboratory, Hiroshima University, 5-8-1 Minato, Takehara, Hiroshima 725-0024, Japan.

Massimo Gabellini, ISPRA, Institute for Environmental Protection and Research, Via di Casalotti, 300, 00166 Roma, Italy.

Kira Gee, GKSS Research Centre, Max-Planck Str. 1, 21502 Geesthacht, Germany.

Bernhard Glaeser, Social Science Research Centre (WZB), Berlin, Germany.

Russel Gorddard, CSIRO Sustainable Ecosystems, Bellenden Street, Crace ACT 2911, Australia.

Maria-José Granjo, Civil Engineering Department, University of Aveiro, Campus Universitário de Santiago, 3810-193 Aveiro, Portugal.

Saskia Hommes, Department of Water Engineering and Management, University of Twente, P.O. Box 217, 7500 AE Enschede, The Netherlands.

Ridwanul Hoque, Faculty of Law, University of Dhaka, Dhaka - 1000, Bangladesh.

Sissel Hovik, Norwegian Institute for Urban and Regional Research (NIBR), Gaustadallen 21, 0349 Oslo, Norway.

Suzanne J.M.H. Hulscher, Department of Water Engineering and Management, University of Twente, P.O. Box 217, 7500 AE Enschede, The Netherlands.

Martin Isæus, AquaBiota Water Research, Svante Arrhenius väg 21A, SE-10405 Stockholm, Sweden.

Svein Jentoft, Department of Economics and Management, Norwegian College of Fishery Science, University of Tromsø, Norway.

Yasuhiro Kamimura, Takehara Fisheries Research Laboratory, Hiroshima University, 5-8-1 Minato, Takehara, Hiroshima 725-0024, Japan.

Andreas Kannen, GKSS Research Centre, Max-Planck Str. 1, 21502 Geesthacht, Germany.

Md. Saiful Karim, Faculty of Law, National University of Singapore, Singapore-259776, Singapore. House # 243, Road # 11, Chandgaon R/A, Chittagong, Bangladesh.

Jan Atle Knutsen, Institute of Marine Research (IMR), Flødevigen Marine Research Station, 4817 His, Norway.

Katharina Licht-Eggert, GKSS Research Centre, Max-Planck Str. 1, 21502 Geesthacht, Germany.

Oddvar Longva, Geological Survey of Norway, 7491 Trondheim, Norway.

Merle Looring, Institute of Geography, University of Tartu, Vanemuise Str. 46, 51014 Tartu, Estonia; Pärnu College Ringi Str. 35, University of Tartu, 80010 Pärnu, Estonia.

Maria Celia Magno, ISPRA, Institute for Environmental Protection and Research, Via di Casalotti, 300, 00166 Roma, Italy.

Robin McCall, Department of Biology, Faculty of Natural Science, University of Guyana, Lot 2, Sixth Street, Cummingslodge, Greater Georgetown, Guyana.

Kathleen McInnes, CSIRO Marine and Atmospheric Research, Private Bag-1, 107-121 Station Street, Aspendale VIC 3195, Australia.

Tom G. Measham, CSIRO Sustainable Ecosystems, Bellenden Street, Crace ACT 2911, Australia.

Isabel Mercatali, ISPRA, Institute for Environmental Protection and Research, Via di Casalotti, 300, 00166 Roma, Italy.

Ophelia Meyer-Engelhard, Ecology Centre of Christian-Albrechts-University Kiel (ÖZK), Germany.

Tanja Michler, Alfred Wegener Institute (AWI), Bremerhaven, Germany.

Even Moland, Centre for Ecological and Evolutionary Synthesis (CEES), University of Oslo, Norway; Institute of Marine Research (IMR), Flødevigen Marine Research Station, 4817 His, Norway.

Craig Morrison, Sydney Coastal Councils Group, Inc., Level 12, 456 Kent Street, GPO Box 1591 Sydney, NSW 2001, Australia

Nibedita Mukherjee, Centre for Ecological Sciences, Indian Institute of Science, Bangalore-560012, India.

Jan P.M. Mulder, Department of Water Engineering and Management, University of Twente, P.O. Box 217, 7500 AE Enschede, The Netherlands; Netherland Centre for Coastal Research (NCK), c/o WL|Delft Hydraulics, P.O. Box 177, 2600 MH Delft, The Netherlands.

Ryoji Naito, Port and Airport Research Institute, PARI, 3-1-1, Nagase, Yokosuka 239-0826, Japan.

Yoshiyuki Nakamura, Port and Airport Research Institute, PARI, 3-1-1, Nagase, Yokosuka 239-0826, Japan.

Quach Thi Khanh Ngoc, Department of Economics and Management, Norwegian College of Fishery Science, University of Tromsø, 9037 Breivika, Tromsø, Norway; Faculty of Economics, Nha Trang University, 02 Nguyen Dinh Chieu, Nha Trang, Vietnam.

Corinna Nunneri, Ecology Centre of Christian-Albrechts-University Kiel (ÖZK), Germany.

Melissa Nursey-Bray, National Centre for Marine Conservation and Resource Sustainability AMC, A specialist Institute of the University of Tasmania, Locked Bag 1370, Newnham Way, Tasmania, Australia, 7248.

Filipa S.B.F. Oliveira, LNEC, Hydraulics and Environment Department, Av. Brasil 101, 1700-066 Lisboa, Portugal.

Stephen Bloye Olsen, The Coastal Resources Center, University of Rhode Island, Narragansett, RI 02882, USA.

Esben Moland Olsen, Institute of Marine Research (IMR), Flødevigen Marine Research Station, 4817 His, Norway.

Henriëtte S. Otter, WL|Deltares, P.O. Box 85467, 3508 AL Utrecht, The Netherlands.

Eivind Oug, NIVA, Branch Office South, Televeien 3, N-4879 Grimstad, Norway.

Alf Ring Pettersen, Institute of Marine Research (IMR), Flødevigen Marine Research Station, 4817 His, Norway; Centre for Marine Resource Management (MAREMA), Norwegian College of Fisheries Science, University of Tromsø, Norway.

Alan H. Pickaver, Coastal & Marine Union (EUCC), International Secretariat, P.O. Box 11232, 2301 EE Leiden, The Netherlands.

Benjamin Preston, CSIRO Marine and Atmospheric Research, Private Bag 1, 107-121 Station Street, Aspendale VIC 3195, Australia.

Peter J. Ricketts, Nipissing University, 100 College Drive, Box 5002, North Bay, Ontario P1B 8L7, Canada.

Eli Rinde, Norwegian Institute for Water Research, Gaustadalléen 21, N-0349 Oslo, Norway.

Elena Romano, ISPRA, Institute for Environmental Protection and Research, Via di Casalotti, 300, 00166 Roma, Italy.

Cristina Segurado-Silva, Civil Engineering Department, University of Aveiro, Campus Universitário de Santiago, 3810-193 Aveiro, Portugal.

Kartik Shanker, Ashoka Trust for Research in Ecology and Environment, No. 659, 5th 'A' Main Road, Hebbal, Bangalore 560 024, Karnataka, India; Centre for Ecological Sciences, Indian Institute of Science, Bangalore 560 012, Karnataka, India.

Jun Shoji, Takehara Fisheries Research Laboratory, Hiroshima University, 5-8-1 Minato, Takehara, Hiroshima 725-0024, Japan.

Mohammad A.L. Siddique, Marine Affairs Program, Dalhousie University, Halifax, Canada; Upazila Fisheries Officer, Ramu, Cox's Bazar, Department of Fisheries, Bangladesh.

Timothy F. Smith, University of the Sunshine Coast, Maroochydore DC 4558, Australia.

Ellen Soldal, The Norwegian University of Life Sciences, P.O. Box 5003, N-1432 Ås, Norway.

Knut Bjørn Stokke, Norwegian Institute for Urban and Regional Research (NIBR), Gaustadalleen 21, 0349 Oslo, Norway.

Josianne G. Støttrup, Technical University of Denmark, National Institute of Aquatic Resources, Coastal Ecology Section, Charlottenlund Castle, DK-2920 Charlottenlund, Denmark.

Sebastian Stragies, Social Science Research Centre (WZB), Berlin, Germany.

Catarina I.C. Vargas, LNEC, Hydraulics and Environment Department, Av. Brasil 101, 1700-066 Lisboa, Portugal.

John P. Volpe, School of Environmental Studies, University of Victoria, British Columbia, Canada.

Wilhelm Windhorst, Ecology Centre of Christian-Albrechts-University Kiel (ÖZK), Germany.

Geoff Withycombe, Sydney Coastal Councils Group Inc., Level 12, 456 Kent Street, GPO Box 1591 Sydney, NSW 2001, Australia.

Referees

Albers, Thorsten, Institute of River and Coastal Engineering, Hamburg University of Technology, Germany

Aps, Robert, Estonian Marine Institute, Estonia

Arbour, Joseph, Department of Fisheries and Oceans, Bedford Institute of Oceanography, Canada

Bachtiar, Ramadian, Bogor Agricultural University, Indonesia

Bailly, Denis, Université de Bretagne Occidentale – CEDEM, France

Bastien-Daigle, Sophie, Université de Moncton, Canada

Bekkby, Trine, Norwegian Institute for Water Research, Norway

Bergamin, Luisa, ICRAM, Italy

Bernabeu, Ana, Universidad de Vigo, Spain

Bhatrasataponkul, Tachanat, Faculty of Marine Technology, Burapha University, Thailand

Bjordal, Åsmund, Institute of Marine Research, Norway

Bosch, Elisabet Roca, The Universitat Politècnica de Catalunya (UPC), Spain

Bruns, Antje, GKSS Research Centre, Germany

Bryceson, Ian, Norwegian University of Life Sciences (UMB), Norway

Campos, Angeline Menina, Marine Science Institute, University of the Philippines, Philippines

Celliers, Louis, Oceanographic Research Institute, South Africa

Chiau, Wen-Yan, National Taiwan Ocean University, Taiwan

Choy, Talia, University of Guyana, Guyana

Christie, Hartvig, Norwegian Institute for Water Research, Norway

Diedrich, Amy, IMEDEA (CSIC-UIB), Institute de Estudios Avanzados, Spain

Dolmer, Per, Technical University of Denmark, Denmark

Erikstad, Lars, Norwegian Institute for Nature Research, Norway

Ervik, Arne, Institute of Marine Research, Norway

Fanning, Lucia M. University of the West Indies, West Indies

Flaaten, Ola, Norwegian College of Fisheries Science, Norway

Fosså, Jan Helge, Institute of Marine Research, Norway

Gjøsæter, Jakob, Institute of Marine Research, Norway

Guidetti, Paolo, University of Salento, Italy

Gulliksen, Bjørn, University of Tromsø, Norway

Hallenstvedt, Abraham, University of Tromsø, Norway

Henocque, Yves, CHARM Project Management Unit Department of Fisheries, Thailand

Hoel, Alf Håkon, Dept. of Political Science, University of Tromsø, Norway

Hommes, Saskia, University of Twente, The Netherlands

Hopkins, Thomas Sawyer, CNR-Institute for the Coastal Marine Environment, Italy

Hovik, Sissel, Norwegian institute for urban and regional research, Norway

Huse, Irene, Institute of Marine Research, Norway

Isaac, Victoria, Universidade Federal do Para, Brazil

Jacobsen, Frank, Directorate of Fisheries, Norway

Jentoft, Svein, Norwegian College of Fisheries Science, Norway

Johannessen, Tore, Institute of Marine Research, Norway

Kannen, Andreas, GKSS Research Centre, Germany

Kerwath, Sven, Marine and Coastal Management, South Africa

Kiousopoulos, John, Technological Educational Institute of Athens, Greece

Klein, Carissa, University of Queensland, Australia

Klungsøyr, Jarle, Institute of Marine Research, Norway

Knutsen, Jan Atle, Institute of Marine Research, Norway

Kotta, Jonne, University of Tartu, Estonia

Kraft, Dietmar, ICBM, Germany

Looring, Merle, University of Tartu, Institute of Geography, Estonia

McCall, Robin S, University of Guyana, Guyana

Mikkelsen, Eirik, Norwegian College of Fisheries Science, Norway

Morales-Nin, Beatriz, IMEDEA, Spain

Mortensen, Lene Buhl, Institute of Marine Research, Norway

Mortensen, Pål Buhl, Institute of Marine Research, Norway

Nursey-Bray, Melissa, Australian Maritime College, Australia

Nygaard, Kari, Norwegian Institute for Water Research, Norway

Næs, Kristoffer, Norwegian Institute for Water Research, Norway

Næsje, Tor F, Norwegian Institute for Nature Research, Norway

Øiestad, Victor, Akvaplan-niva, Spain

Ostrom, Elinor, Workshop in Political Theory and Policy Analysis, Indiana University, USA

Pettersen, Alf Ring, Institute of Marine Research, Norway

Rice, Jeep, AFSC's Auke Bay Laboratory, USA

Sandberg, Audun, Bodø University College, Norway

Sandberg, Jan Henrik, The Norwegian Fishermen's Association, Norway

Sandersen, Håkan T, Bodø University College, Norway

Sardá, Rafael, Consejo Superior de Investigaciones Científicas, Spain

Skaare, Bent, Norwegian Institute for Water Research, Norway

Smith, Tim, CSIRO, Australia

Steen, Henning, Institute of Marine Research, Norway

Stojanovic, Timothy, Cardiff University, UK

Stokke, Knut Bjørn, Norwegian institute for urban and regional research, Norway

Sunnanå, Knut, Institute of Marine Research, Norway

Søvik, Guldborg, Institute of Marine Research, Norway

Thorarinsdóttir, Gudrun, Marine Research Institute, Iceland

Uglem, Ingebrigt, Norwegian Institute for Nature Research, Norway

Vatn, Arild, Norwegian University of Life Sciences, Norway

Vølstad, Jon Helge, Institute of Marine Research, Norway

Yamashita, Yoh, Kyoto University, Japan

Preface

Coastal waters around the globe suffer from strain due to a wide range of human activities. The situation calls for a holistic approach, combining expertise from nature science and social science, to reach a balanced and sustainable development of the coastal zone. The International Symposium on Integrated Coastal Zone Management took place in Arendal, Norway, between 11 and 14 June 2007. The main objective of the Symposium, 'Integrated Coastal Zone Management', was to present current knowledge and to address issues on advice and management related to the coastal zone. This international multidisciplinary conference intended to promote science and integration of knowledge for the sustainable management of coastal resources. It provided a venue for scientists, engineers, managers and policy makers to discuss recent advances and innovative ideas, share experiences and develop networks. A total of 167 persons (including 19 students) from 36 countries participated in the symposium (Australia, Austria, Bangladesh, Barbados, Belgium, Brazil, Canada, Chile, Denmark, Estonia, EU, France, Germany, Greece, Guyana, Iceland, India, Indonesia, Iran, Italy, Japan, Latvia, Norway, Philippines, Poland, Portugal, Russia, Singapore, South Africa, Spain, Sri Lanka, Taiwan, Thailand, the Netherlands, the UK and the USA). During the Symposium, a total of 133 presentations (8 keynotes, 55 oral and 70 posters) addressed issues within the following four themes:

- Coastal habitats
- Impacts on coastal systems
- Integrated coastal zone management (ICZM)
- Coastal governance

To develop sustainable utilization of coastal resources, the major challenge facing us is to manage human activities, including conserving significant coastal resources such as tropical reefs, mangroves and sea grass. Internationally (e.g. Convention on Biodiversity) and within the EU (e.g. EC Habitat Directive), there are agreements and new legislation that address the issues related to habitat degradation and the sustainability of coastal resources. An important tool to apply to this is spatial planning that draws upon the new tools provided by geographic information systems (GIS). In particular, the ability to capture information on human activities, the protection provided for resources and the abundance and occurrence of unique resources. Spatial planning helps address the scramble that occurs for space in the coastal zone. This approach should also draw upon the fisher's knowledge of the area as well, since they often know a great deal about where essential fish habitats are located. Several recommendations about habitats that need to be included in future management were stated as:

- The importance of habitat integrity for the maintenance of diversity, productivity and fisheries.

- Importance of the evaluation of degradation and the restoration possibilities.
- Coastal zones receive impact from different areas, so the study scale cannot be local, i.e. general circulation patterns must be considered.
- The importance to take advantage of long-time series, or start building them.
- Need to develop network of observations, mapping of resources, inclusion of social values.

Among the central messages of the theme was the challenge of doing integrated coastal management in a responsible and open manner. There are so many different goals that participants in an ICZM process are pursuing that finding effective ways of doing the 'I' (integration) part of ICZM is a very demanding task, leading as well to the potential for increasing conflicts among participants. One core issue concerned protection of coastal zones. It was emphasized that protection is not necessarily in opposition to use. Protection may also be a prerequisite for use at it is a prerequisite for sustaining resources like fish stocks. Nevertheless, making trade-offs between conflicting uses and between protection and use is a demanding task. As an example, small and poor coastal communities see tourism as a way to increase the income of the residents (as well as the tax base). Tourism can be eco-friendly, but when done rapidly for immediate gain, can result in degradation of environmental conditions in coastal waters as well as in coastal ecosystems themselves. Similarly, to protect the biodiversity of coastal waters, local fishers may be restricted in pursuing commercial fishing, even though this pushes long-term residents out of a job. The situation for the poorest of the poor is that these persons are more likely than others to pay a very large share of the costs of developing coastal areas. Those who benefit from knowing that biodiversity is preserved rarely need to earn their living from fishing or have not established small huts along the shore for living. While tourism frequently creates jobs and can be done in a way that enhances the overall socio-ecological coastal system, if the interests of poor residents of a region are not well represented in the planning of new policies, they may be the ones who end up losing livelihoods and long-term links to a local community.

Governance was keynote defined as encompassing the values policies, laws and institutions by which a society defines a course of action or addresses a set of issues. Governance probes the fundamental goals, the institutional processes and the structures that are the basis for planning and decision making. It spans the formal and informal arrangements, institutions and values that structure and influence:

- How a resource or an environment is utilized
- How problems and opportunities are evaluated and analyzed
- What behavior is deemed acceptable or forbidden
- What rules and sanctions are applied to affect how natural resources are distributed and used

The processes of governance are expressed through the institutions and arrangements of markets, government and civil society. During the conference, it was pointed out that coastal governance is informed by science but is only sometimes science-driven. It was stressed repeatedly that coastal governance makes it imperative to integrate information and knowledge from both the social and the natural sciences. The diversity of contexts from which the speakers drew in their chapters underscored the crucial importance of the condition of the ecosystem, the preexisting traditions of governance and the spatial scale of a project or program in determining how best to tailor the processes of governance to a specific place and set realistic goals within a given time period. Several speakers identified and reaffirmed broad, universally applicable principles that are emerging as useful in guiding coastal governance at a time of accelerating global change. They emphasized the need to set realistic goals and to structure initiatives to overcome the widening 'implementation gap' between issue analysis and planning and the effective implementation of a plan of action directed at selected social and environmental issues. Many urged that those funding and practicing and evaluating coastal governance initiatives accept the diversity and the complexity that is a defining characteristic of both coastal systems and their governance. In addition, the importance of putting people and ecology together and on how difficult it is to achieve was also discussed. Universities were criticized for keeping people and ecology apart by compartmentalizing them into widely separate disciplines, often in different faculties or schools. This lack of interdisciplinarity makes it difficult to bring about a true understanding of the nature of interaction within complex ecosystems, of which humans are an integral part. Overcoming this challenge is essential for understanding dynamic systems.

We hope the present book contributes with new insight into this fast-moving field, highlights the multidisciplinarity required, the complexity of issues at stake and the major task ahead of integrating the sciences with policy.

Erlend Moksness
Einar Dahl
Josianne Støttrup

Participants at the conference held in Ardenal, Norway in June 2007.

Chapter 1
State of Fear or State of Oblivion? What Coastal Zones Are Telling Us about Global Change and Why We Need Integrated Coastal and Ocean Management on a Global Scale

Peter J. Ricketts

Abstract

Globally, our coastal zones and oceans are providing significant evidence of the reality of climatic change and global warming. Yet we are still a long way from approaching an effective regime for managing ocean and coastal resources. This chapter outlines on a global scale the ways by which oceans and coasts are exhibiting the effects of climate change and the potential impacts on huge concentrations of the world's population. Canada is used as an example of a country that has made advances in ocean and coastal management over the past decade, but not nearly enough considering the level of risk for resources, environments and people. The chapter challenges scientists and researchers to step beyond academic boundaries and contribute in a serious manner to the public debate about global climate change, its impacts on oceans and coasts and the implications for public policy. Finally, the author calls for the United Nations to establish a Commission on Ocean and Coastal Management to bring these critical issues to the global political agenda, and to provide the necessary support to regional, national and sub-national efforts to manage the world's oceans and coasts on a sustainable basis within the context of global climate change.

1.1 Prelude

> There is something fascinating about science. One gets such wholesale returns of conjecture out of such a trifling investment of fact.
>
> (Mark Twain, *Life on the Mississippi*, 1883)

Mark Twain's famous caustic quotation about the foibles of scientific theories and projections is used by Michael Crichton at the beginning of his 2004 *New York Times* best-seller novel *State of Fear* (Crichton, 2004). In this novel, Crichton uses his significant talents as a writer of commercial fiction to take on those who believe that global environmental change, and global warming in particular, is being caused significantly by human activities. In addition to challenging the scientific basis of global warming, Crichton launches a vigorous attack on key environmental principles such as sustainable development and the precautionary approach. He even goes as far as to equate global warming with past discredited scientific theories such as eugenics where pseudoscience and politics were so intertwined that they had enormous negative impacts on the world.

So what has all this to do with integrated coastal zone management (ICZM)? Why should the broad interdisciplinary community of scientists, engineers, managers, policy makers and communities that live, work and study in our coastal zones and oceans around the world be concerned with a work of commercial fiction? Surely, those of us who are scientists and serious researchers should not need to concern ourselves with such trivia. In fact, nothing could be further from the truth. After all, not many coastal and ocean scientists have had a book that has made it to the *New York Times* best-seller list, and like it or not the public mood around issues of immense importance like global change can be heavily influenced by popular, commercial fiction. While being a work of fiction, Crichton writes in the world of 'faction' whereby elements of fact are mixed in with a good story to create a work of popular appeal. In a similar fashion to Dan Brown's blockbuster *The Da Vinci Code* (Brown, 2003), readers can easily get confused between what is fact and what is fiction, and the inclusion of references, figures, tables, data and in the case of *State of Fear* an annotated bibliography and a detailed Author's Message section (Crichton, 2004) add to the air of authenticity for the entire work.

Popular literature has an impact on perceptions and opinions, and as scientists we must look at how we can communicate our knowledge, expertise and insights to a broader community, and not just to our academic and scientific peers. This is a dilemma and a significant challenge because we have been trained to speak to and be judged by our scientific peers, and not those who we consider to be outside of that favoured group.

1.2 State of fear or state of oblivion?

Globally, our coastal zones and oceans are providing significant evidence of the reality of climatic change and global warming. In many ways, our oceans and coasts are among the natural systems that are most vulnerable to global climatic change. Yet we are still a long way from approaching an effective regime for managing ocean and coastal resources. Despite mounting scientific and anecdotal evidence to the contrary, there is still a great deal of scepticism amongst the general public, politicians and decision-makers about the reality of global climate change and about the sensitivity of the world's oceans and coasts to such change. While the science of global warming has become increasingly overwhelming, the management response has been less than stellar. True, we have made significant advances in ocean and coastal management over the past decades, but not nearly enough considering the level of risk for resources, environments and people throughout the world. As scientists and researchers we have a responsibility to step beyond the academic boundaries of our expertise and contribute in a serious manner to the public debate about global climate change, its impacts on oceans and coasts and the need to manage human actions and activities accordingly. This begs the question of whether environmentalists are actually creating a 'state of fear' about the impacts of human activity on global climate, or are politicians and decision-makers living in a state of oblivion about the human potential to inflict significant changes on the global climatic system. If the latter is the case, then what are the implications for our planet?

It might be suggested that it would be more appropriate to describe our current state as one of 'denial' rather than 'oblivion'. However, given the actual climatic changes that are occurring and the current and potential impacts of those changes on global oceans and coasts, 'oblivion' may not be too strong a word to use. The *American Heritage Dictionary* provides one definition of the word 'oblivion' as 'official disregard'. So while world leaders may not be oblivious to the issues and concerns about our oceans and coasts, they are certainly disregarding the urgent need to take immediate and effective action to reverse the trends and reduce the detrimental impacts of the present global condition. Arguably, they are living in a state of oblivion!

For ocean and coastal management, the implications of global climate change are especially significant because of the great potential to radically alter the sustainability and habitability of oceans and coasts around the world. The question is how we get that message out to the general public and politicians alike, and how do we get them to take the actions necessary.

1.3 Reaching out

The backdrop for this book was a gathering of scientists, engineers, managers and policy makers in Norway in June 2007 to discuss the importance of ICZM through the auspices of the International Council for the Exploration of the Seas (ICES). In the country of Gro Harlem Brundtland, arguably the 'birthplace' of the concept of sustainable development, it was particularly appropriate to consider the importance of reaching out to the broader global community and the efforts to promote an integrated approach to manage the world's oceans and coasts. Indeed, the very purpose of Bruntland's milestone work that resulted from her World Commission on Environment and Development, the 1987 report *Our Common Future* (Brundtland, 1987), was to do just that – to speak to the general public, to the people of the world and so to their politicians and governments. Brundtland challenged all of us to raise the level of concern about how we are managing our planet, and what it means for future generations and, indeed, the future of the planet itself as a safe and sustainable home for humankind.

Building on Brundtland's challenge, the 1992 World Conference on Economy and Development in Rio de Janeiro developed the *Agenda 21* statement, which remains as a clarion call to action on how to change the way in which we use and exploit our environment and natural resources. This is essential so that not only we may live in harmony with a planet that will sustain us into the future, but also, and perhaps this is the most radical aspect of *Agenda 21*, we will actually be able to sustain the planet for its future survival. Within *Agenda 21*, Chapter 17 (UN, 1992) dealt specifically with oceans, coastal and estuarine environments. This chapter has inspired numerous governmental and non-governmental efforts to promote integrated coastal zone management and ocean management, including the World Bank's

Noordwijk Guidelines for Integrated Coastal Zone Management (World Bank, 1993) and the Organisation for Economic Co-operation and Development's *Recommendations on Integrated Coastal Zone Management* (OECD, 1993).

Many authors have attempted to bring the importance of nurturing and cherishing our oceans and coasts to a more general audience, including Rachel Carson's pioneering work *The Sea Around Us* (Carson, 1951) and more recent works such as *The Living Beach* by the renowned Canadian author Silver Donald Cameron (Cameron, 1998). Also, global warming is an increasingly popular theme for writers, such as Tim Flannery's highly acclaimed *The Weather Makers* (Flannery, 2006), and former US Vice President Al Gore has been promoting action on global warming through his documentary film and book *An Inconvenient Truth* (Gore, 2006). That book did reach number one on the *New York Times* best-seller list in June 2006, and with the documentary winning an Oscar in 2007, Gore exemplifies the kinds of impacts that individuals can have when they shed the limitations of their traditional roles and speak more directly about issues of importance in a language that can be widely understood. The battle for the hearts and minds of the general public is well and truly in full swing.

In North America during 2006 and 2007, a series of spectacular and disturbing trends in climate (successive record average high temperatures), weather patterns (hurricanes, tornadoes and severe storms) and environmental events (droughts, fires and floods) have turned public opinion back towards the environment and climate change in particular, even though these events in and of themselves may have nothing to do with global warming. In Canada, a new conservative federal government elected in January 2006 with a very pro-oil industry agenda and a strong anti-Kyoto stand has been forced to backtrack on its inherent indifference to the environment (and global warming in particular) in response to the astonishing and rapid changes in public opinion. In the US, the George W. Bush administration will not go down in the history books as being environmentally friendly, but it is to the credit of many of the individual states (helped and perhaps lead by California) that the US has moved further towards meeting its Kyoto targets than Canada. Again, public reaction has been a significant force behind these regional and local level initiatives and has had a demonstrated impact on a national and global level, despite the indifference and often outright hostility of President Bush and his administration. The election of Barack Obama as president will hopefully be accompanied by a dramatic change in the way the US has addressed issues of global climate change over the past 8 years.

Perhaps most importantly, a series of major reports over the past year from the Intergovernmental Panel on Climate Change (IPCC) has strengthened the scientific basis for global warming and largely offset political and industrial efforts to undermine public confidence in the science behind climate change and produce an electorate of global warming sceptics (IPCC, 2007a). As the *Summary for Policy Makers* states:

> Warming of the climate system is unequivocal, as is now evident from observations of increases in global average air and ocean temperatures, widespread melting of snow and ice, and rising global average sea level.
>
> (IPCC, 2007b)

Each of these IPCC reports has been released with much public fanfare, and they have moved the debate from one where the science of human-induced global warming was just a theory to one where there is widespread acceptance of some significant level of proof. While detractors might accuse the members of the IPCC of selectively looking at the scientific evidence that supports their pre-convictions, those who know the IPCC and its highly political and bureaucratic nature (being essentially a creature of the UN) understand that it is a very conservative body and has taken a long time to get to the point where it was willing to claim direct evidence demonstrating anthropogenic causes of global warming.

1.4 What are oceans and coasts telling us about global warming?

The debate about global warming is extremely important for ocean and coastal management. Within the IPCC's integrated framework for climate change (as depicted in Figure 1.1), oceans and coasts form a significant component of the human and natural systems response to climate change. With over 23% of the world's population living within a 100 km distance from a coast and at elevations of less than 100 m above sea level (including 23 of the world's

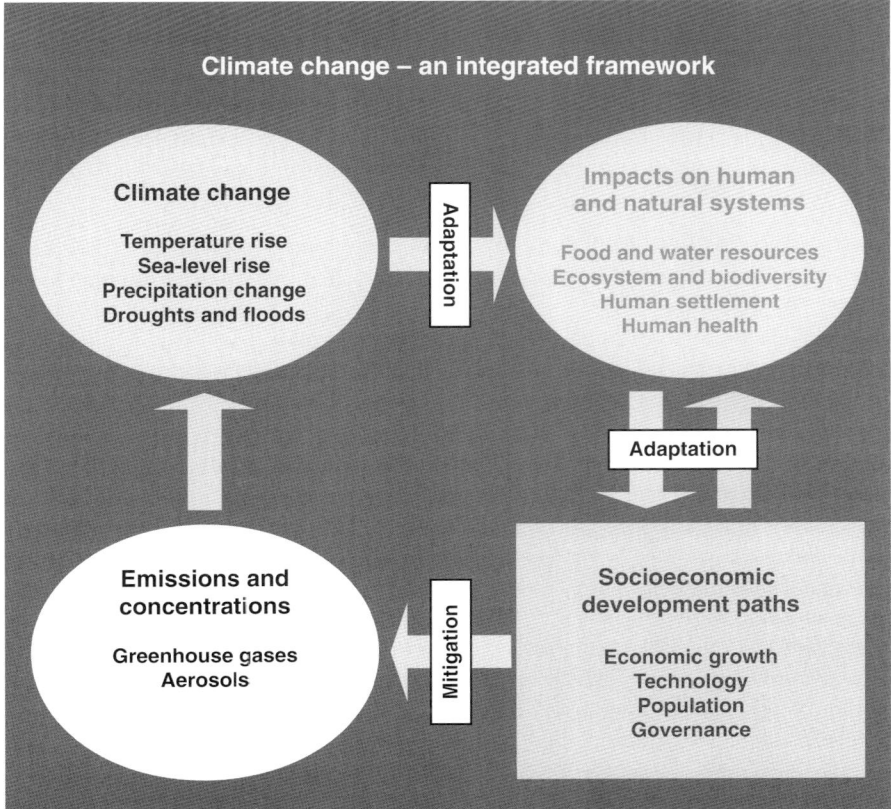

Figure 1.1 IPCC integrated framework for climate change.

39 cities with populations over 5 million), the potential coastal impacts of climate change on global populations are enormous. Those of us who have been researching this field over the past 35 years have long argued the importance of coastal zone management in order to manage effectively the rapid coastal population growth compared with non-coastal regions of the world, and consequently the increased hazards that are faced by those expanding coastal communities as well as the accelerated destruction and overexploitation of coastal environments and resources. Global warming considerably exacerbates those threats and the exposure to risks faced by those coastal populations. Indeed, some of the greatest potential adverse impacts of global warming are faced by coastal environments and communities.

In reviewing the areas of concern around the globe that are being experienced, observed and measured within the context of global warming, it is difficult not to be alarmed by what our oceans and coasts are telling us about the state of our planet's environment. The cumulative impacts of climate change are now being felt in ways that can actually be measured with some high degree of accuracy. These changes and trends, if continued unabated, predict dire consequences for planet Earth and we would be foolish not to recognise them and take action while we still can.

1.4.1 Loss of polar ice sheets

Thinning of the polar ice sheets is being measured at astonishing rates in recent years, with predictions of major loss of ice cover over the next 20 years in both the Arctic and Antarctic. The 2005 Arctic Climate Impact Assessment Report (Symon et al., 2005) estimates that summer ice in the Arctic has decreased by almost 27% in the past 50 years, that sea ice loss has increased by 20% over the past 30 years, and that the sea ice is thinning at a rate of 7–9% per decade. Some implications of this include the following:

- The opening up of the Arctic Ocean to year-round marine transportation through the Northwest Passage with the associated impacts of increased marine shipping and pollution.
- Potential changes in the Arctic gyre and currents that will impact life and feeding cycles for polar marine species and those animals that rely on them for food.
- Huge impacts on marine mammals and migratory species, such as polar bears, who use the ice for feeding.
- Higher shoreline erosion rates due to increased exposure of coastlines to longer fetches, stronger wave action and nearshore currents.
- Increased sub-aerial erosion of coastal cliffs and bluffs and subsidence of coastal lowlands due to the melting permafrost.
- Wholesale changes to nearshore sediment transportation and deposition regimen along the Arctic coasts.
- Major impacts on coastal plain settlements and infrastructure due to increased average temperatures, higher precipitation, permafrost melting, and changes in migratory patters of marine and terrestrial animals upon which the inhabitants depend for their food and livelihood.

Indeed, monitoring of Arctic sea-ice break-up in 2007 has indicated a significant increase in the rate of melting, and predictions of an ice-free Northwest Passage have now been accelerated to 2030 and are being revised on an alarmingly consistent basis.

Table 1.1 Sources of sea-level rise 1961–2003.

Source of sea-level rise	Rate of sea-level rise (mm per year)	
	1961–2003	1993–2003
Thermal expansion	0.42 ± 0.12	1.6 ± 0.5
Glaciers and ice caps	0.50 ± 0.18	0.77 ± 0.22
Greenland ice sheet	0.05 ± 0.12	0.21 ± 0.07
Antarctic ice sheet	0.14 ± 0.41	0.21 ± 0.35
Sum of individual climate contributions to sea-level rise	1.1 ± 0.5	2.8 ± 0.7
Observed total sea-level rise	1.8 ± 0.5[a]	3.1 ± 0.7[a]
Difference (observed minus sum of estimated climate contributions)	0.7 ± 0.7	0.3 ± 1.0

Source: IPCC, 2007b, p. 7.
[a] Data prior to 1993 are from tide gauges and after 1993 are from satellite altimetry.

1.4.2 Increased sea-level rise

Accelerated sea-level rise associated with increased ocean water temperatures and polar ice cap melting is one of the most widely known impacts of global warming. As Table 1.1 shows, the 2007 IPCC report states that sea-level rise has accelerated since 1993 and is increasingly explained by climate change. Sea-level rise is accompanied by associated risks of flooding and inundation of low-lying coastal lands, and increased susceptibility to coastal storms, tsunamis and other extreme events. While not the result of global warming, the Indian Ocean tsunami of 2004 brought home the devastating power of the ocean and the high level of danger to which coastal communities in low-lying coastal zones are exposed. The comparatively low level of damage and loss of life along those coastlines that had retained their natural mangrove forests provide an example of the importance of retaining and restoring mangroves and other environments (such as coral reefs) that dissipate wave energy in order to provide natural protection against such hazards as tsunamis. However, sea-level rise makes it increasingly difficult for such natural environments to maintain their capacity to protect coastlines and low-lying coastal plains in their present location. The inevitable trend is for coastlines to retreat landwards as sea-level rise increases the erosive power and the surge effect of waves, tides and currents.

1.4.3 Impacts of changing ocean currents

Temperature changes and influxes of cold water from melting polar ice have the potential to impact the nature and location of global ocean currents. The North Atlantic is especially susceptible here given the potential of the Gulf Stream to change its flow and its temperature characteristics due to increased influxes of cold water from the Arctic Ocean and Greenland ice sheets. With the increased melting of the ice sheets and sea ice, the Labrador Current could become much stronger with the effect of pushing the Gulf Stream southwards in the northwest Atlantic. The potential impacts on the climate of Western Europe could be significant, with a worst-case scenario suggesting that the Gulf Stream could be shut off from reaching the eastern side of the Atlantic (Flannery, 2006).

1.4.4 Increases in ocean acidity

The German Advisory Council on Global Change (Schubert et al., 2006) has measured increases in the acidity of oceans and the possibility that the oceans are reaching some kind of limit in their ability to absorb increased carbon dioxide from the atmosphere. The potential impacts on marine environments and species that are sensitive to water acidity levels are enormous. Scientists are measuring detrimental effects of acidification on coral reefs and other species that rely upon the building of calcite shells, the long-term growth and reproduction of squid and certain fishes, and the potential for an imbalance in phytoplankton growth resulting in an imbalance in the oceans and a loss of biodiversity (Flannery 2006; Schubert et al., 2006; IPCC, 2007c).

1.4.5 Deterioration of coral reefs

A combination of increased water temperatures, rising sea levels and increasing ocean acidity will have a serious impact on coral reefs, especially those in sub-tropical coastal areas. Coral reefs are being identified as under increasing threat (Schubert et al., 2006), and any acceleration of coral reef destruction resulting from climate change will have serious and immediate impacts. Not only would this mean the destruction of some of the most highly productive ecosystems in the world, but also the impacts on the coastal zones that are protected by coral reefs would be significant. Small island nations are particularly at risk and are actively planning for the impacts of climate change, with some of them already considering the possibility of complete inundation and eventual evacuation (Tompkins et al., 2005; IPCC, 2007c).

1.4.6 Changes to living marine species

Already reeling from major collapses of commercial marine species, the environmental changes resulting from climate change could result in further collapse of marine animal populations, both commercial and non-commercial. Continued use of and improvement in commercial harvesting techniques will render critical stocks and habitats unsustainable as the changes in oceanic environmental conditions exert increased pressure on the ability of marine species to procreate. While overfishing is still identified as the primary reason for marine stock collapse, the impacts of climate change are accelerating an already catastrophic deterioration of marine ecosystems and loss of biodiversity. Recent studies even project that all commercial fish and seafood species will collapse by 2048 unless serious mitigation efforts are commenced (Worm et al., 2006).

1.4.7 Increased hazardousness of coastal zones

As global oceanic changes are reflected in the increased exposure of coasts to storms, flooding, erosion and tsunamis, coastal zones will become increasingly hazardous places to live. Higher rates of flooding, erosion, damage to human infrastructure and loss of life will result in increased risks for coastal populations and infrastructure across the globe. The IPCC states that:

> Many millions of people are projected to be flooded every year due to sea-level rise by the 2080s. Those densely populated and low-lying areas where adaptive capacity is relatively low, and which already face other challenges such as tropical storms or local coastal subsidence, are especially at risk. The numbers affected will be largest in the mega-deltas of Asia and Africa while small islands are especially vulnerable.
>
> (IPCC, 2007c)

Nicholls et al. (2007) provide an excellent assessment of the state of the world's coastal systems in relation to climate change. Table 1.2 summarises their assessment and presents a daunting overview of the challenges for global ocean and coastal management in an environment of climate change and global warming. Finally, the Working Group on Climate, Oceans and Security under the Strategic Oceans Planning to 2006–2016 initiative of the Global

Table 1.2 Climate drivers and biophysical effects on coastal systems.

Climate driver (Trend[a])	Main physical and ecosystem effects on coastal systems
CO_2 concentration (↑)	Increased CO_2 fertilisation; decreased seawater pH (or 'ocean acidification') negatively impacting coral reefs and other pH-sensitive organisms
Sea surface temperature (↑, R)	Increased stratification/changed circulation; reduced incidence of sea ice at higher latitudes; increased coral bleaching and mortality; poleward species migration; increased algal blooms
Sea level (↑, R)	Inundation, flood and storm damage; erosion; saltwater intrusion; rising water tables/impeded drainage; wetland loss (and change)
Storm intensity (↑, R)	Increased extreme water levels and wave heights; increased episodic erosion, storm damage, risk of flooding and defence failure
Storm frequency (?, R)	Altered surges and storm waves and hence risk of storm damage and flooding
Storm track (?, R)	
Wave climate (?, R)	Altered wave conditions, including swell; altered patterns of erosion and accretion; reorientation of beach planform
Run-off (R)	Altered flood risk in coastal lowlands; altered water quality/salinity; altered fluvial sediment supply; altered circulation and nutrient supply
Air temperature (↑, R)	Increased permafrost melting; increased surface, through, and ground water flow; increased sub-aerial erosion (slumping, gullying, erosion, etc.) along Arctic coastlines

Source: After Nicholls et al. (2007).
[a] Trend: ↑, increase; ?, uncertain; R, regional variability.

Forum on Oceans, Coasts, and Islands (http://www.globaloceans.org/planning/index.html – accessed 28 February 2008) predicts the following regional impacts of climate change on the coastal areas of the world:

- Africa – by the end of the twenty-first century, sea-level rise is projected to affect low-lying coastal areas, with further degradation of mangroves and coral reefs; the cost of adaptation to sea-level rise estimated to be at least 5–10% of gross domestic product.
- Asia – predicted significant sea-level rise will result in greater risks of flooding and seawater intrusion; loss of coral reefs is estimated at 24% in the next 10 years and 30% within 30 years.
- Australia/New Zealand – continued coastal development and population growth will lead to higher risks from sea-level rise.
- Europe – sea-level rise will cause a loss of up to 20% of wetlands resulting in increased risk of flash floods, coastal flooding and coastal erosion; coastal flooding predicted to negatively impact up to 1.6 million people annually.
- Latin America – sea-level rise will negatively impact Mesoamerican coral reefs; increased risk of flooding in low-lying coastal areas due to sea-level rise.
- North America – continued coastal development and population growth will lead to higher risks from tropical storms.
- Polar regions – potential navigable northern sea routes for shipping.
- Small islands – sea-level rise may lead to increased storm surge, erosion and inundation, leading to threats to vital infrastructure; deterioration of coastal conditions, including coral bleaching and beach erosion.

1.5 The global response

So what are we doing about all of this? In January 2006, the Global Forum on Oceans, Coasts and Islands met in Paris for its third forum. This conference followed on from the Ocean Policy Summit held in Lisbon in 2005. The Global Forum is the nearest thing we have to a world organisation for the sustainable management of oceans and coasts. Funded through the United Nations Educational, Scientific and Cultural Organization (UNESCO) and the United Nations Environment Programme (UNEP), the Global Forum brings together a wide range of governments, groups, non-governmental organisations (NGOs), and individuals from around the world to facilitate collaboration and cooperation, to promote integrated coastal and ocean management (ICOM) and to encourage the expansion of efforts to protect, conserve and sustain ocean and coastal resources. The Global Forum collaborated with the Third World Ocean Network conference in Boulogne-sur-Mer (also in January 2006), which is linking aquariums, museums and marine tourism to educate the public about the plight of the oceans. The Third Global Forum, *Third Global Conference on Oceans, Coasts, and Islands: Moving the Global Oceans Agenda Forward*, which was attended by 403 delegates representing 78 countries, identified some critical issues for ICOM globally, and they are worth thinking about (Cicin-Sain et al., 2006a, b):

- The targets of eliminating illegal, unreported and unregulated fishing, and fishing overcapacity by 2004 and 2005 have not yet been met, and 75% of fish stocks are classified

by Food and Agriculture Organization (FAO) of the United Nations as fully exploited, overexploited or depleted.
- In many of the poorest countries, a cycle of extreme poverty coupled with excessive exploitation of the environment needed for survival still prevails, contributing to marine pollution and excessive resource depletion.
- The goal of establishing representative networks of marine protected areas by 2012 will not be met until 2085 at the present rate of designation, although some countries, such as Australia, Palau, the Cook Islands and Costa Rica, have made significant progress in establishing marine protected areas with a view towards protecting marine and coastal biodiversity.
- Although half of the world's 43 small island developing state, as well as a number of other countries, have adopted ecosystem-based management and coastal and ocean management programs, no international organisation is responsible for tracking progress in the establishment of these program. In addition, there is no regular collection of information on the social and economic well-being of coastal communities.
- Sixty states have initiated national plans of action to address land-based sources of marine pollution under the Global Programme of Action for the Protection of the Marine Environment from Land-Based Activities, which accounts for 80% of marine pollution.
- According to the UN Millennium Project, urgent action is needed to achieve the Millennium Development Goals (MDGs), but it remains to be seen if sufficient action has occurred to move closer towards their achievement. Much effort has been put into the development of indicators to measure achievement towards the targets and goals, but little emphasis has been placed on oceans, coasts and coastal populations.
- One area where tangible progress is being made is in the protection of marine biodiversity and networks of marine protected areas (MPAs). Implementation of the biodiversity and MPA goals established by the 2002 World Summit on Sustainable Development (WSSD) has benefited from the connection to the International Convention on Biological Diversity and its Secretariat, which provide a well-organised structure and process for advancing biodiversity around the world. The large majority (71%) of nations reporting to the Convention on Biological Diversity indicate that they have designated MPAs and/or have plans to improve existing MPAs; while 51% report that development of an MPA system or network is underway, and 29% report that an MPA system or network is already in place.
- There are now more than 700 ICZM initiatives in more than 90 nations around the world, but there are no standardised evaluation criteria for measuring their performance in achieving the goals established by the WSSD and MDGs, although there are efforts towards this direction.

While there is a huge variation in the level, commitment and effectiveness of formal ICZM programs and initiatives around the world, nonetheless progress is being made even if it is painfully slow and lagging far behind the management needs of the world's oceans and coasts. Sadly, these efforts might be more acceptable under a relatively stable global environment, but in the context of global change and increasing global warming they are woefully and pitifully inadequate. The commitment to ICZM on a global level is simply not there, and Canada is an example of the good, the bad and the ugly when it comes to ICZM.

1.6 Canada's experience in ICOM

As illustrated in Figure 1.2, Canada has a huge coastal zone and ocean area incorporating the longest coastline (243 797 km) and the second largest contiguous continental shelf area (2 877 623 km^2, but the total area exceeds 5.5 million square kilometers if one includes the 0–12 nautical mile territorial sea) of any country in the world. Despite these facts of global significance, Canada's experience in ICOM[1] has been characterised by periods of impressive advances separated by long periods of relative inaction.

A recent special issue of the journal *Coastal Management* has assessed the state of ocean and coastal management in Canada (Ricketts and Harrison, 2007a). Canada's marine and freshwater coasts include three oceans – Atlantic, Pacific and Arctic; the Gulf of St. Lawrence and the Great Lakes; and important seas, bays and straits, including the Beaufort Sea, Labrador Sea, Hudson Bay and the Davis Strait (Ricketts and Harrison, 2007b). There is a huge diversity of physical, oceanographic and biological characteristics, as well as significant socioeconomic dependence upon coastal and ocean resources and transportation. Because Canada has a relatively small population for its size, it is often assumed that there are few people living along the coast and hence little incentive for the government to be concerned about managing coastal and ocean resources. However, Manson (2005) shows that a growing number of Canadians are living in the coastal zone, and consequently are becoming increasingly exposed to coastal hazards (such as sea-level rise, erosion, flooding, earthquakes and tsunamis) and are putting increased pressures on the environmental and resource capacities of Canada's coasts and oceans. The sustainable development and management of Canada's coasts and oceans is an issue of national and international importance, and this has been reflected by increasing government interest in ICOM. Furthermore, Canada's coastal zones have been and continue to be severely impacted by environmental change.

Indeed, major crises in coastal economic development have been instrumental in influencing the development of ICOM in Canada. In 1992, Atlantic cod stocks became the first of a series of commercial fish stock collapses that sent a shiver throughout coastal communities in Canada. The impacts on government were enormous, as for the first time a coastal resource issue became a major political headache, and the public became aware of how predictions of a stock collapse by fisheries scientists had long been ignored, indeed stifled, by the responsible government agency for over 10 years. It is important here to note the brave and pioneering work of Dr Ransom Myers, whose untimely death from cancer in 2007 marked a huge loss to the global scientific community and the cause for sustainable fisheries management.

Before moving to Dalhousie University, Dr Myers was one of a number of fisheries scientists at the Department of Fisheries and Oceans (DFO) who raised the alarm about the reproductive future of the Atlantic cod stocks and called for a reduction in commercial fishing and a change to the way in which DFO was managing fish stocks and establishing

---

[1] In Canada, the term integrated coastal and ocean management (ICOM) is used rather than ICZM because of the fact that the involvement of the federal government (through the Department of Fisheries and Oceans) is primarily through its authority for the oceans, whereas responsibility for the coastal zones involves the jurisdictions of the provincial governments.

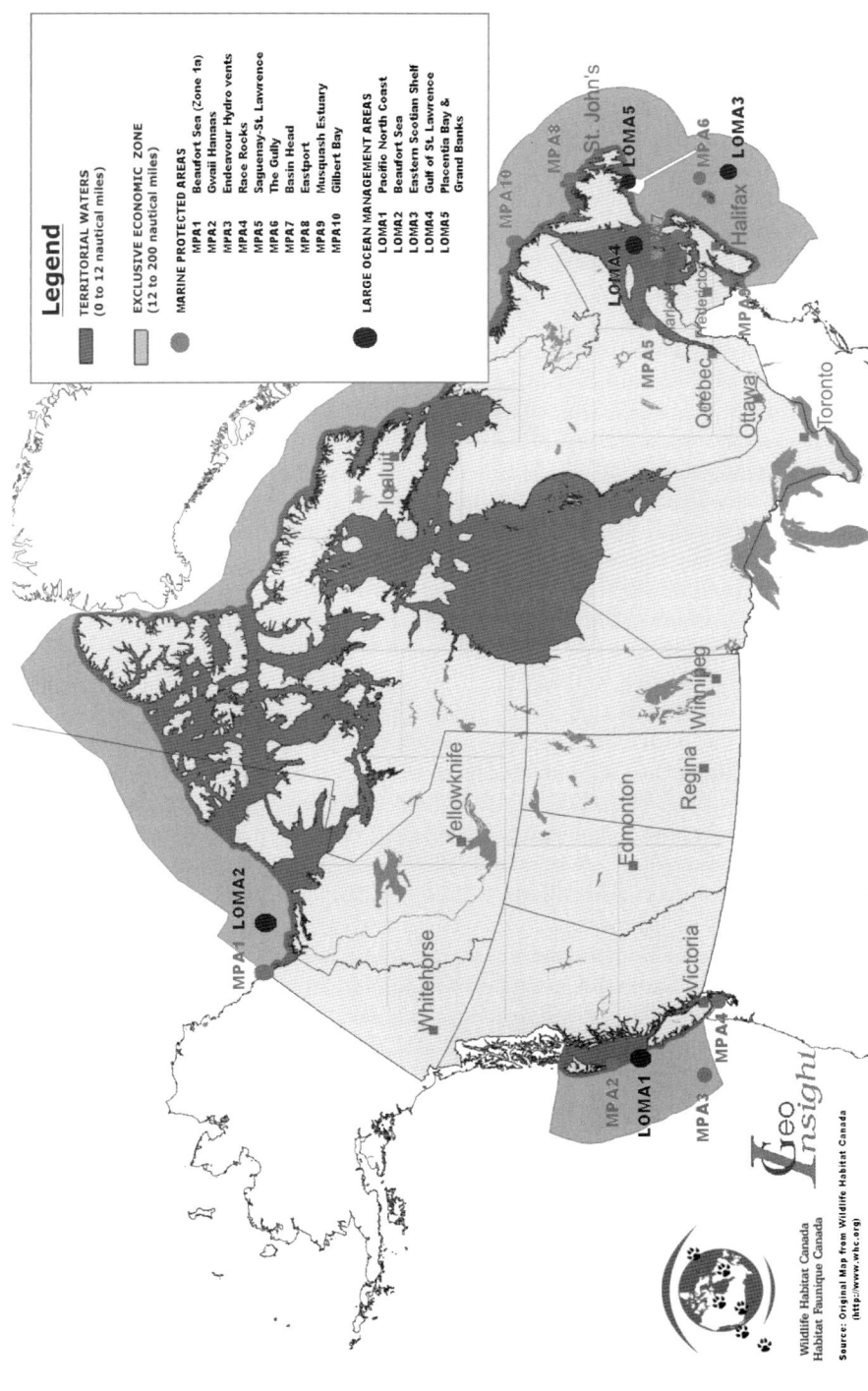

Figure 1.2 Canada's marine protection and management areas.

total allowable catch limits. Dr Myers's research was effectively shut down and he was forced to leave his position to seek academic freedom at a university. Later as a marine biologist at Dalhousie University and holder of the Killam Chair in Ocean Studies, Dr Myers went on to discover that 90% of the world's sharks, bluefish tuna, swordfish, cod and other big predatory fish had been serially stripped from the seas by industrialised fishing since the early 1950s. In October 2005, *Fortune* magazine declared him one of the top ten people in the world to watch, listing him between the then US Senator Barack Obama and the cofounders of Google.

1.6.1 The Canada Oceans Act

The collapse of the cod fishery helped promote government action, albeit slowly, towards ICOM, starting with the passage of the Oceans Act in 1997 (Government of Canada, 1996), and followed some 6 years later with the implementation of the oceans action plan (OAP) in 2004. Along a similar time scale, the Kyoto Accord was opened for signature in December 1997. However, partly due to the opposition of key provincial governments (which control much of the resource production in the country) at the time, the Government of Canada only ratified it in 2003, thus belatedly commencing the development of a plan of action to reduce carbon emissions in accordance with Kyoto targets. Although not linked at the time, ICOM and global warming are now very much related in terms of the urgency of action required to reduce greenhouse gas emissions and prepare for and, where possible, mitigate against the dramatic impacts on our oceans and coasts.

Without doubt, the Oceans Act was the most significant and hopeful development in Canadian coastal and ocean management to date. This clear expression of political will, missing so long from the ICOM equation in Canada, was finally a reality (Ricketts and Harrison, 2007b). The Oceans Act included three parts covering sovereign rights, a national oceans management strategy and the consolidation of federal legislation on oceans (see Figure 1.3), and the DFO was designated as the lead federal agency for ICOM in Canada. With this legislation, Canada effectively adopted the provisions of the United Nations Convention on the Law of the Sea (UNCLOS) (UN, 1982) including the establishment of a 200 nautical mile exclusive economic zone. The Act also incorporated three key principles of sustainable development, integrated management and the precautionary approach, and is the first comprehensive oceans legislation of its kind in the world. Within the context of the Oceans Act, DFO was legislated to lead and facilitate the development and implementation of a national strategy for the management of Canada's estuarine, coastal and marine ecosystems.

By 1998, DFO had begun to put flesh on the framework of the Oceans Act by releasing a review document entitled *Toward Canada's Oceans Strategy* (DFO, 1998). This proposed Oceans Strategy was to be built on the three principles of the Act and would focus on the goals of replacing the existing fragmented approach to oceans management with a collaborative, integrated approach; expanding working partnerships among oceans stakeholders and increase their responsibility and accountability; optimising the economic potential of our oceans while ensuring their conservation and sustainability; and positioning Canada as a world leader in oceans management. However, in the years following, there was little actual progress towards the development of an integrated approach to coastal and ocean

> - **PART I: Sovereign Rights and Jurisdiction**
>   - EEZ and Contiguous Zone
>   - RATIFICATION OF UNCLOS
>
> - **PART II: National Oceans Management Strategy**
>   - Sustainable Development
>   - Integrated Management of Ocean and Marine Activities and Resources
>   - Common Heritage of Canada's Three Oceans
>   - Ecosystem Approach to Maintain Biodiversity
>   - Precautionary Approach
>   - Economic Diversification and Wealth Generation
>   - National Strategy for Management of Estuarine, Coastal, and Marine Ecosystems
>   - DFO as Lead Agency – mandate to collaborate – federal, provincial, First Nations, coastal communities, etc.
>
> - **PART III: Consolidation of Fedreal Legislation**

Figure 1.3 Key elements of the Canada Oceans Act 1997.

management. Despite the passage of the Oceans Act, the development of coastal and ocean management in Canada moved slowly, sporadically, and with more talk than walk. Most activity occurred within the various resource sectors, and at the provincial and community levels.

It was not until 2002 that some momentum developed with the long anticipated publication by DFO of *Canada's Oceans Strategy* (DFO, 2002), followed by Canada's ratification of the UNCLOS at the end of 2003. Then in February 2004, the federal government announced its intention to develop an OAP. Finally, some 7 years after the Oceans Act the government was proposing some action to implement its content. This fact was emphasised in 2005 by the Auditor General of Canada who released his findings of the actions taken by the Government of Canada to implement the Oceans Act (Auditor General of Canada, 2005). The report found that following the passing of the Oceans Act, oceans ceased to be a government priority and that 'after 8 years, the promise of the *Oceans Act* is unfulfilled'. However, despite lamenting the glacially slow progress in implementing the Oceans Act, the report did recognise the progress now being made by DFO under the OAP.

1.6.2 The oceans action plan

Phase I of the OAP is composed of a number of initiatives grouped under four pillars: international leadership, sovereignty and security; integrated oceans management for sustainable development; health of the oceans; and ocean science and technology. These pillars constitute the framework for action under Phase 1 of the OAP, as illustrated in Figure 1.4.

> **Oceans Action Plan: The Four Pillars**
>
> I. **International Leadership, Sovereignty, and Security**
>   - Arctic Marine Strategic Plan
>   - Regional Cooperation – e.g. Gulf of Maine
>   - Continental Shelf sovereignty
>
> II. **Integrated Oceans Management for Sustainable Development**
>   - Ecosystem-based Management
>   - Five pilot Large Ocean Management Areas (LOMAs)
>
> III. **Health of the Oceans**
>   - Ballast water discharge regulations re. invasive species
>   - Enhanced marine pollution surveillance and enforcement on continental shelf
>   - Marine Protected Area Strategy – network of MPAs
>
> IV. **Ocean Science and Technology**
>   - Marine and Ocean Industry Technology Roadmap
>   - Oceans Science and Technology Partnership
>   - Placentia Bay Technology Demonstration Platform

Figure 1.4 Four pillars of Canada's oceans action plan.

1.6.2.1   Pillar 1: international leadership, sovereignty and security

The initial passage of the Oceans Act in 1997 was followed shortly by Canada declaring its exclusive economic zone and ratifying the United Nations Law of the Sea Convention (UNCLOS III). Under the OAP, Canada will continue to play a leadership role in international oceans bodies (such as UNCLOS; Global Forum on Oceans) and will work closely on oceans stewardship with the United States and Mexico under the 2005 *Security and Prosperity Partnership of North America*. The OAP provides specific support for collaboration with the US in the Gulf of Maine, and to the Arctic Marine Strategic Plan, which was adopted by the Arctic Council in November 2004. Under this pillar, the OAP will also emphasise both the significant problem of overfishing in the Northwest Atlantic Fisheries Organization's (NAFO) regulatory area and elsewhere in the world's oceans, and the quest to confirm our sovereign rights over the full extent of the continental shelf (UNCLOS, art. 76).

1.6.2.2   Pillar 2: integrated oceans management for sustainable development

Integrated management (IM), based on an ecosystems approach, is one of the key requirements of the Oceans Act. The IM pillar is a key component of the OAP through the establishment of five large ocean management areas (LOMAs) – Placentia Bay and the Grand Banks, Scotian Shelf, Gulf of St. Lawrence, Pacific North Coast and the Beaufort Sea (see Figure 1.2). The LOMA is at the core of one of the specifically Canadian elements of the approach to ocean and coastal management under the Oceans Act. This Canadian approach is one of starting out in the ocean (where the federal government has clear jurisdiction) and then moving landwards to bring in the provincial governments, municipal governments and other players and stakeholders in the coastal zone, to bring the LOMAs onshore. This is quite the opposite from traditional approaches to coastal and ocean management, such as in the US, where they have started with the coastal zone (and often a fairly narrow definition of that zone) and are now trying to incorporate broader ocean management approaches into the existing CZM structure. Because of Canada's difficulties in getting the federal and provincial governments to agree on a national approach to CZM, the federal government has chosen to move ahead in the area where it has sole jurisdiction (i.e. the ocean) and work with the provinces and key stakeholders to bring in the coastal zone as the IM process develops. This is occurring more quickly in some areas than in others (Guénette and Alder, 2007).

1.6.2.3   Pillar 3: health of the oceans

Under this pillar, Transport Canada will develop appropriate regulations for ballast water discharge in order to reduce the risk of aquatic invasive species, and will enhance its pollution surveillance on the East Coast by doubling the number of its patrols and coordinating closely with Radarsat information. An amendment to the Migratory Birds Convention Act will make the enforcement of marine pollution cases more effective. A significant component of this pillar is the implementation of the *MPA strategy* under which Fisheries and Oceans Canada, Parks Canada and Environment Canada will work towards establishing a coordinating network of MPAs. As shown in Figure 1.2, some ten areas either have been designated or are slated for designation as MPAs under this pillar of the OAP.

1.6.2.4   Pillar 4: ocean science and technology

A number of ocean technology networks already exist across Canada which link together ocean researchers in various domains, ocean-related institutions and technology developers. Under this pillar, the 'oceans science and technology partnership' (OSTP) has been created to build on existing synergy and to lay the ground for more extensive cooperation in the future. Also, support is being given to the Placentia Bay Technology Demonstration Platform, which will use state-of-the-art systems to provide stakeholder access to data and information in support of effective management and sustainable development of coastal and ocean areas. On Canada's west coast, a number of university-based initiatives (primarily through the University of Victoria in British Columbia) are pushing forward the boundaries of knowledge about ICOM. These include VENUS, a joint Canada–US project that will create the largest cable-linked seafloor observatory in the world, and the larger scale NEPTUNE project that will cover the entire Juan de Fuca plate.

With the time and budget for Phase 1 of the OAP completed, the development of Phase 2 was underway when something happened that is part of the democratic process. In January 2006, the election of a minority Conservative government following 13 years of successive Liberal governments brought a considerable chill to the development and implementation of climate change policy and the Oceans Act in particular. All new governments go through a period of review and reflection of the policies that they inherit from the preceding government, especially after such a long period of relative stability in the governing party. The Conservative government has moved considerably from its initial hostility to climate change, especially as embodied in the Kyoto Accord, and has started to embrace the need for the control of greenhouse gases albeit at a different rate and intensity to those agreed under Kyoto. Hopefully, there will be a similar shift in terms of ICOM and Canada's government will move forward to the second phase of implementing its responsibilities under the Oceans Act.

Canada is now well placed to develop and implement a structure and process of IM of coastal and ocean resources in a manner that is most effective and appropriate for the political, legal and constitutional circumstances of the country. The development of the next phase of Canada's OAP is critical to determining whether or not Canada is up to the challenge, and whether Canada will continue to take a global leadership role in ICOM. After a long period of gestation and then an accelerated period of progressive action, it would be tragic if Canada now slipped back from its current position of global leadership. Unfortunately, at the time of writing the jury is still out on this!

1.7 Promoting ICOM to a broader audience

Recently, public opinion has turned once again towards highlighting the environment, and 'recovering politicians' such as former US Vice President Al Gore are promoting action on global warming. Given this resurgence in public support for environmental issues and the need to address them on a global scale, what future lies for ICOM and what role should scientists and researchers play in this movement to mobilise action to address these critical issues?

As scientists of oceans and coasts, we have perhaps the greatest potential to educate the public about the reality of global warming and the importance of recognising the contribution that we as humans are having on global change. Also, global change provides a powerful argument for ICOM given that the world's coastal zones are among the most susceptible to global climate change, be it human induced or otherwise. One might have thought that the major environmental crises that have affected coastal zones over the past 10–15 years would have made a compelling case for ICOM on their own. Such events as the collapse of the northern groundfisheries, increased intensity of hurricanes and cyclones, the Indian Ocean tsunami, the melting of permafrost and the Arctic and Antarctic ice sheets, and the numerous signs of global sea-level rise all have major implications for the world's coastal zones. Yet we still do not have consensus about the need for an effective, global regime for ICOM. Why is that?

The answer lies in the singular failure to translate our scientific knowledge about the threats to oceans and coasts from our current unsustainable approach into a message that can be recognised and embraced by the general public and politicians alike. Those of us who are counted as scientists and researchers have consistently failed to take it upon ourselves to develop a clear and understandable case for the management of our oceans and coasts. We have chosen, instead, to prefer to talk amongst ourselves in a language that inspires great minds and intellect, but says little of meaning to elected politicians and decision-makers and even less to the average citizen. There are many successful efforts underway to change this, but we must do more to ensure that the scientific message gets out to the public, politicians and decision-makers.

One example of a concerted effort to move the public and political ICOM agenda forward is the Coastal Zone Canada Association (CZCA), which was established in 1993 to help bridge the communication gaps between scientists, coastal communities, decision-makers, politicians and the general public. Through its seven biennial Coastal Zone Canada (CZC) conferences to date, the CZCA has attempted to translate the science of coastal management into advice for policy and calls for action. The CZC conferences have a broad and varied mix of participants and have all been designed to stimulate interactive dialogue and understanding across the spectrum of ocean and coastal stakeholder groups.

The CZCA has worked to create a meaningful dialogue between the wide range of stakeholders in the coastal zone. Since its creation in 1993, the Association has seen some success in influencing Canadian public policy on ICOM, and provides one example for other scientific organisations to use their collective expertise to influence public policy. The CZCA is not a lobbying organisation, but by creating real dialogue, developing conference statements and outputs that are directed towards influencing public policy, and through the influence and actions of individual members, the CZCA has been able to be an influence towards developing more effective ICOM policy in Canada. Table 1.3 provides a summary of the outcomes of the CZC conferences till 2006, and indicates some of the progress made towards moving ICOM to a higher priority within the public policy agenda in Canada. In particular, the *Rimouski Declaration* from CZC 1996 was influential in the development and content of the Oceans Act, while the most recent *Tuk Declaration* from CZC 2006 identified many concerns that have subsequently been addressed by the federal government in relation to resource management, climate change impacts and security in Canada's Arctic.

However, despite these advances, we have failed to influence public opinion in a sustained and significant way and remain susceptible to the swings of public support for serious action on environmental issues. In promoting marine research, organisations such as ICES must also engage in the broader debate of influencing public opinion and public policy to promote ICOM. When we come together in scientific gatherings, we should always have as one objective some kind of statement or message to bring to policy makers and the general public. We cannot simply continue to talk to ourselves as scientists and researchers and we must be prepared to engage in the broader debate that we have traditionally left to others. We have a responsibility to take our knowledge and our expertise beyond the limits of our scientific community and into the realm of public debate.

In their analysis of the importance of communications in accomplishing a global marine conservation strategy, Adabia et al. (2004) conclude that the key is to communicate globally and advocate regionally and locally. This involves a multi-scaled approach to marine conservation communications, requiring international, national and local campaigns tailored to reach target audiences in priority regions of the world. The integration and communication of climate and ocean science and policy to governments, decision-makers, civil society and the public is crucial for both the developed world and the developing world in order to build support for the necessary mitigation and adaptation measures, as well as the protection, conservation and sustainable management of our oceans and coasts.

Table 1.3 Outputs of Coastal Zone Canada (CZC) conferences 1994–2006.

CZC year	Location	Problems and issues	Outputs and solutions
CZC 1994	Halifax, NS	Lack of national policy discussion and development on ICOM in Canada Empowerment of communities and stakeholders	Call for action: recommendations on addressing poverty, empowerment, rights of indigenous peoples, integrating traditional knowledge into science
CZC 1996	Rimouski, QC	Lack of national ICOM policy in Canada Lack of follow-up on Agenda 21	Rimouski Declaration on Canada Oceans Policy Call for Canada Oceans Act Declaration for international action
CZC 1998	Victoria, BC	Lack of capacity for communities and stakeholders to participate in ICOM Need to involve youth	ICOM toolkit for communities Statement on youth action and perspectives
CZC 2000	Saint John, NB	Lack of capacity to measure progress in ICOM	Baseline 2000 Beyond 2000 Agenda for ICOM development ICOM toolkit
CZC 2002	Hamilton, ON	Getting freshwater and saltwater communities together Managing transboundary waters	Hamilton statement on managing shared waters Resource kit for managing transboundary coastal ecosystems
CZC 2004	St. John's, NL	Lack of implementation of ICOM components of the Canada Oceans Act	Conference statement on Canada OAP
CZC 2006	Tuktoyaktuk, NWT	Problems affecting Canada's northern coasts and communities Impacts of climate change on Arctic coastal zone	Tuk declaration Call for effective ocean and coastal strategy for the Arctic under OAP Creation of a northern CZC organisation

ICOM, integrated coastal and ocean management; OAP, ocean action plan.

1.8 Who speaks for the oceans?

There is a great need for a concerted global effort for sustainable management and protection of our oceans and coasts. The movement for managing the world's large marine ecosystems (LMEs) is an important force towards global ICOM. But what about the rest of the oceans, most of which is largely unknown? Who speaks for the oceans?

The sad truth is that nobody speaks for the oceans in a truly coherent and effective manner, and even less so for including oceans and coasts together as an essential step towards global sustainability. Defying Ocean's End is one initiative that presents a valuable agenda for global action on sustainable ocean and coastal management (Glover and Earle, 2004). Focusing on six key areas – ocean governance, marine protected areas, communications, the land–ocean interface, fisheries reform and aquaculture, and global science, the Defying Ocean's End initiative presents an agenda for action to build upon existing global initiatives in order to bring about a 'sea change' in awareness, action and behaviour towards the oceans. We need to link these with other global efforts, such as the Global Forum on Islands, Oceans and Coasts (including their new Strategic Oceans Planning Initiative), the Intergovernmental Oceanographic Commission and the World Ocean Network, but above all we need action and we need it sooner rather than later. The time has come for the United Nations to establish a Commission on Ocean and Coastal Management in order to bring ICOM to the global political agenda and to provide the necessary support to regional, national and sub-national efforts to manage the world's ocean and coastal resources on a sustainable basis. The aforementioned relative success in the area of biodiversity can be directly linked to support structures provided under the Convention on Biological Diversity. Furthermore, it is time to join forces with the IPCC in order to highlight the enormous implications of climate change on oceans and coasts, and to start taking serious efforts to reverse the path to oblivion and mitigate the impacts on coastal populations. If ICOM was not important enough to have as a UN global initiative before the issue of human-induced climate change emerged, then surely it is now. As Gordon Moore and Sylvia Earle state:

> Our collective challenge is nothing less than the creation of a framework and steps for a practical agenda of global action – including coasts and impacts – to safeguard the ocean for generations to come.
>
> (Glover and Earle, 2004)

Do we live in a state of fear or a state of oblivion? Or do we want to live in a state of awareness, reason and responsibility? Do we even think it our responsibility as scientists to do anything about it? The challenge of this chapter is that it is our responsibility and that we must care enough to take action collectively. The reality is that we must speak for the oceans and coasts – because if not us, then who?

References

Adabia, R., Day, B., Knowlton, N., McCosker, J.E., Baron, N., Katoppo, A. and Hough, H. (2004) Defying ocean's end through the power of communications. In: Glover, L.K. and Earle, S.A. (eds). *Defying Ocean's End: An Agenda for Action*. Island Press, Washington, DC, pp. 183–196.

Auditor General of Canada (2005) *Report of the Commissioner of the Environment and Sustainable Development, Chapter 1, Fisheries and Oceans Canada – Canada's Oceans Management Strategy*. Office of the Auditor General of Canada, Minister of Public Works, Ottawa.

Brown, D. (2003) *The Da Vinci Code*. Doubleday, New York, 454 pp.

Brundtland, G.H. (1987) Our common future. *Report of the World Commission on Environment and Development*. Oxford University Press, Oxford.

Cameron, S.D. (1998) *The Living Beach*. Macmillan, Toronto.

Carson, R. 1951. *The Sea Around Us*. Oxford University Press, New York.

Cicin-Sain, B., Vandeweerd, V., Bernal, P.A., Williams, L.C., Balgos, M.C. and Barbiere, J. (eds) (2006a) *Meeting the Commitments on Oceans, Coasts, and Small Island Developing States Made at the 2002 World Summit on Sustainable Development: How Well Are We Doing?* Co-Chairs' Report – Volume 1. Third Global Conference on Oceans, Coasts and Islands, UNESCO, Paris, June.

Cicin-Sain, B., Vandeweerd, V., Bernal, P.A., Williams, L.C., Balgos, M.C. and Barbiere, J. (eds.) (2006b) *Reports from the Third Global Conference on Oceans, Coasts, and Islands: Moving the Global Oceans Agenda Forward*. Co-Chairs' Report – Volume 2. Third Global Conference on Oceans, Coasts, and Islands, UNESCO, Paris, June.

Crichton, M. (2004) *State of Fear*. HarperCollins, New York, 640 pp.

DFO (1998) *Toward Canada's Ocean Strategy*. Department of Fisheries and Oceans, Ottawa.

DFO (2002) *Canada's Oceans Strategy: Our Oceans, Our Future*. Department of Fisheries and Oceans, Ottawa.

Flannery, T. (2006) *The Weather Makers*. HarperCollins, New York, 274 pp.

Glover, L.K. and Earle, S.A. (eds) (2004) *Defying Ocean's End: An Agenda for Action*. Island Press, Washington, DC, 283 pp.

Gore, A. (2006) *An Inconvenient Truth*. Rodale, New York, 325 pp.

Government of Canada (1996) *Canada Oceans Act, RSC 1996: Bill C-26*, Chapter 31, 2nd Session, 35th Parliament, 45, Elizabeth 2.

Guénette, S. and Alder, J. (2007) Lessons from marine protected areas and integrated ocean management initiatives in Canada. *Coastal Management* **35**, 51–78.

IPCC (2007a) Summary for policymakers. In: Solomon, S., Qin, D., Manning, M., Chen, Z., Marquis, M., Averyt, K.B., Tignor, M. and Miller, H.L. (eds). *Climate Change 2007: The Physical Science Basis of Climate Change. Contribution of Working Group I to the Fourth Assessment Report of the Intergovernmental Panel on Climate Change*. Cambridge University Press, Cambridge.

IPCC (2007b) Summary for policymakers. In: Solomon, S., Qin, D., Manning, M., Chen, Z., Marquis, M., Averyt, K.B., Tignor, M. and Miller, H.L. (eds). *Climate Change 2007: The Physical Science Basis. Contribution of Working Group I to the Fourth Assessment Report of the Intergovernmental Panel on Climate Change*. Cambridge University Press, Cambridge.

IPCC (2007c) Summary for policymakers. In: Parry M.L., Canziani, O.F., Palutikof, J.P., van der Linden, P.J. and Hanson, C.E. (eds). *Climate Change 2007: Impacts, Adaptation and Vulnerability. Contribution of Working Group II to the Fourth Assessment Report of the Intergovernmental Panel on Climate Change*. Cambridge University Press, Cambridge, pp. 7–22.

Manson, G.K. (2005) On the coastal populations of Canada and the world. *Proceedings of the Canadian Coastal Conference 2005*. Canadian Coastal Science and Engineering Association, Ottawa, Canada.

Nicholls, R.J., Wong, P.P., Burkett, V.R., Codignotto, J.O., Hay, J.E., McLean, R.F., Ragoonaden, S. and Woodroffe, C.D. (2007) Coastal systems and low-lying areas. In: Parry, M.L., Canziani, O.F., Palutikof, J.P., van der Linden, P.J. and Hanson, C.E. (eds). *Climate Change 2007: Impacts, Adaptation and Vulnerability. Contribution of Working Group II to the Fourth Assessment Report of the Intergovernmental Panel on Climate Change*. Cambridge University Press, Cambridge, pp. 315–356.

OECD (1993) *Coastal Zone Management: Integrated Policies*. Organization for Economic Cooperation and Development, Paris.

Ricketts, P. and Harrison, P. (eds) (2007a) *Coastal and Ocean Management in Canada: Moving into the 21st Century, Coastal Management*, Theme Issue, Vol. 35, No. 1. Taylor & Francis, Philadelphia, 162 pp.

Ricketts, P. and Harrison, P. (2007b) Coastal and ocean management in Canada: moving into the 21st century. *Coastal Management* **35**, 5–22.

Schubert, R., Schellnhuber, H-J., Buchmann, N., Epiney, A., Griesshammer, R., Kulessa, M., Messner, D., Rahmstorf, S. and Schmid, J. 2006. *The Future Oceans – Warming Up, Rising High, Turning Sour*. Special Report, German Advisory Council on Climate Change, Berlin, 110 pp.

Symon, C., Arris, L. and Heal, B. (eds) (2005) *Arctic Climate Impact Assessment – Scientific Report, Arctic Climate Impact Assessment*. Cambridge University Press, Cambridge, 1046 pp.

Tompkins, E.L., Nicholson-Cole, S.A., Hurlston, L., Boyd, E., Brooks Hodge, G., Clarke, J., Gray, G., Trotz, N. and Varlack, L. (2005) *Surviving Climate Change in Small Islands: A Guidebook*. Tyndall Centre for Climate Change Research, School of Environmental Sciences, University of East Anglia, Norwich, 128 pp.

UN (1982) *United Nations Convention on the Law of the Sea*. United Nations, New York.

UN (1992) *Protection of the Oceans, All Kinds of Seas, Including Enclosed and Semi-Enclosed Seas, and Coastal Areas and the Protection, Rational Use and Development of Their Living Resources*. United Nations, New York, Chapter 17, Agenda 21.

World Bank (1993) *Noordwijk Guidelines for Integrated Coastal Zone Management*. World Coast Conference, Noordwijk, The Netherlands.

Worm, B., Barbier, E.B., Beaumont, N., Duffy, J.E., Folke, C., Halpern, B.S., Jackson, J.B.C., Lotze, H.K., Micheli, F., Palumbi, S.R., Sala, E., Selkoe, K.A., Stachowicz, J.J. and Watson, R. (2006) Impacts of biodiversity loss on ocean ecosystem services. *Science* **314**, 787–790.

Section 1
Coastal Habitats

Chapter 2
The Challenge towards Sustainable Utilisation of Coastal Fish Resources

Josianne G. Støttrup

Abstract

The marine coastal environment harbours some of the richest marine biodiversity, but is under threat from increasing and often conflicting pressures. Major threats include global warming, destructive land uses, industrial pollution and eutrophication. The challenge for this generation is to establish practices for the sustainable utilisation of the resources that are adaptable to changing baselines. In particular, progress on integrated coastal zone management (ICZM) objectives is needed to manage human activities to enable the preservation of coastal fish stocks, sustain coastal and recreational fishery and to preserve those species' contribution to a healthy ecosystem. The concept of ICZM has been introduced in most European countries, but in practice, the fragmented administrative and management systems do not provide a framework for communication between authorities. Spatial planning, using geographic information system maps, has gained popularity as an ICZM tool, but its increasing use has uncovered the urgent need to develop habitat maps for coastal resources for their protection or sustainable exploitation.

2.1 Introduction

Coastal and estuarine marine areas are highly productive areas with high biodiversity. Shellfish are directly exploited from the coastal areas (MacKenzie, 2004) or produced in extensive aquaculture, where the natural phytoplankton product is utilised (Kennedy and Roberts, 1999). Both commercial and recreational fisheries of most species rely on fish recruits produced in nursery areas. They are highly diverse from the structurally complex tropical reefs and mangroves, temperate seagrass meadows, kelp forests and shellfish reefs to less structurally complex habitats such as mud flats and sandy shores. These coastal systems also provide many services such as nutrient cycling, natural barriers to coastal erosion, pollution filters, water quality control and recreational uses. Tropical habitats have been the subject of several reviews aimed at examining the extent of habitat loss in tropical regions such as coral reefs (Pandolfi et al., 2003) and mangroves (Alongi, 2002). Increasing awareness and concern about the degradation of natural habitats such as seagrasses and shellfish reefs is reflected in a number of reviews on the subjects (Shepherd et al., 1989; Short and Wyllie-Echeverria, 1996; Duarte, 2002; Steneck et al., 2002; Thompson et al., 2002; Lotze et al., 2006).

In 2001, an estimated 19% of the European population inhabited the coastal zone (Carreau and Gallego, 2006), 10% of Europe's coastlines are already artificial and less than 15% of the European coastline was in 'good' condition (EEA, 2006). Continuing degradation and fragmentation of coastal habitats pose a serious threat to biodiversity and future resource availability (EEA, 2006; Airoldi and Beck, 2007). Human activities often result in loss of structural complexity, but as they cannot be measured in terms of lost habitat their impact may be overlooked. Since many of these activities have taken place over decades or centuries, attempts to study human impacts may be confounded by shifting baselines.

The chapter focuses on progress and obstacles concerning integrated coastal zone management (ICZM) objectives to manage human activities to enable the preservation of fish stocks for sustainable coastal and recreational fisheries, preserve that species' contribution to a healthy ecosystem and sustain biodiversity within a system with shifting baselines.

2.2 Coastal juvenile fish habitats

Both the structurally complex rocky coastlines and mudflats or sandy coastlines are important nursery areas for several fish species. Seventy-five per cent of the world's fish species important for commercial or recreational fishery are dependent on the coastal zone or coastal waters at some stage in their life cycle (Chambers, 1992). Flatfishes, in particular, utilise the coastal areas during the summer growth seasons during the first 2–3 years and spend a smaller or larger portion of the summer months in the coastal zone (Gibson et al., 2002). The most environmentally sensitive coastal habitats are those in which a resource (such as flatfish nursery areas or eelgrass habitats) is mainly restricted to a few sites or spatially restricted in size (Parrish et al., 1997; Dean et al., 2000; Le Pape et al., 2003). Important sole, *Solea solea* (L.), nursery grounds in the Bay of Biscay, France, were shown to be highly restricted, shallow, muddy, estuarine areas (Le Pape et al., 2003). The factors controlling the suitability of these nursery areas were the sediment preference of the sole juveniles, which coincided with an ample food supply, and transport mechanisms during the planktonic stage that supply the nursery grounds with larvae. A positive relationship between nursery areas size and average year-class strength was found in that study.

The size of nursery habitats in other regions was also demonstrated closely related to recruitment for sole (Rijnsdorp et al., 1992) as well as for other flatfish species such as plaice (*Pleuronectes platessa* L.), flounder (*Platichthys flesus* L.), dab (*Limanda limanda* L.), turbot (*Psetta maxima* L.) and brill (*Scophthalmus rhombus* L.) (Gibson, 1994; van der Veer et al., 2000). The size of the habitat for these species was thus influenced by the depth distribution of the juveniles and the bathymetry of the nursery areas, in addition to sediment type and transport processes. These factors determining the spatial extent of essential habitats for juvenile fish are useful for ICZM. Mapping these areas is not only an important task, but also a complicated one. Geographic information system (GIS) maps on essential fish habitats (EFHs) can be used to identify where anthropogenic activity may have serious consequences for a particular stock. In some cases, these areas may already be protected from human activities as they coincide with Natura 2000 sites, but without maps that identify those areas vital for fish recruitment, they will not be ensured the necessary protection (ICES, 2007). Although it may be relatively easy to set boundaries for important habitats of species with well-defined habitat requirements such as nursery grounds for plaice, flounder, turbot and sole, essential habitats for species, such as cod (*Gadus morhua* L.) or dab that have a more general distribution, may be more difficult to map (ICES, 2004).

Small local populations may be particularly vulnerable to adverse local system changes. Distinct populations of coastal cod have been demonstrated at short distances from each other, and do not interbreed despite ample opportunity for mixing (Knutsen et al., 2003). In many cases, spawning occurs within the fjords and the pelagic offspring are largely retained within the fjord due to hydrodynamic processes preventing egg or larval drift among fjords (Knutsen et al., 2007). Since these systems depend on local recruitment, both spawning and nursery areas need to be preserved to ensure effective management. Further, this adds a new challenge to traditional fisheries management to manage these small-scale stocks to prevent overfishing.

Seagrass beds are under decline worldwide due to anthropogenic pressures (Shepherd et al., 1989). Tropical habitats in particular have been well studied, and goods and services provided by seagrass meadows are described in several studies (Parrish, 1989; Robertson and Blaber, 1992; Blaber, 1997). Apart from providing important habitats for fish, they play an important role in sustaining biological diversity. Seagrasses play an important role in dissipating wave energy and anchoring sediments. In temperate regions, the role of seagrass meadows and other important coastal ecosystems have been the topic of a few studies (Mathieson et al., 2000; Nagelkerken and van der Velde, 2002, 2004). Further research on habitat utilisation by fishes in different types of temperate coastal habitats is needed for a better understanding of the processes that structure the fish community and to evaluate the importance of these habitats for fish species.

The concept of EFHs was introduced through the Sustainable Fisheries Act in the US in the mid-1990s. EFH is defined as 'those waters and substrate necessary to fish for spawning, breeding, feeding or growth to maturity'. This concept was reviewed in Europe by the ICES (International Council for the Exploration of the Sea) Working Group on Fish Ecology in 2003, where nursery grounds were identified among the list of EFHs (ICES, 2003). Since most of these nursery grounds are coastal and major impacts of human activities concentrate in coastal areas, it should be logical that priority should be given to coastal nursery grounds. However, few studies have examined the importance of nursery grounds for fish stocks and much of the work towards mapping or preserving coastal habitats is fragmentary. Furthermore, focus on commercial species by fishery managers and red-listed species by nature conservation manager draws away attention from fish species important to recreational fishery or biodiversity in general. The challenge here is to ensure nursery grounds for commercially important species and other species to continue to support a sustainable fishery as well as protect and support biodiversity.

Human impacts on coastal habitats (seagrass habitats, sandy coasts, mud flats or other) may be a slow degradation process that does not result in a spatial loss of habitats but nevertheless decreases their value. Examples include dredged oyster reefs or mussel beds that result in mudflats (Beck et al., 2001; Dolmer, 2002), the removal of large boulders for harbour construction (Støttrup, 1999) or the regular shore nourishment for coastal protection purposes (Støttrup et al., 2005, 2007). These activities may result in less structurally complex habitats in the case of dredging activities, or more unstable habitats. Often the result is a different kind of habitat, which may also be of some value but with a different function. This type of human impact may be difficult to measure and is often not treated as a loss. Documentation of the negative impacts may be difficult to provide in those cases where the impact is a gradual change of structurally complex habitats to less complex ones. The changes occurring may be a slow process of increasing habitat fragmentation or the impact may cover a geographically broader area than expected, effecting possible reference sites as much as impact sites (Støttrup et al., 2005, 2007). Further, activities that have been taking place over several decades may be difficult to terminate without considerable evidence regarding their relative negative impact. The proof of burden lies on the scientists' ability to prove negative impacts from human activities that may have been conducted over several decades and with the added problem of shifting baselines.

2.3 Spatial planning and fishing activity

The preferred tool for ICZM in many coastal systems is GIS maps that map activities (pipelines, marine windmills, harbours) and protected areas (Natura 2000 sites, Ramsar areas) or maps indicating densities and occurrence of bottom fauna and flora (seagrass meadows, mussel beds, distribution and occurrence of fish of different age groups). GIS maps are implemented for prioritising areas for particular activities. Coastal zone planning therefore requires GIS maps for all activities, areas and resources. This was, for example, the case in a Danish fjord where GIS maps were used to improve the sustainable production of cultured mussels (Dolmer and Geitner, 2004). Areas were mapped that were unavailable for production of mussels due to harbours, pollution or restricted for mussel production due to conservation or for recreational purposes. The maps also included important fishing grounds for wild mussels and other fisheries. In most cases, as was also demonstrated in the Danish fjord example, maps on fishing or recreational activities were unavailable and had to be constructed.

Increasing spatial planning will inevitably increase conflicts with activities which have historically enjoyed free access to these resources. Historically, fishermen may fish in all marine areas with few restrictions. The severe decline of many fish stocks in recent decades, partly or largely due to overfishing, has resulted in novel fisheries restrictive measures in addition to traditional catch restrictions. Spatial restrictions for fishery are steadily increasing worldwide. The introduction of marine protected areas (MPA) (Dayton et al., 2000) includes an increasing number of no-take zones (Fletcher, 2004) and seasonal 'boxes' such as the North Sea plaice box and the Baltic cod box, where fishery is not allowed or prohibited during certain months (http://ec.europa.eu/fisheries/press_corner/press_releases/archives/com05/com05_46_en.htm, accessed April 2008; http://www.mmm.fi/en/index/frontpage/Fishing,_game_reindeer/Fisheriesindustry/Fisheries_control/action_plan_2008.html, accessed April 2008). Other spatial restrictions are introduced to ensure safety, such as within and around windmill farms, along pipelines or in the proximity of aquaculture cages. In most cases, maritime space is requested by private enterprise for a specific activity and the fishermen need to defend their rights for fishing within that particular area. Unless they can make a good case in demonstrating the area economically important for their fishery, and in the absence of objections from other sectors, yet another area is closed for fishing. In this way, spatial restriction of fishery evolves in a random fashion and may or may not result in sub-optimal fishing opportunity for fishermen, impacting the economic/socioeconomic value of that activity. A challenge is to change this process to map the economically important areas for fishery and reserve these areas for this activity as suggested by the ICES Working Group on ICZM (ICES, 2007). It would require the development of a framework for sustainable fishery with fishing rights for specific areas taking into consideration fish population dynamics within that ecosystem, and perhaps allow stakeholders involvement in management, such as area-based co-management. An example of such an approach is to be found in Sweden in their so-called co-management initiatives where commercial and recreational fishermen met with local and regional governments, NGOs and other stakeholders within an area to discuss problems and define management objectives for problems as they arise (ICES, 2007).

Spatial planning impacts both commercial and recreational fishery. Not all leisure and recreational fishermen are organised, and they may not have a strong organisation that can collectively ensure their interests. In the scramble for limited coastal space, lack of maps that reserve areas for recreational fishery can be a source of constant conflict with various new initiatives that seek permission for alternative activities. The objections of a few leisure or recreational fishers may be overruled in the face of strong economic interests or strong organisations. Few valuations of the recreational fisheries have been conducted despite its popularity, pressure on the resources and social/cultural or economic value. Increasing resource depletion and increasing conflicts with commercial fishers may threaten the future of this activity (Sutinen and Johnston, 2003). They suggested integration of recreational sector into fishery management to resolve some of these conflicts and treats. However, coastal recreational fishery is also threatened by effects from anthropogenic activity within the coastal zone; this is the case in Denmark with deteriorating water quality due to pollution and eutrophication (Toivonen et al., 2004). As in this case for commercial fishermen, mapping important recreational fishing areas may help preserve these small cultural clusters of society (Hansen, 1996) and ensure their inclusion in coastal zone planning.

2.4 Conclusions and perspectives

Recognition of the impact of spatial planning within ICZM prompted mapping of resources and activities, but these maps should include all resources including EFHs and activities. Evaluations of the importance of different resources and activities for ecosystems, economics and society are important components of ICZM. Methodology for impact studies for activities which have taken place over decades need to be developed to encompass the concept of changing baselines. Top-down management of local resources should change towards co-management to increase accountability. Communication should be prioritised especially between users (stakeholders, citizens, etc.), scientists and managers and possibly enhanced through area-based co-management initiatives. Finally, ICZM is not a one-on-one decision, but a continuous process. Such a process requires flexibility to allow for adjustment of policies, perceptions and incorporation of new information and data.

References

Airoldi, L. and Beck, M.W. (2007) Loss, status and tends for coastal marine habitats of Europe. In: Gibson, R.N., Atkinson, R.J.A., Gordon, J.D.M. (eds). *Oceanography and Marine Biology*. An Annual Review, vol. 45. Taylor & Francis, UK, pp. 345–405.

Alongi, D.M. (2002) Present state and future of the world's mangrove forests. *Environmental Conservation* **29**, 331–349.

Beck, M.W., Heck, K.L., Able, K.W., Childers, D.L., Eggleston, D.B., Gillanders, B.M., Halpern, B., Hays, C.G., Hoschino, K., Minello, T.J., Orth, R.J., Scheridan, P.F. and Weinstein, M.P. (2001) The identification, conservation, and management of estuarine and marine nurseries for fish and invertebrates. *BioScience* **51**, 633–641.

Blaber, S.J.M. (1997) *Fish and Fisheries of Tropical Estuaries*. Chapman and Hall, London.

Carreau, P.R. and Gallego, F.J. (2006) *EU25 Coastal Zone Population Estimates from the Disaggregated Population Density Data 2001*. European Commission, DG Joint Research Centre.

Chambers, J.R. (1992) Coastal degradation and fishery declines in the U.S. In: Stroud, R.H. (ed.). *Stemming the Tide of Coastal Fish Habitat Loss*. National Coalition for Marine Conservation, Savannah, Georgia, pp. 45–51.

Dayton, P., Sala, E., Tegner, M. and Thrush, S. (2000) Marine reserves: parks, baselines, and fishery enhancement. *Bulletin of Marine Science* **66**, 617–635.

Dean, T.A., Haldorson, L., Laur, D.R., Jewett, S.C. and Blanchard, A. (2000) The distribution of nearshore fishes in kelp and eelgrass communities in Prince William Sound, Alaska: associations with vegetation and physical habitat characteristics. *Environmental Biology of Fishes* **57**, 271–287.

Dolmer, P. (2002) Mussel dredging: impact on epifauna in Limfjorden, Denmark. *Journal of Shellfish Research* **21**, 529–537.

Dolmer, P. and Geitner, K. (2004) Integrated Coastal Zone Management of cultures and fishery of mussels in Limfjorden, Denmark. International Council for the Exploration of the Sea (ICES) CM 2004/V:07, 9 pp.

Duarte, C.M. (2002) The future of seagrass meadows. *Environmental Conservation* **29**, 192–206.

EEA (2006) The changing faces of Europe's coastal areas. European Environmental Agency Report 6, 91 pp.

Fletcher, K.M. (2004) Legal challenges and issues for no-take zones. *American Fisheries Society Symposium* **42**, 267–273.

Gibson, R.N. (1994) Impact of habitat quality and quantity on the recruitment of juvenile flatfishes. *Netherlands Journal of Sea Research* **32**, 191–206.

Gibson, R.N., Robb, L., Wennhage, H. and Burrows, M.T. (2002) Ontogenetic changes in depth distribution of juvenile flatfishes in relation to predation risk and temperature on a shallow-water nursery ground. *Marine Ecology Progress Series* **229**, 233–244.

Hansen, K.M. (1996) Kampen om Limfjorden. DFU-report 7–96, 114 pp.

ICES (2003) Report of the Working Group on Fish Ecology. ICES CM 2003/G:04, 110 pp.

ICES (2004) Report of the Working Group on Fish Ecology. ICES CM 2004/G:09, 257 pp.

ICES (2007) Report of the Working Group on Integrated Coastal Zone Management. ICES CM 2007/MHC:09, 69 pp.

Kennedy, R.J. and Roberts, D. (1999) A survey of the current status of the flat oyster *Ostrea edulis* in Strangford Lough, Northern Ireland, with a view to the restoration of its oyster beds. *Biology and Environment: Proceedings of the Royal Irish Academy* **99**, 79–88.

Knutsen, H., Jorde, P.E., André, C. and Stenseth, N.C. (2003) Fine-scaled geographical population structuring in a highly mobile marine species: the Atlantic cod. *Molecular Ecology* **12**, 385–394.

Knutsen, H., Olsen, E.M., Ciannelli, L., Espeland, S.H., Knutsen, J.A., Simonsen, J.H., Skreslet, S. and Stenseth, N.C. (2007) Egg distribution, bottom topography and small-scale cod population structure in a coastal marine system. *Marine Ecology Progress Series* **333**, 249–255.

Le Pape, O., Chauvet, F., Mahévas, S., Lazure, P., Guérault, D. and Désaunay, Y. (2003) Quantitative description of the habitat suitabililty for the juvenile common sole (*Solea solea* L.) in the Bay of Biscay (France) and the contribution of different habitats to the adult population. *Journal of Sea Research* **50**, 139–150.

Lotze, H.K., Lenihan, H.S., Bourque, B.J., Bradbury, R.H., Cooke, R.G., Kay, M.C., Kidwell, S.M., Kirby, M.X., Peterson, C.H. and Jackson, J.B.C. (2006) Depletion, degradation, and recovery potential of estuaries and coastal seas. *Science* **312**, 1806–1809.

MacKenzie, C.L. Jr. (2004) The slow development of culture (aquaculture) of shellfish and finfish in estuaries of eastern North America and its future. *Journal of Shellfish Research* **23**, 639–640.

Mathieson, S., Catrijsse, A., Costa, M.J., Drake, P., Elliott, M., Gardner, J. and Marchand, J. (2000) Fish assemblages of European tidal marshes: a comparison based on species, families and functional guilds. *Marine Ecology Progress Series* **204**, 225–242.

Nagelkerken, I. and van der Velde, G. (2002) Do non-estuarine mangroves harbour higher densities of juvenile fish than adjacent shallow-water and coral reef habitats in Curacao (Netherlands Antilles)? *Marine Ecology Progress Series* **245**, 191–204.

Nagelkerken, I. and van der Velde, G. (2004) A comparison of fish communities of subtidal seagrass beds and sandy seabeds in 13 marine embayments of a Caribbean island, based on species, families, size distribution and functional groups. *Journal of Sea Research* **52**, 127–147.

Pandolfi, J.M., Bradbury, R.H., Sala, E., Hughes, T.P., Bjorndal, K.A., Cooke, R.G., McArdle, D., McClenachan, L., Newman, M.J.H., Paredes, G., Warner, R.R. and Jackson, J.B.C. (2003) Global trajectories of the long-term decline of coral reef ecosystems. *Science* **301**, 955–958.

Parrish, F.A., De Martini, E.E. and Ellis, D. (1997) Nursery habitat in relation to production of juvenile snapper, *Pristipomoides filamentosus* in the Hawaiian Archipelago. *Fishery Bulletin* **95**, 137–149.

Parrish, J.D. (1989) Fish communities of interacting shallow-water habitats in tropical oceanic regions. *Marine Ecology Progress Series* **58**, 143–160.

Rijnsdorp, A.D., Van Beek, F.A., Flatman, S., Millner, R.M., Giret, M. and De Clerck, R. (1992) Recruitment of sole stocks, *Solea solea* (L.), in the northeast Atlantic. *Netherlands Journal of Sea Research* **29**, 173–192.

Robertson, A.I. and Blaber, S.J.M. (1992) Plankton, epibenthos and fish communities. In: Robertson, A.I. and Alongi, D.M. (eds). *Tropical Mangrove Ecosystems. Coastal and Estuarine Studies 41*. American Geophysical Union, Washington, DC, pp. 173–224.

Shepherd, S.A., McComb, A.J., Bulthuis, D.A., Neverauskas, V., Steffensen, D.A. and West, R. (1989) Decline of seagrasses. In: Larkum, A.W.D., McComb, A.J. and Shepherd, S.A. (eds). *Biology of Seagrasses*. Elsevier, Amsterdam, pp. 346–393.

Short, F.T. and Wyllie-Echeverria, S. (1996) Natural and human-induced disturbance of seagrasses. *Environmental Conservation* **23**, 17–27.

Steneck, R.S., Graham, M.H., Bourque, B.J., Corbett, D., Erlandson, J.M., Estes, J.A. and Tegner, M.J. (2002) Kelp forest ecosystems: biodiversity, stability, resilience and future. *Environmental Conservation* **29**, 436–459.

Støttrup, J.G. (ed.) (1999) Mapping reefs, stone and gravel extraction and hard-bottom fishery, and description of methods for scientific studies on reefs or hard-bottom substrates (in Danish with English summary). DFU-report 63–99, 63 pp.

Støttrup, J.G., Dolmer, P., Røjbek, M., Nielsen, E., Ingvardsen, S., Laustrup, C. and Sørensen, S.R. (2005) Coastal nourishment and sustainable fishery (in Danish with English summary). DFU-report 156–05, 48 pp.

Støttrup, J.G., Dolmer, P., Røjbek, M., Nielsen, E., Ingvardsen, S., Sørensen, P. and Sørensen, S.R. (2007) Coastal nourishment and coastal ecology – an evaluation of near-shore nourishment near Fjaltring (in Danish with English summary). DFU-report 171–07, 46 pp.

Sutinen, J.G. and Johnston, R.J. (2003) Angling management organizations: integrating the recreational sector into fishery management. *Marine Policy* **27**, 471–487.

Thompson, R.C., Crowe, T.P. and Hawkins, S.J. (2002) Rocky intertidal communities: past environmental changes, present status and predictions for the next 24 years. *Environmental Conservation* **29**, 168–191.

Toivonen, A.-L., Roth, E., Navrud, S., Gudbergsson, G., Appelblad, H., Bengtsson, B. and Tuunainen, P. (2004) The economic value of recreational fisheries in Nordic countries. *Fisheries Management and Ecology* **11**, 1–14.

van der Veer, H.W., Geffen, A.J. and Witte, J.I.J. (2000) Exceptionally strong year classes in plaice *Pleuronectes platessa*: are they generated during the pelagic stage only, or also in the juvenile stage? *Marine Ecology Progress Series* **199**, 255–262.

Chapter 3
Evaluating the Geomorphologic Stability of an Estuarine Sandy Beach

Filipa S.B.F. Oliveira and Catarina I.C. Vargas

Abstract

The hydrodynamics and sediment dynamics of a low-energy estuarine beach was analysed. The beach is subjected to local wind waves generated in an area of restricted fetch, wake waves generated by catamarans and a semi-diurnal meso-tidal regime. Only storm events of short duration modify the beach profile, which, once the normal hydrodynamic conditions are restored, naturally recovers its initial shape. The beach is characterised by having a steep upper slope until mean sea level, followed, seaward, by a low-gradient terrace, and a bimodal sediment distribution. To investigate the hydro-sedimentologic beach behaviour, wind, topo-hydrographic and sedimentologic data were used as input and validation of process-based numerical modelling. Wave generation and transformation in an area of restricted fetch, nearshore circulation, sediment transport and morphological evolution were simulated for average annual conditions and storm conditions. The statistical analysis of a 6-year wind data series allowed to derive the average wind regime, based on which the average annual wave regime and the average annual longshore sediment budget, 14.5×10^3 m^3 year^{-1}, were calculated. The contribution of the individual components of the representative wave regime, discretised by directional sector and height class of incidence, was evaluated as well as the spatial distribution of the longshore sediment transport in the active part of the beach profile. The characteristics of the wave groups generated by the passage of catamarans at different speeds were estimated and their action on the beach morphology was simulated. Although when the catamarans travel at 20 knots speed the average annual wake wave energy dissipated at the beach is 2.5 times higher than the average annual wind wave energy, the erosion effect on the beach profile is still not relevant for the present traffic. In opposition, short-duration storm events generate the formation of an erosion scarp at the upper part of the beach face. The numerical modelling of this phenomenon allowed to acknowledge on the protective effect that the low-gradient terrace has on the beach face.

3.1 Introduction

Low-energy beaches are generally not studied despite their heritage value but the recent increase of recreational use enhances the need to understand the sediment dynamics of these particular coastal systems to execute suitable preservation, protection or even rehabilitation interventions.

The distinct concepts of low-energy beach and fetch-limited beach were discussed by Jackson et al. (2002) and Goodfellow and Stephenson (2005). Jackson et al. (2002) suggested that the term low energy should be used in locations where (1) non-storm significant wave heights are minimal (<0.25 m), (2) significant wave heights during strong onshore winds are low (<0.50 m), (3) beach face widths are narrow (<20 m in micro-tidal environments) and (4) morphologic features include those inherited from higher energy events. Low-energy beaches are located in sheltered and/or fetch-limited environments and have little cross-shore sediment exchange. The present case study fits into both concepts: it is a low-energy estuarine beach sheltered from oceanic waves and it is dependent on local wind and area of fetch for wave generation–propagation.

Nordstrom and Jackson (1992) observed two general types of profile response in low-energy beaches, each associated with one of the two directional components of the sediment transport, alongshore and cross-shore. In response to storm waves, the sediment is removed from the upper foreshore and deposited in the lower foreshore changing the profile to a concave upward profile. During post-storm recovery, sediment is transported up to the foreshore resulting in restoration of the previous foreshore slope. The other type of response, result of longshore, rather than cross-shore, sediment transport, results in the vertical displacement of the entire foreshore while the slope is maintained. In the present case study, only the first type of profile modification was observed in the last decade.

Longshore transport rates on estuarine beaches have been inferred from sediment accumulations and dredging volumes or calculated over the long term, yielding annual rates within the range 17×10^3 to 47×10^3 m^3 year^{-1} for energetic beaches in the West and East Coast of USA, but lower rates ($<10 \times 10^3$ m^3 year^{-1}) for more sheltered beaches (Nordstrom et al., 2003). Longshore sediment transport rates can be determined using field measurements on the basis of techniques like sediment tracers, short-term impoundment or streamer traps, all of which have advantages and limitations (Wang et al., 1998). Nordstrom et al. (2003) estimated the longshore sediment transport rate on a micro-tidal estuarine beach in Great South Bay, USA, during two dyed sand tracer experiments using a temporal sampling method. These rates were 3.1 and 6.5 times greater than those predicted by three existing empirical formulations using standard coefficients. The authors attributed these greater rates to the concentration of sediment transport in the energetic swash zone under plunging breakers. In fact, the application of the existing formulas implies that a greater proportion of sediment is transported in the surf zone, which is not the case of estuarine beaches.

The aim of this study was to evaluate short- and medium-term geomorphologic stability of a low-energy and fetch-limited estuarine sandy beach through the quantification of the natural modelling agents and numerical simulation and verification of the beach response.

3.2 Material and methods

The Tagus estuary, of about 320 km^2, extends from the Tagus mouth until a location named Vila Franca de Xira, where the salinity is negligible (at normal hydrological conditions). The estuary has two main parts: the channel, deep and narrow, with main alignment ENE–WSW;

and the internal estuary, shallow and large, with main alignment NNE–SSW (Figure 3.1). The internal estuary has a maximum width of 15 km in the alignment of the predominant wind, favouring the establishment of a sandy beach, named Alfeite, in the south bank of the internal estuary.

Alfeite beach, with main alignment WNW–ESE, is characterised by having a steep upper slope (average 0.15) until mean sea level (MSL), followed, seaward, by a low-gradient terrace of approximately 800 m width. The sediment grain size distribution is bimodal: coarse sand grains ($D_{50} = 1.26$ mm, where D_{50} is the mean grain diameter) at the upper beach and finer grains ($D_{50} = 0.33$ mm) at the low-gradient terrace. Both the steep upper beach and the low-gradient terrace are engaged in the surf zone at different states of the tide, therefore, the beach responds as a reflective beach and as a dissipative beach. The Tagus estuary is subjected to a semi-diurnal astronomic tidal cycle and a meso-tidal regime, according to the classification established by Davies (1964). At Alfeite beach, the average tidal range at spring tide is 3.2 m and at neap tide is 1.5 m. The alongshore extension of the beach is about 2.5 km.

Freire (1999) studied the evolution of the coastline at Alfeite beach since 1849, through the comparison of surveys of different date, and concluded that the main modifications occurred at the ESE beach extreme and in the beach main alignment. Initially, the coastline presented a rectilinear shape, which, in 1979, had changed into a slightly curved line. The present configuration of Alfeite beach is maintained since the survey of 1979. The present study also included the collection, treatment and analysis of field data: topo-hydrographic and sedimentologic surveys were performed to characterise the morphology and sediment size sorting. Several topo-hydrographic surveys performed along cross-shore transects, covering the full alongshore extension of the beach, were compared with surveys performed by Freire (1999) in the last decade. The results allowed us to conclude on the three-dimensional morphological stability of the beach. The cross-shore profiles A, B and C, which locations are presented in Figure 3.1, are shown in Figure 3.2. Four sedimentologic surveys, numbered (1) to (4) from the terrace towards onshore, were performed along each transect: (1) located at the terrace, (2) located at the bottom of the beach face, where the beach slope changes, (3) located at the lower face region and (4) at the beach face. From the analysis of these surveys, the two sedimentologic parameters mean grain diameter, D_{50}, and geometrical deviation or spreading, $\sigma = (D_{84}/D_{16})^{1/2}$ (where D_{84} and D_{16} are the sizes for which 84% and 16% by weight of the material is finer), were calculated (Figure 3.2).

The hydrodynamic modelling agents of this coastal system are wind waves, wake waves, generated by catamarans that circulate in front of the beach, and the tide. The numerical wave generation–propagation model SWAN (Simulating WAves Nearshore) (Booij et al., 1996) was applied in the estuary to estimate the wave field generated by the components of the representative average annual wind wave regime with incidence at the beach. The waves generated by the passage of the catamarans were also estimated. Once these results were obtained, the beach hydrodynamics and sediment dynamics were analysed along both components, longshore and cross-shore, at different time scales, medium- and short-term, respectively.

In order to extrapolate the sedimentologic characteristics for the entire profile, two theoretical grain size distributions were considered: the reference distribution, in which a linear interpolation of the above parameters was applied for the positions between (1) and

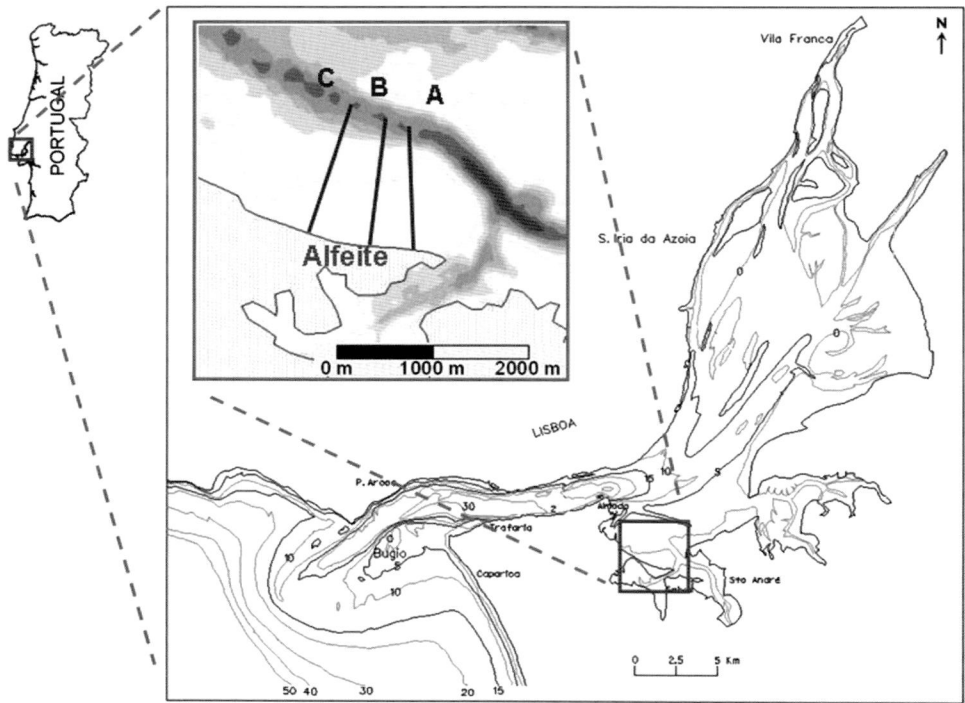

Figure 3.1 Location of the study area.

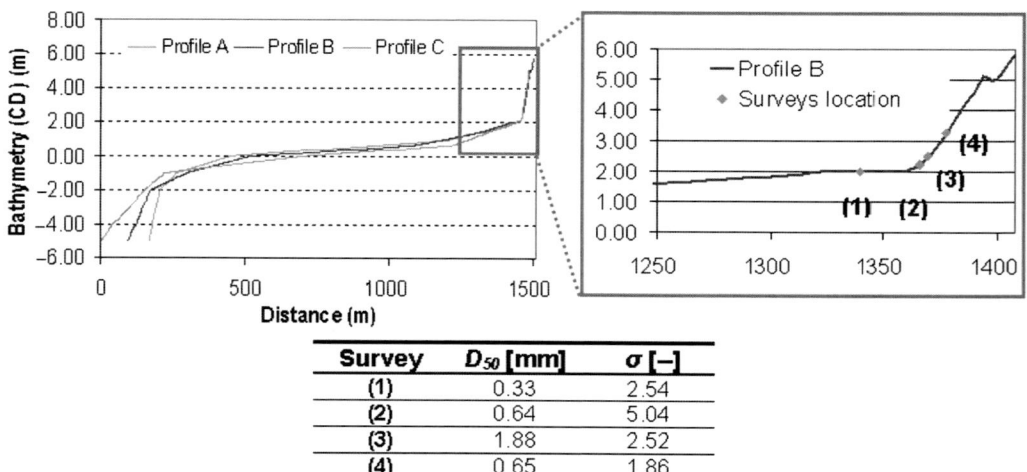

Survey	D_{50} [mm]	σ [–]
(1)	0.33	2.54
(2)	0.64	5.04
(3)	1.88	2.52
(4)	0.65	1.86

Figure 3.2 Geometry of the beach profiles A, B and C; detail of profile B with location of the sedimentologic surveys (1) to (4); and sedimentologic parameters, D_{50} and σ, for each survey. (To see this figure in colour, please see colour plate 1.)

(4), except in (2) and (3) where the respective values were considered; and the alternative distribution, in which a linear interpolation of both parameters was considered between each consecutive pair of positions. In both distributions, the sedimentologic parameters offshore (1) and onshore (4) were considered constant and equal to the values obtained in positions (1) and (4), respectively.

The alongshore wave-induced current field and sediment transport were calculated along the beach profiles using the numerical model LITDRIFT (DHI, 2005) and validated on the basis of testing the physical parameters that, being sensitive to the local conditions, bring uncertainty to the calculations. These were the wave theory used to estimate the wave kinematics, the theoretical grain size distribution along the beach profile and the tidal amplitude. The beach morphological response to wave series generated by catamarans and maritime storm events was also evaluated through the simulation of the short-term cross-shore beach morphodynamics, using the numerical model LITPROF (DHI, 2005).

The numerical study of the sediment dynamics was developed on the basis of the application of the two above-mentioned modules of an integrated modelling system for littoral processes and coastline kinetics (LITPACK) of commercial license. The estimation of the sediment transport (in both modes, suspension and bed load) under waves and currents, common to all the modules, was based on a deterministic approach. The longshore dynamics was simulated on the basis of static morphological conditions, whereas the cross-shore dynamics was simulated on the basis of dynamic morphological conditions. The wave transformation processes considered were shoaling, refraction, directional dispersion and dissipation through breaking. The physical processes considered as contributing to the sediment transport outside the surf zone, where the energy loss occurs predominantly at the near-bed wave boundary layer, were Lagrangian drift, wave vertical asymmetry and streaming (or near-bed current). Inside the surf zone, where the energy loss occurs at the near-bed wave boundary layer but mostly at the water surface due to wave breaking, the considered contributing processes to the sediment transport were the same processes acting outside the surf zone plus the flux of the surface roller and the undertow.

3.3 Results and discussion

3.3.1 *Wave action*

3.3.1.1 Wind waves

Wind data (parameters intensity and direction) of a 6-year series (of four daily records), from 1999 to 2004, locally recorded (at Lisboa/Gago Coutinho meteorological station) was statistically treated to obtain the representative average annual wind regime (Figure 3.3). This regime, considered reliable since only 10.8% of the totally expected records were either missing or not valid, was used as forcing agent to estimate the average annual wave regime in front of the beach through a numerical model of wave generation–propagation. Since the internal estuary has the natural geomorphologic required conditions (enough depth and fetch length), the SWAN model was applied to estimate the wave field generated by the 13 components of the representative average annual wind regime, each associated with a frequency of occurrence (Table 3.1), that have incidence at the beach, main alignment

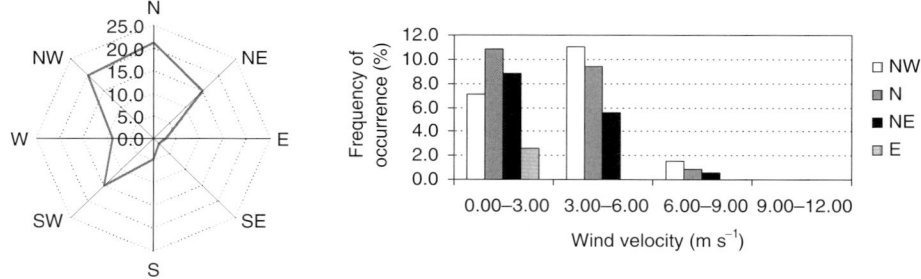

Figure 3.3 Representative average annual wind regime.

of which is WNW–ESE. These components, incoming from sectors NW, N, NE and E, correspond to 58% of the total regime. Their velocity intensity was derived as being the average for each of the classes (0–3), (3–6), (6–9) and (9–12), for each octant.

The result of the 13 simulations, performed at MSL, allowed the characterisation of the wind wave regime, based on the wave parameters such as root-mean-square height, H_{rms}, mean period, T_m, and direction. The comparison of these parameters before the surf zone reveals a great uniformity of the wave action alongshore. At the entrance of profile B (Figure 3.4), at 7.2 m below MSL, the most energetic sector with incidence at the beach is normal to the shoreline; however, there is still a significant amount of energy, incoming from the octants NW, NE and E, that reaches the beach obliquely because the representative waves are short (maximum and minimum wavelength of about 6.9 and 0.6 m). It is therefore expected that the wave action upon the beach is capable of mobilizing and transporting a small volume of sediments.

Table 3.1 Components of the representative average annual wind regime with incidence at the beach.

	Wind parameters	
Direction*	Velocity (m s^{-1})	Occurrence (%)
N	2.17	10.83
N	4.60	9.44
N	7.17	0.90
N	10.42	0.02
NE	2.04	8.85
NE	4.57	5.52
NE	7.27	0.55
NW	2.27	7.09
NW	4.75	11.09
NW	7.26	1.51
NW	10.42	0.02
E	2.14	2.57
E	5.00	0.02

*Octant bisecting line.

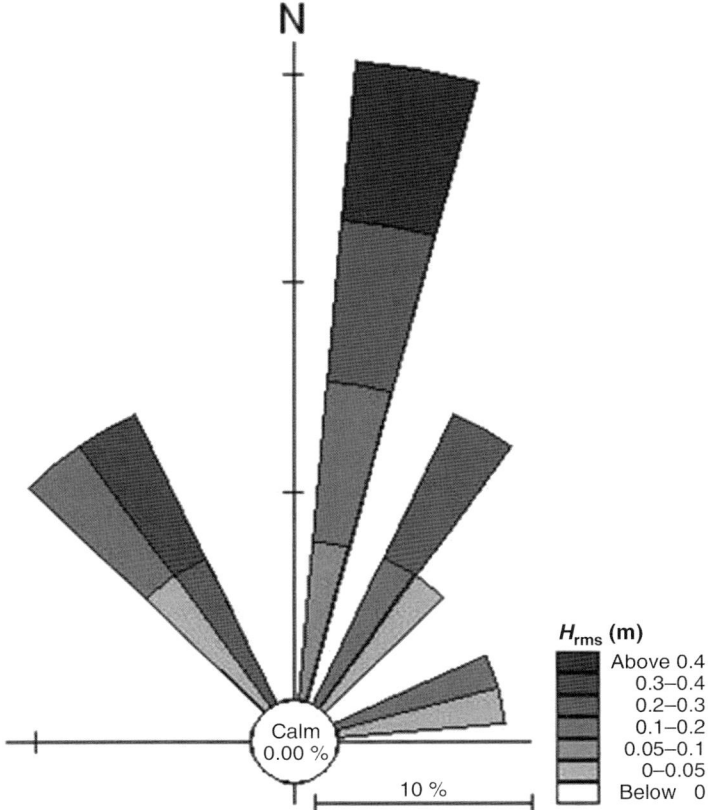

Figure 3.4 Representative average annual wave regime (H_{rms} along directional sectors of 10°).

3.3.1.2 Wake waves

Besides wind waves, Alfeite beach is also subjected to waves generated by catamarans that navigate in the internal estuary, between Terreiro do Paço and Seixal (at north and south banks, respectively). In front of the beach, the catamarans follow a route nearly parallel to the shoreline. This stretch of the route is located at approximately 1500 m distance from the shoreline and over a depth that varies between 7 and 12 m at MSL. The impact of wake waves on the beach sediment transport and morphodynamics and its importance relatively to the impact of wind waves was evaluated.

In what concerns the longshore sediment dynamics, it was concluded that the passage of the catamarans does not play a significant role because the catamarans transit is performed in both directions the same number of times, making the resultant of the longshore sediment transport induced by waves null.

In what concerns the transversal sediment dynamics, the impact of the catamarans depends on the characteristics of the waves generated by them. In general, wake waves depend on the boat velocity, on the geometry of the boat, on the distance from the route to the shoreline and on the route's depth. Based on the formulation suggested by

Blaauw et al. (1984), the maximum wave height, H_{max}, of the wave group generated by the passage of the catamarans was calculated for three different velocities, 10, 15 and 20 knots (5.44, 8.17 and 10.89 m s^{-1}), depending on the distance to the route. The results show that the catamaran speed has a major effect on the amplitude of the water surface fluctuations generated and that at about 500 m from the sailing line the maximum wave height stabilises (Figure 3.5a).

Based on the average annual wind wave regime and on the present traffic of catamarans, an estimate of the ratio between the wake energy and the wind wave energy was calculated depending on the H_{rms} of the wave train generated by the passage of the catamarans (Figure 3.5b). The results show that for a catamaran speed of 10, 15 and 20 knots the average annual wake energy that reaches the beach, for the present catamarans traffic, is about 5, 52 and 250% of the average annual wind wave energy that reaches the beach. Since the service velocity of the catamarans is 20 knots, it can be concluded that the impact of the catamarans traffic in Alfeite beach is very important when compared with the wind action.

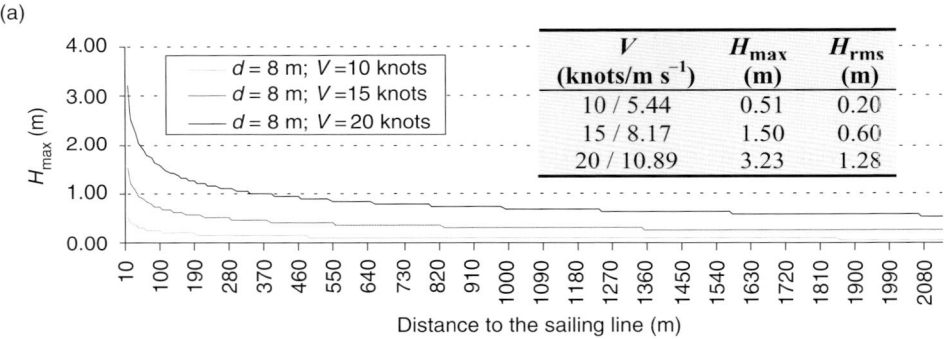

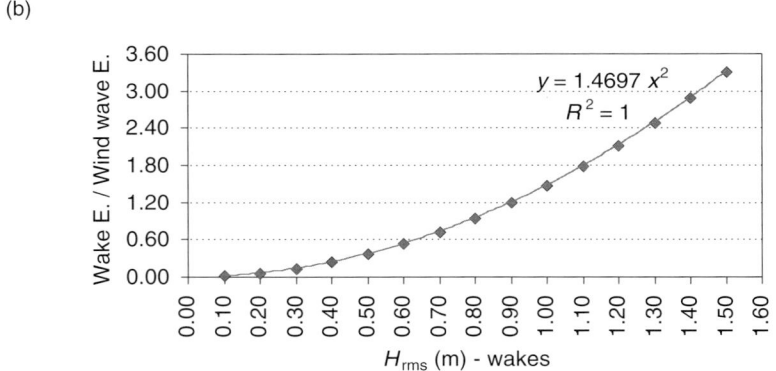

Figure 3.5 (a) Wave heights generated by the passage of catamarans and (b) ratio between the energy imparted by catamaran wakes and the energy imparted by wind waves versus H_{rms} of the wave train generated by the catamarans.

3.3.2 Sediment dynamics

3.3.2.1 Longshore dynamics

The average annual longshore sediment transport, induced by the representative average annual wave regime described in Section 3.3.1, was estimated assuming the following conditions: (a) sea surface elevation described by the semi-empirical wave theory suggested by Doering and Bowen (1995), (b) theoretical grain size reference distribution as described in Section 3.2 and (c) variation of the sea level in agreement with the tidal regime described in Section 3.2. The result is an average annual longshore sediment transport of 14.5×10^3 m^3 year^{-1}, with net sediment transport approximately null in agreement with the observed stability of the shoreline configuration. This value is of the same order of magnitude as values mentioned in the literature for estuarine and assigned to fetch-restriction beaches (Nordstrom et al., 2003).

The integrated analysis of the wave regime, at the initial position of profile B (Figure 3.4), and the sediment budget, for the total transport along profile B (Figure 3.6a), both discretised as function of the incident wave directional sector and class of H_{rms}, shows that the sector N5°–N15°, despite having the largest frequency of occurrence of waves and waves with higher H_{rms} (Figure 3.4), is not the sector that gives the largest contribution to the longshore sediment transport because its direction of incidence relative to the shoreline is nearly normal. In opposition, the two western sectors of incidence contribute largely to the transport directed to east (Figure 3.6a) because their angle of incidence relative to the shoreline is about 45° (the most effective direction in generating longshore transport).

Because the sea level variation correspondent to the semi-diurnal meso-tidal regime was considered in the calculations, the active part of the beach (where the significant sediment transport occurs) for profile B extends along a cross-shore length of 500 m (Figure 3.6b). When the sea level variations are not taken into account (as commonly happens in a great number of studies), that is, the sea level is considered constant, at MSL, the extension of the active part of the beach is only about 100 m. The variation of the sea level causes enlargement and flattening of the curve of the cross-shore distribution of the longshore transport. In this case, the transport peak is located in the lower part of the beach face, in the vicinity of the changing slope zone, in a zone with about 50 m, but there is also significant sediment transport in the low-gradient terrace (Figure 3.6b).

The validation of the longshore transport can be performed through experimental field methodologies, evaluation of the evolution of the shoreline and the vegetation line, comparison with results of empirical formulations and testing the influence of the numerical model parameters that carry uncertainty into the results. Experimental field methodologies like sand tracers and sand traps would not be successfully applied in this particular coastal environment due to its low-sediment dynamics. The analysis of the vegetation line and shoreline evolution through recent aerial photographs (Valente et al., 2006) does not reveal expressive variations to conclude on the longshore transport and therefore validate the numerical estimation. The large disagreements between results of empirical formulations and field experimentation previously obtained for this type of coastal environment (Nordstrom et al., 2003; Silva et al., 2006) indicate that the estimations obtained with empirical formulations are as much reliable as the characteristics of the study area are closer to the conditions which were in the genesis of the formulations derivation. Thus,

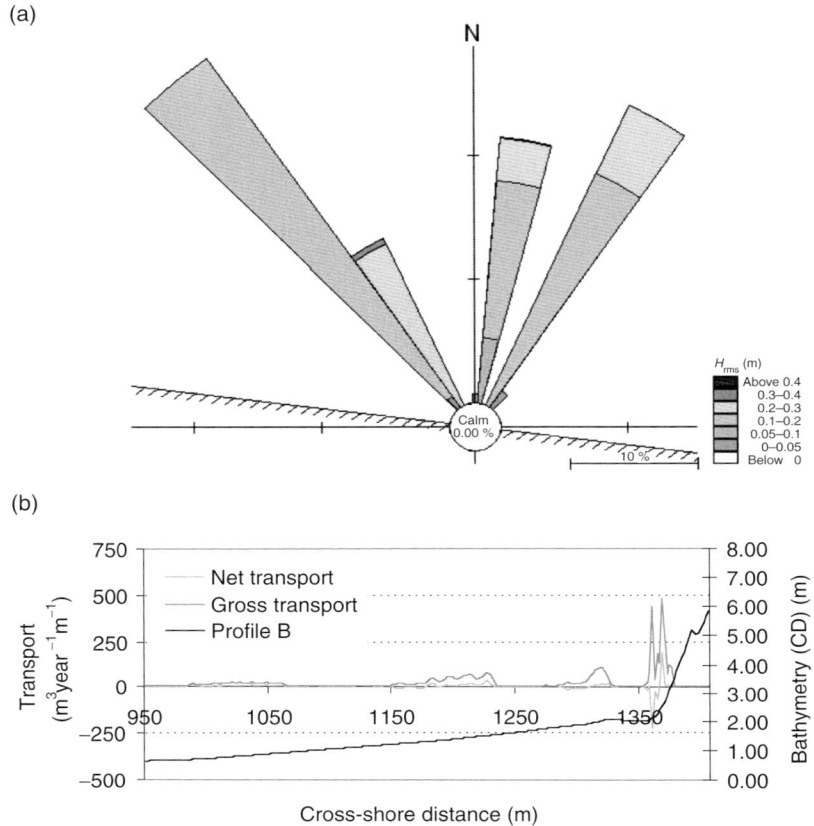

Figure 3.6 Longshore sediment budget at profile B: (a) discretisation per H_{rms} class and 10° directional sector and (b) cross-shore distribution in the active profile. (To see this figure in colour, please see colour plate 2.)

the validation of the longshore transport obtained with the deterministic model is done through the testing of the uncertainty of the parameters that are sensitive to the local conditions.

The wave theory that describes the sea surface elevation and the wave kinematics was made variable in the numerical simulations. The theories tested were Stokes first and fifth order (Fenton, 1985), Cnoidal third order (Fenton, 1990), Vocoidal (Swart, 1982), Isobe and Horikawa (Isobe and Horikawa, 1982) and Doering and Bowen (Doering and Bowen, 1995). Their impact on the longshore transport was evaluated through the comparison of the results (Table 3.2). The results show a large variety of values, meaning that the wave theory has a significant impact in the longshore transport calculations. In fact, the longshore transport varies between 9.6×10^3 m^3 year^{-1}, obtained with the Vocoidal theory (34% of the value obtained with Dowering and Bowen wave theory), and 15.2×10^3 m^3 year^{-1}, obtained with the Isobe and Horikawa wave theory (5% higher than the value obtained with Dowering and Bowen wave theory).

Table 3.2 Longshore sediment transport estimated with six wave theories.

Wave theory	Total longshore transport (m^3 year^{-1})
Stokes first order	12 145
Stokes fifth order	12 366
Cnoidal third order	13 360
Vocoidal	9 601
Isobe and Horikawa (1982)	15 203
Doering and Bowen (1995)	14 540

The Dowering and Bowen wave theory was chosen to be the reference theory because not only it is valid for a wide range of depths but it also accounts for the larger number of processes of wave deformation, such as horizontal asymmetry (skewness) and vertical asymmetry of the sea surface elevation.

The impact of the theoretical grain size distribution along the beach profile on the longshore transport was evaluated on the basis of the results of the numerical simulations performed considering the reference distribution and the alternative distribution, both described in Section 3.2. The longshore transport considering the alternative theoretical grain size distribution was 16.1×10^3 m^3 year^{-1}, which corresponds to an increase of 11% over the transport considering the reference distribution.

The sea level variations generate the displacement of the different hydrodynamic zones, shoaling zone, surf zone and swash zone. Hence, and due to the geometric characteristics of the beach profile, it is expected that if the sea level is assumed constant, at MSL, as commonly happens in a great number of studies, not only the active part of the profile is considerably shorter (as described before) but also the sediment transport is miscalculated. The longshore sediment transport obtained considering the sea level constant, at MSL, was 12.3×10^3 m^3 year^{-1}, which corresponds to a reduction of bout 15% of the transport considering the semi-diurnal meso-tidal regime of tidal amplitude 2.40 m.

3.3.2.2 Cross-shore dynamics

The evaluation of the impact of the passage of the catamarans on the beach morphological stability was performed for the three navigation velocities 10, 15 and 20 knots. The action of the wave trains generated by the catamarans, with H_{rms} 0.2, 0.6 and 1.28 m (Section 3.3.1), was tested at mean low sea level, MSL and mean high sea level. The results obtained reveal that the sediment transport increases with the catamaran speed independently of the sea level; however, the volume of sediment mobilised and consequently the profile changes are insignificant even for the fastest catamaran (Figure 3.7).

At mean low sea level, results with notation (2) in Figure 3.8, the energy dissipation occurs smoothly as the wave decay occurs over the entire low-gradient terrace. For this case, the active part of the beach is larger than at mean high sea level, results with notation (1) in Figure 3.8, when the wave decay occurs localised on the beach face, because the very smooth gradient of the terrace guaranties the maintenance of the wave shape during its progression towards the shore.

Maritime storm events generated by extreme meteorological conditions of wind and atmospheric pressure have strong impact on the beach profile, which rapidly recovers

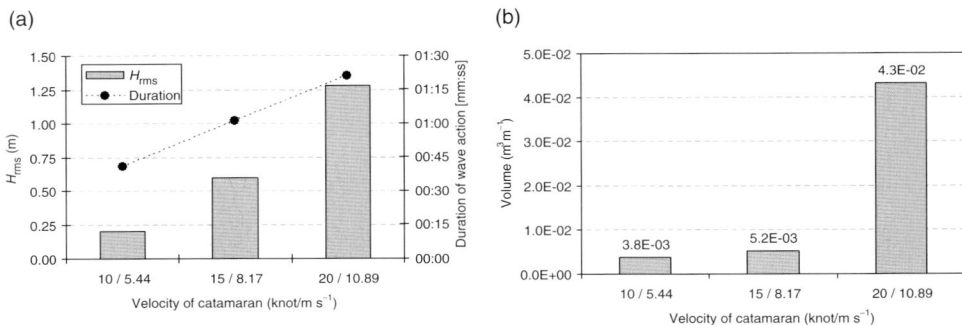

Figure 3.7 Effect of the catamaran in the hydrodynamics (a) and sediment transport (b) of the beach at mean high sea level.

to its previous shape under normal wave conditions. The maritime storm episode which occurred on the 29 January, 2006, was observed, simulated and analysed. Between 14:00 and 22:00 hh:mm of that day, the wind blew from NE (direction of maximum fetch) with speed between 5 and 10.8 m s^{-1} (data recorded at Lisboa/Gago Coutinho meteorological station). Based on previous climatologic studies (IM, 1994) and on the analysis of the wind series from 1999 to 2004, this episode can be classified as exceptional for the study area. The wave series in front of the beach was obtained on the basis of the same methodology applied to obtain the representative average wind wave regime.

The storm event generated an erosion scarp at the beach face (Figure 3.9). The short-term morphological evolution of the beach profile under this storm event was simulated and analysed. Although the volume of sediment mobilised in the beach face of the numerical profile, 9.34 m^{3}m^{-1}, was approximate, the erosion scarp was located below the scarp observed in the field. This can be explained because the storm surge was not considered since no data were available. Another discrepancy between the observed and the numerical profile concerns the formation of a bar in the lower part of the beach face in the numerical profile resultant from the accumulation of the sand extracted from the upper part of the beach face (Figure 3.9). In the field, this feature was not exhibited. This might be explained

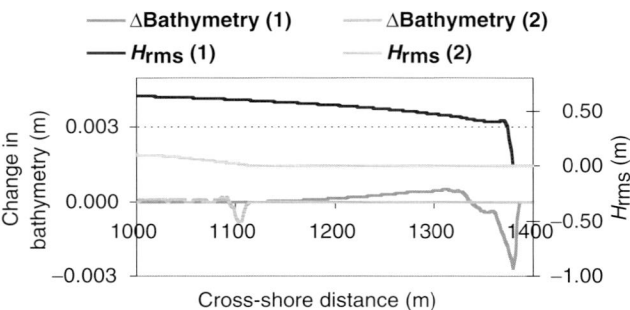

Figure 3.8 Wave propagation and morphological variation along profile B due to the passage of a catamaran at speed 20 knots, at mean high sea level (1) and mean low sea level (2). (To see this figure in colour, please see colour plate 3.)

(a)

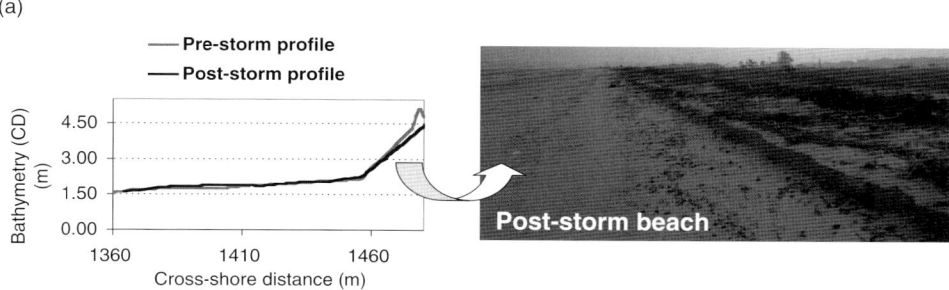

(b)

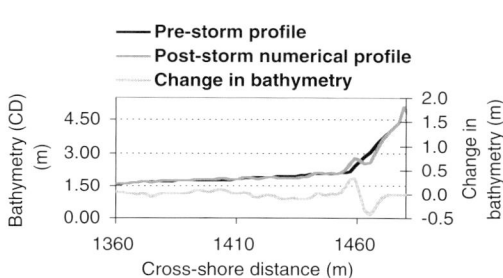

Figure 3.9 Topo-hydrographic surveys of the pre- and post-storm profile C (a) and the numerical profile C (b). (To see this figure in colour, please see colour plate 4.)

because the dynamics of the swash zone under storm conditions is very high and was not considered in the simulation.

3.4 Conclusions

The nearshore sediment dynamics is induced by locally generated waves, wind and wake waves, and the tide.

The geomorphology is characteristic of a low-energy beach: a low-gradient terrace until MSL followed by a steep and narrow beach face. The sediment grain size distribution is bimodal: coarser sand grain at the upper beach and finer grain at the terrace. The comparison of the topo-hydrographic survey with others from previous studies indicates a three-dimensional geomorphologic stability of the beach.

Waves incoming from north, the most frequent directional sector of incidence, had no significant impact on the longshore transport. The sectors NW and NE had higher capacity to generate longshore transport due to their degree of obliquity relative to the shoreline. Wave theory, tidal amplitude and theoretical grain size distribution were parameters that carried uncertainty into the longshore transport estimation by decreasing order of importance (34, 15 and 11% of the estimated value).

The simulation of wave action generated by catamarans on the beach profile reveals that the cross-shore transport is insignificant for the actual traffic. However, if the catamarans pass in front of the beach at the service speed (20 knots), the average annual wake energy that reaches the beach is about 2.5 times the average annual wind wave energy that reaches the beach.

The topo-hydrographic surveys revealed that modifications in the beach profile occurred only in connection with storm events. After these episodes, the beach presents an erosion scarp at the upper part of the beach face. The morphodynamics of the beach that occurred during the last episode of this type was simulated numerically. Despite the good approximation of the volume mobilised, the location of the scarp was not correctly simulated because storm surge was not considered (no measurements were available). Besides, the complexity of the three-dimensional processes that dominated the swash zone was not included in the model, and this might explain why the numerical results exhibit the formation of a bar, which in situ was flattened by the uprush and downrush flows.

The study revealed that the sediment transport, alongshore and cross-shore, in Alfeite beach was higher at the higher sea levels (above MSL), when the beach face was acted and that, since the tidal regime was meso-tidal, the low-gradient terrace had the effect of beach protection.

Neglecting (a) the higher dimension material and litter observed at the base of the beach face and (b) the very fine sediments with aggregation capacity observed in the terrace should have resulted in a slight overestimation of the sediment transport. Jackson et al. (2002) pointed out the necessity of obtaining a relation between the wave energy and the density, position and volume of the litter to identify its importance as morphological agent. The morphodynamics of Alfeite beach is not a scale-down version of the morphodynamics of other beaches located at the mouth of the estuary, fully exposed to the much more energetic wave regime of the Atlantic Ocean, mostly due to the different geomorphologic characteristics. For this reason, future research effort should be done to develop a morphodynamic model suitable to beaches where cohesive and non-cohesive sediments coexist.

Acknowledgements

The authors wish to thank colleagues, fieldwork assistants, and those who contributed to obtaining the data, the Portuguese Research Foundation (FCT) for financing the project BERNA (POCTI/CTA/45431/2002), and the Fundação Calouste Gulbenkian for financing the participation in this international scientific meeting.

References

Blaauw, H., De Groot, M., Van Der Knaap, F. and Pilarczyk, K. (1984) Design of Bank Protection for Inland Navigation Fairways. *Proceedings of Flexible Armour Revetments Incorporating Geotextiles Conference*, Institute of Civil Engineers, London.

Booij, N.R., Holthuijsen, L.H. and Ris, R.C. (1996) The SWAN Wave Model for Shallow Water. *Proceedings of International Conference on Coastal Engineering*, Orlando, pp. 668–676.

Davies, J.L. (1964) A morphogenic approach to world shorelines. *Zeitschrift für Geomorphology* **8** (Mortensen Sonderheft), 127–142.

DHI (2005) *Litpack – Noncohesive Sediment Transport in Currents and Waves*. User Guide, Danish Hydraulic Institute, Denmark.

Doering, J.C. and Bowen, A.J. (1995) Parametrization of orbital velocity asymmetries of shoaling and breaking waves using bispectral analysis. *Coastal Engineering* **26**, 15–33.

Fenton, J. (1985) A fifth-order Stokes theory for steady waves. *Journal of Coastal, Port, Waterway and Ocean Engineering, ASCE* **111**, 216–234.

Fenton, J. (1990) Non-linear wave theories. In: Mehaute, B. Le and Hanes, D.M. (ed.). *The Sea, Vol. 9: Ocean Engineering Science, Part A*, Wiley, New York.

Freire, P.M.S. (1999) *Morphological and Sedimentary Evolution of Estuarine Banks (Tagus Estuary, Portugal)* (in Portuguese). PhD dissertation, Faculdade de Ciências da Universidade de Lisboa, Lisboa, Portugal, pp. 320.

Goodfellow, B.W. and Stephenson, W.J. (2005) Beach morphodynamics in a strong-wind bay: a low-energy environment? *Marine Geology* **214**, 101–116.

IM (1994) *Ten-Day Normal Values Correspondent to the Period 1961–1990 – Region of Ribatejo and West* (in Portuguese). Instituto de Meteorologia, Lisboa.

Isobe, M. and Horikawa, K. (1982) Study on water particle velocities of shoaling and breaking waves. *Coastal Engineering in Japan* **25**, 109–123.

Jackson, N.L., Nordstrom, K.F., Eliot, I. and Masselink, G. (2002) 'Low energy' sandy beaches in marine and estuarine environments: a review. *Geomorphology* **48**, 147–162.

Nordstrom, K.F. and Jackson, N.L. (1992) Two dimensional change on sandy beaches in estuaries. *Zeits für Geomorphologie* **36**, 465–478.

Nordstrom, K.F., Jackson, N.L., Allen, J.R. and Sherman, D.J. (2003) Longshore sediment transport rates on a microtidal estuarine beach. *Journal of Waterway, Port, Coastal and Ocean Engineering* **129**, 1–4.

Silva, A., Taborda, R., Rodrigues, A., Duarte, J. and Cascalho, J. (2006) Longshore Sand Transport Estimates: Results from the Comporta Beach Tracer Experiment. *Proceedings of 5th Symposium on the Atlantic Iberian Margin*, Aveiro, Portugal, pp. 201–202.

Swart, D.H. (1982) The nature and analysis of random waves in shallow water. CSIR Research Report 388/2. Stellenbosch, South Africa.

Valente, C., Freire, P. and Taborda, R. (2006) Mesoscale morphological evolution of Alfeite estuarine beach (in Portuguese). *Livro de Resumos do VII Congresso Nacional de Geologia* **2**, 437–440.

Wang, P., Kraus, N.C. and Davis, R.A. (1998) Total longshore sediment transport rate in the surf zone: field measurements and empirical predictions. *Journal of Coastal Research* **14** (1), 269–282.

Chapter 4
Macro-Algae Habitat as Fish Nursery: Evaluation of Function as Predation Refuge

Jun Shoji, Yasuhiro Kamimura and Chiaki Fujiki

Abstract

Vegetated habitats in shallow waters such as macro-algae bed have been suggested to be important nursery of fish. However, few studies exist on the quantitative evaluation of the functions of vegetation in nursery habitats. In the present study, predation experiments using mesocosms were done to test the hypothesis that habitat complexity provided by macro-algae (*Sargassum* spp.) affects vulnerability of red sea bream (*Pagrus major*) juveniles to piscivorous fish predators. Juvenile behavior was video-recorded in two structurally different habitats: vegetated (with macro-algae) and unvegetated in 500-L tanks. Juvenile red sea bream were associated with the macro-algae in the vegetated tank. Predation experiments were conducted with 30 red sea bream juveniles (31.9 mm in body length) exposed to two piscivorous fish predators, Chinese sea bass, *Lateolabrax* sp. (260.4 mm), for 6 h in 1000-L tanks. Predation rate was significantly lower in the vegetated tank (0.03 fish predator^{-1} h^{-1}) than in the unvegetated tank (0.24 fish predator^{-1} h^{-1}). Present results indicate that habitat complexity provided by macro-algae reduces vulnerability of juvenile red sea bream to predation by piscivorous fish predators by serving as physical and/or visual barriers and limiting the predator's ability to pursue and capture prey.

4.1 Introduction

The Seto Inland Sea (Figure 4.1) is one of the most productive estuaries around the world (Okaichi et al., 1996). The maximum tidal amplitude is as high as about 5 m. A total of 2192 fish species has been recorded in the Inland Sea. Recent annual fish catch has decreased to less than 50% of the maximum annual catch in the mid-1980s (Figure 4.2). Environmental changes in the fish nursery caused by reclamation of coastal zone as well as overfishing are considered to have damaged fish population of the Inland Sea. Total reclamation for recent 100 years amounted to ca. 30 000 ha, which approximates to 20% of the shallow waters with depth <20 m of the Inland Sea (Figure 4.2). About 46% of the total coastline (6760 km) has been changed from natural to artificial (Okaichi et al., 1996; Anon., 2005). The area of shallow waters, which is important spawning grounds and/or nurseries for fish, has been consistently decreasing in recent years.

Macro-Algae Habitat as Fish Nursery: Evaluation of Function as Predation Refuge 51

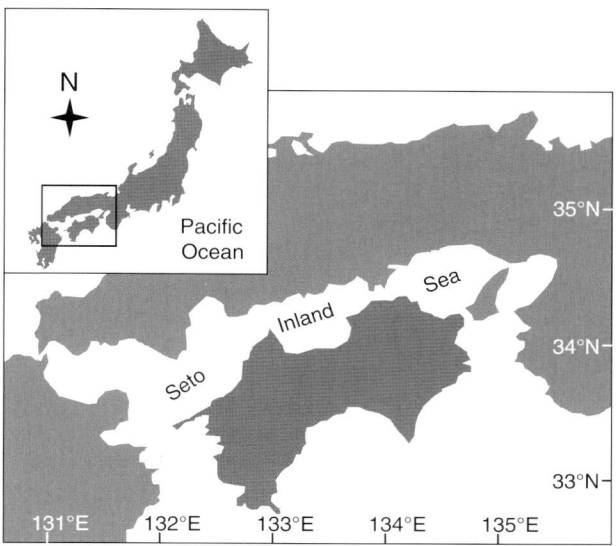

Figure 4.1 Map of the Seto Inland Sea, southwestern Japan.

Among the fish habitats in the shallow waters, macro-algae bed has been suggested to contribute to high fish production, since it serves as feeding ground and/or predation refuge during the early life stages of fish (Beck et al., 2001; Sano et al., 2008). However, information on fish nursery functions of macro-algae beds is limited, while several studies have demonstrated such functions for seagrass beds (Main, 1987; Orth, 1992; Rooker et al., 1998; Shoji et al., 2007). Quantitative evaluation of the functions of macro-algae beds as fish nursery is indispensable for future conservation and management of a variety of fish habitat in the shallow waters.

Predation is one of the most important factors for mortality of young fish, since it prevails throughout all the early life stages (Houde, 1987). Fishes, which experience habitat shifts in their early life, are exposed to large changes in biotic and abiotic environmental

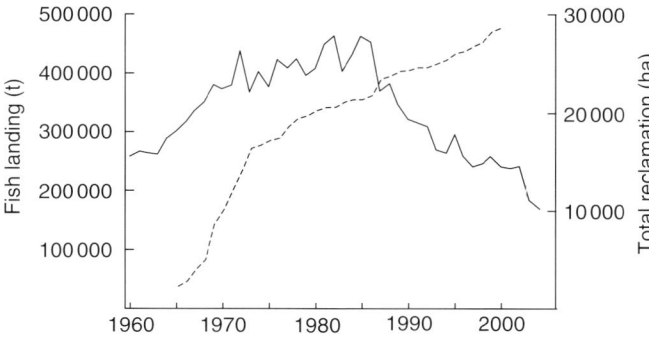

Figure 4.2 Annual fish catch (from 1960 to 2004: solid line) and total reclamation (from 1965 to 2000) in the Seto Inland Sea.

conditions. Successful migrations into their nursery areas can assure high survival rates, while migration into a habitat with inappropriate conditions (e.g. high predator abundance) leads to high mortality.

In the present study, experiments in mesocosm were conducted in order to test the hypothesis that habitat complexity reduces predation mortality of juvenile fish. The effect of physical complexity on juvenile survival was detected by (1) examining association behavior of juveniles with macro-algae and (2) comparing predation rates by piscivorous fish predator between vegetated and unvegetated habitat conditions. Red sea bream (*Pagrus major*) juveniles were used for the experiments. Red sea bream is widely distributed in the coastal waters of Japan. Larvae and early juveniles migrate from offshore waters into shallow coastal habitats and inhabit in/around seagrass and macro-algae beds (Azeta et al., 1980; Tanaka and Matsumiya, 1986).

4.2 Materials and methods

4.2.1 Experimental fish

Red sea bream larvae and juveniles were hatched from naturally spawned eggs in the Hiroshima Prefecture Sea-Farming Center and fed with rotifers *Brachionus plicatilis*, *Artemia* spp. and artificial dry pellets in 200 000-L tanks for 30 days after hatching. About 1000 juveniles were transferred into 500-L tanks for additional rearing and were fed only dry pellets for 4 weeks prior to the experiments. Water temperature ranged between 18.5 and 21.9°C. Mean (standard deviation, SD) body length of the juveniles used for the experiments was 31.9 (2.9) mm.

4.2.2 Association experiment

In order to see if the red sea bream juveniles associate with macro-algae, their behavior was observed in two 500-L tanks prior to predation experiment. Bottom of the tanks was divided into four areas (Figure 4.3). Macro-algae (*Sargassum* spp.) were set on one (section 1) of the four sections at a density of about 3200 g m^{-2}, which approximates the density in natural macro-algae beds around the Takehara Fisheries Research Laboratory, Hiroshima University (240–4340 g m^{-2} during winter 2006 through spring 2007 – Shoji and Kamimura, unpublished data) in one tank (vegetated) and no macro-algae in another

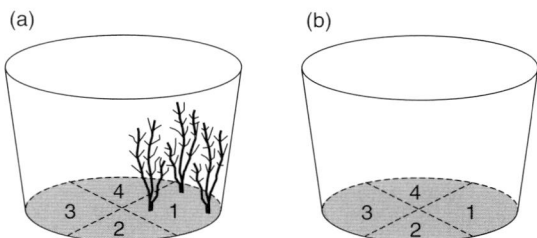

Figure 4.3 Schematic drawings of tanks used for the association experiments: (a) vegetated and (b) unvegetated.

tank (unvegetated as control). Thirty red sea bream were introduced all together using a 2-L plastic cup into the middle of each tank. In order to avoid possible effects of observers, fish distribution was recorded using a remote-controlled video camera (CCD-TRV126, Sony). The number of fish in each section was counted for each tank at 5, 15, 30, 60, 120, 180, 240, 300 and 360 min after the start of the experiment. Association ratio (Masuda and Tsukamoto, 2000) in the section 1 (number of fish in section 1 divided by total number of fish) was calculated. This value would be 1.0 if all fish are associated with the macro-algae and 0.25 with a random distribution. Single trial was made for each observation. Since the observation was repeated nine times using same fish, the difference in the ratio between the two treatments was compared by Chi-square test for only the first observation. Air and water were not supplied to the tanks. No fish died during the experiment.

4.2.3 Predation experiment

Predation rate of the red sea bream juveniles by piscivorous fish predators was compared between vegetated and unvegetated 1000-L tanks. In the vegetated tanks, macro-algae were set at the same density as in the 'Association Experiment' on the 1/4 area of the tank bottom. No macro-algae were set in the unvegetated tanks. Chinese sea bass (*Lateolabrax* sp.) were used as predator for the experiment. Thirty red sea bream juveniles together with two Chinese sea bass were introduced into each tank. Mean (SD) body length of the predators was 264.4 mm (2.2) and 256.3 mm (3.4) in vegetated and unvegetated conditions, respectively ($n = 12$ for each condition). Predators and preys were not reused. Number of red sea bream predated on by the Chinese sea bass for 6 h was counted at the end of the experiment. Predation rate was expressed as the number of juveniles predated on by a predator per hour (number of fish predated predator^{-1} h^{-1}) for each trial. Six replicates were made for each condition. The difference in the predation ratio between the two treatments was compared by Mann–Whitney U test. No air and water were supplied in the tanks during the experiment.

4.3 Results

The association ratio ranged between 0.6 and 0.75 in the vegetated tank and was higher than that in the unvegetated tank (0.15–0.3) throughout the experiment (Figure 4.4). The difference in the ratio between the two treatments at first observation was significant (Chi-square test, d.f. = 3, $p = 0.023$).

Mean ($\pm$SD) predation rate was eight times higher in the unvegetated tanks (0.24 ± 0.14 fish predator^{-1} h^{-1}) than in the vegetated tanks (0.03 ± 0.04 fish predator^{-1} h^{-1}) (Figure 4.5). There was a significant difference in the predation rate between the two conditions (Mann–Whitney U test, $p = 0.020$).

4.4 Discussion

Present experiments demonstrate that macro-algae bed contributes to reducing vulnerability of red sea bream juvenile to predation by piscivorous fish. Structural complexity of a habitat

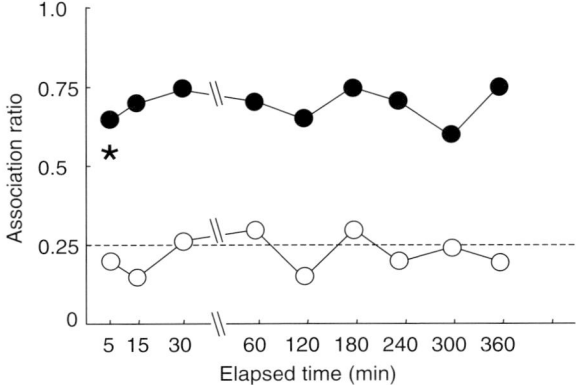

Figure 4.4 Association rates of red sea bream juveniles on each observation in vegetated (closed circles) and unvegetated (open circles) tanks. Asterisk shows a significant difference between the two treatments at first observation (Chi-square test, $p = 0.023$). Dotted line shows limit where fish show a random distribution.

produced by vegetation such as macro-algae as well as seagrass has been suggested as an important determinant for survival of fish early life stages (Stoner and Lewis, 1985; Tanaka and Matsumiya, 1986; Main, 1987; Orth, 1992; Rooker et al., 1998). In a previous study (Shoji et al., 2007), the predation rate of juvenile red sea bream was shown to be lower in vegetated (with seagrass) tank than in unvegetated tank. From this study, it was concluded that habitat complexity provided by the seagrass decreased predation rate through two processes: (1) association with the seagrass decreased encounter rate with predator and (2) seagrass served as physical and/or visual barriers between predator and prey and limited the predator's ability to pursue and capture prey. The present results indicate that macro-algae bed also serves as predation refuge for red sea bream juveniles.

Association behavior with plants and objects such as rocks and floating objects has been observed also in many marine and freshwater fishes and considered to be adaptive behavior for successful feeding, avoidance from predators, migration, school formation and

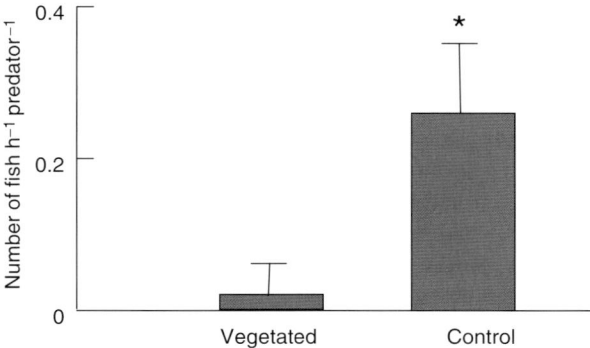

Figure 4.5 Predation rates (number of red sea bream juveniles predated on predator^{-1} h^{-1}) in vegetated and unvegetated tanks. Asterisk shows a significant difference between the two treatments (Mann–Whitney U test, $p = 0.020$).

maintenance (Gooding and Magnuson, 1967; Hunter and Mitchell, 1967; Kingsford, 1993; Masuda and Tsukamoto, 2000). In early life stages of some seagrass-associated fishes, positive correlations between habitat complexity and fish density have been reported in natural environments (Holt et al., 1983; Orth, 1992; Sogard and Olla, 1993). Previous field observations showed that red sea bream juveniles inhabit macro-algae and seagrass beds for feeding (Tanaka and Matsumiya, 1986). Association of red sea bream juveniles with macro-algae, with their cruising activity reduced, would decrease the encounter rate with predators, while swimming at a distance from macro-algae would increase the probability to be observed by visual predators (Bailey and Houde, 1989).

The evidence of low predation rates in the vegetated condition in the present experiments indicates that habitat complexity provided by macro-algae led to lower predation rate of red sea bream juveniles. Efficiency of prey detection by the Chinese sea bass would have decreased because of the lower visibility of red sea bream juveniles provided by macro-algae in the vegetated tanks. In addition, foraging behavior of the predator would also have been restricted due to the decreased attack speed and mobility in the macro-algae area, resulting in lower capture efficiency. Previous laboratory and field experiments (Ryer, 1988; Rooker et al., 1998) also show that the physical complexity of habitats alters the vulnerability of larval and juvenile fishes to fish predators through affecting the predator's ability to pursue and capture prey. Variation in seagrass blade width has been reported to influence foraging efficiency of fish predator (Ryer, 1988). Higher survival ratios were reported in juvenile red drum (*Sciaenops ocellatus*) exposed to predation by piscivorous fish in seagrass habitat than in unvegetated habitat (Rooker et al., 1998).

We conclude that macro-algae beds, as well as seagrass beds (Shoji et al., 2007), serve as predation refuge of red sea bream juveniles through serving as physical and/or visual barriers and limiting the predator's ability to pursue and capture prey. Conservation of macro-algae bed in coastal waters would be important for conservation and sustainable exploitation of coastal fisheries resources such as red sea bream, which inhabit macro-algae bed during their early life stages. Further study on functions of macro-algae bed as nursery for a variety of fish species would be needed for integrated conservation and management of coastal fisheries resources.

Acknowledgments

We thank S. Tanaka and K. Hirakawa, Hiroshima Prefecture Sea-Farming Center, for providing red sea bream and S. Iwasaki and staff of the Takehara Fisheries Research Laboratory for supporting the experiments.

References

Anon. (2005) *Environmental Conservation of the Seto Inland Sea*. Association for the Environmental Conservation of the Seto Inland Sea, Kobe, Japan, p. 101.

Azeta, M., Ikemoto, R. and Azuma, M. (1980) Distribution and growth of demersal 0-age red sea bream, *Pagrus major*, in Shijiki Bay. *Bulletin of the Seikai National Fisheries Research Institute*, **54**, 259–278.

Bailey, K.M. and Houde, E.D. (1989) Predation on eggs and larvae of marine fishes and the recruitment problem. *Advances in Marine Biology* **25**, 1–83.

Beck, M.W., Heck, K.L., Able, K.W., Childers, D.L., Eggleston, D.B., Gillanders, B.M., Halpern, B., Hays, C.G., Hoshino, K., Minello, T.J., Orth, R.J., Sheridan, P.F. and Weinstein, M.P. (2001) The identification, conservation, and management of estuarine and marine nurseries for fish and invertebrates. *BioScience* **51**, 633–641.

Gooding, R.M. and Magnuson, J.J. (1967) Ecological significance of a drifting object to pelagic fishes. *Pacific Science* **21**, 486–497.

Holt, S.A., Kitting, C.L. and Arnold, C.R. (1983) Distribution of young red drum (*Sciaenops ocellatus*) among different sea-grass meadows. *Transactions of the American Fisheries Society* **112**, 267–271.

Houde, E.D. (1987) Fish early life dynamics and recruitment variability. *American Fisheries Society Symposium* **2**, 17–29.

Hunter, J.R. and Mitchell, C.T. (1967) Association of fishes with flotsam in the offshore waters of central America. *Fishery Bulletin* **66**, 13–29.

Kingsford, M.J. (1993) Biotic and abiotic structure in the pelagic environment: importance to small fish. *Bulletin of Marine Science* **53**, 393–415.

Main, K.L. (1987) Predator avoidance in seagrass meadows: prey behavior, microhabitat selection, and cryptic coloration. *Ecology* **68**, 170–180.

Masuda, R. and Tsukamoto, K. (2000) Onset of association behavior in striped jack, *Pseudocaranx dentex*, in relation to floating objects. *Fishery Bulletin* **98**, 864–869.

Okaichi, T., Komori, S. and Nakanishi, H. (1996) *Biological Resources and Environment of the Seto Inland Sea*. Kiseisha-koseikaku, Tokyo, 272 pp.

Orth, R.J. (1992) A perspective on plant-animal interactions in seagrass: physical and biological determinants influencing plant and animal abundance. In: John, D.M., Hawkins, S.J. and Price, J.H. (eds). *Plant-Animal Interactions in the Marine Benthos*. Claredon Press, Oxford, pp. 147–164.

Rooker, J.R., Holt, G.J. and Holt, S.A. (1998) Vulnerability of newly settled red drum (*Sciaenops ocellatus*) to predatory fish: is early life survival enhanced by seagrass meadows? *Marine Biology* **131**, 145–151.

Ryer, C.H. (1988) Pipefish foraging: effects of fish size, prey size and altered habitat complexity. *Marine Ecology Progress Series* **48**, 37–45.

Sano, M., Nakamura, Y., Shibuno, T. and Horinouchi, M. (2008) Are seagrass and mangroves in the tropics nursery habitats for many fish species? *Bulletin of the Japanese Society of Scientific Fisheries* **74**, 93–96.

Shoji, J., Sakiyama, K., Hori, M., Yoshida, G. and Hamaguchi, M. (2007) Seagrass habitat reduces vulnerability of red sea bream *Pagrus major* juveniles to piscivorous fish predator. *Fisheries Science* **73**, 1281–1285.

Sogard, S.M. and Olla, B.L. (1993) The influence of predator presence on utilization of artificial seagrass habitats by juvenile walleye pollock, *Theragra chalcogramma*. *Environmental Biology of Fishes* **37**, 57–65.

Stoner, A.W. and Lewis, F.G. III. (1985) The influence of quantitative and qualitative aspects of habitat complexity in tropical sea-grass meadows. *Journal of Experimental Marine Biology and Ecology* **94**, 19–40.

Tanaka, M. and Matsumiya, Y. (eds) (1986) *Sea Farming Technology of Red Sea Bream*. Kiseisha-koseikaku, Tokyo, 170 pp.

Chapter 5
Predictive Probability Modelling of Marine Habitats – A Case Study from the West Coast of Norway

Ellen Soldal, Trine Bekkby, Eli Rinde, Vegar Bakkestuen, Lars Erikstad, Oddvar Longva and Martin Isæus

Abstract

There is a general lack of knowledge on the distribution of coastal habitats in relation to geophysical factors. Additionally, the factors used to describe the physical variability of the coastal habitats are rarely measured or quantified in an objective and reproducible manner. By developing geographic information system-based models of geophysical factors and analysing presence and absence data with respect to these factors, we are able to calculate and map the probability of finding habitats or species. Here, we present a case study modelling the probability of finding channelled wrack, *Pelvetia canaliculata*. We used generalised additive models to build statistical models of the predictor–response relationships and the Akaike information criterion to compare the models. The best statistical model was used to calculate and map the probability of finding channelled wrack within an area on the west coast of Norway. This approach has been applied to other habitats and species and provides a useful tool for a cost-effective planning, mapping and monitoring of habitats and species considered to be especially important or vulnerable. Probability maps provide users with the possibility of applying the precautionary principle approach or of reducing the focal area. Using the precautionary principle, you include all possible habitats for the given species (i.e. also including areas with low probabilities) in your mapping, monitoring or research activity. However, if you prefer guaranteed findings of the habitat, you will be interested in identifying areas with high probabilities and reduce the focal area.

5.1 Introduction

Several studies have documented that depth (most often as an indicator of light level), terrain structures (e.g. slope), physical factors (e.g. wave exposure) and other environmental characteristics (e.g. desiccation, seabed substrate) are important for the distribution of marine habitats and species (e.g. Connel, 1961; Paine, 1974; Whittick, 1983; Thomas, 1986; Lein et al., 1987; Kautsky and Van der Maarel, 1990; Kiirikki, 1996; Ruuskanen et al., 1999). Concerning algal communities in particular, studies have documented relationships between wave exposure and distribution, density and size of the communities (e.g. Subrahmanyan,

1960; Crothers, 1976; Sjøtun and Fredriksen, 1995; Sjøtun et al., 1998; Bruntse et al., 1999; Bekkby et al., 2002; Isæus, 2004).

In the last few years, several distribution models have been developed (e.g. Guisan and Zimmermann, 2000; Bekkby et al., 2002; Elith et al., 2006; Bekkby et al., 2008a,b). Such models contribute to a better understanding of the factors and processes structuring the distribution and composition of marine habitats, and are increasingly used by managers, e.g. as part of the Norwegian mapping programmes on marine biodiversity (Rinde et al., 2006) and the management of the Gullmarsfjord, Sweden (Bekkby and Rosenberg, 2006). However, good models require good input data, which are often unavailable. Particularly are the factors needed to describe the physical variability in the coastal zone rarely measured or quantified in an objective and reproducible manner (e.g. Norton et al., 1977; Schultze et al., 1990; Christie et al., 2003; Norderhaug, 2004). Numerical models based on adequate input data may be used to develop probability maps and contribute with knowledge on possible habitat distribution in areas where such information has not been collected in the field.

This chapter provides insight into predictive probability modelling of marine habitats, using the channelled wrack, *Pelvetia canaliculata* (L.) Dcne. et Thur., as a case study. Channelled wrack is an algae living in the littoral zone (algal zonation summarized in White, 2003). Its density decreases with increasing wave exposure (Subrahmanyan, 1960; White, 2003), desiccation and heat stress (Schonbeck and Norton, 1978). Hence, the channelled wrack presence was expected to be related to geographic information system (GIS)-modelled wave exposure and solar radiation. As slope influences the wave exposure and the area suitable for channelled wrack growth (as simple geometrical effect), this factor was also included in the analyses.

The geophysical data used in our study were quantitative, objectively defined (through GIS modelling) and aggregated over time (buffering for extraordinary events). The model selection technique Akaike information criterion (AIC) was used to select the best statistical model, which further was used to develop a predictive probability map for channelled wrack within the selected area.

5.2 Methods

5.2.1 Study site and field sample characteristics

The fieldwork was carried out in Sandøy municipality (62°N 6°E), Møre and Romsdal, Norway, during the period 25 May to 1 June 2004. This area is typical for the central west coast of Norway, with small islands, underwater shallows and rocks and high tidal amplitude (about 1.80 m). In all, we visited 98 stations. The stations were selected to represent the variability in wave exposure, solar radiation conditions and terrain within the study area. Channelled wrack occurrence was defined through recordings of presences and absences at each station. At 64 stations, channelled wrack was present (presence defined as any coverage above zero within a 0.5 m wide transect going perpendicular to the water line), at 34 stations it was absent. By confining the study to areas of similar bedrock types, we have reduced the effect of boulder size and the extent of microhabitats due to boulders. Also, the sampling locations had similar tidal amplitudes, hence leaving out possible effects of tidal exposure.

5.2.2 Predictor variables – GIS models

The predictor dataset consists of digitally modelled information on wave exposure, slope and solar radiation. These are all digital models integrated in ArcView 3.3.

- Wave exposure modelling – uses data on fetch (distance to nearest shore; island or coast), wind strength and direction (i.e. 16 directions) in modelling wave exposure at a grid resolution of 25 m (Isæus, 2004). This wave exposure model has been validated in the Stockholm archipelago (Isæus, 2004) and has been applied in the Norwegian mapping programmes on marine biodiversity (Rinde et al., 2006), in Swedish research projects (Eriksson et al., 2004; Sandström et al., 2005) and in Finland (unpublished data). The channelled wrack is sedentary and perennial, and the distribution is considered to be a result of environmental conditions integrated over several years (Thomas, 1986). Hence, the model is based on wind statistics averaged over 10 years.
- Slope (10 m grid resolution) was modelled from an interpolated digital elevation model (DEM) based on data purchased from the Norwegian Mapping Authority (41–85 depth/height recordings per 100 m^2 at the selected stations). The DEM and hence the derived slope model includes both terrestrial and marine data, hence providing the necessary spatial resolution of the littoral zone where channelled wrack is distributed.
- The modelled solar radiation index (10 m grid resolution) provides information on the solar radiation relative to the optimal slope and aspect conditions. Hence, the calculation is based on slope and aspect derived from the interpolated DEM. The index has highest values where the slope and aspect conditions are optimal for solar radiation (Parker's index, Parker, 1988; see Økland, 1990; Økland and Eilertsen, 1993).

5.2.3 Analyses – generalised additive model and AIC model selection

The main aim of this study is to quantify the influence of the most important factors for the distribution of channelled wrack, and to use this information to build a predictive probability model for the algae. We analysed the statistical influence of wave exposure, slope, solar radiation and the exposure–slope interaction to channelled wrack occurrence using generalised additive models (GAM) (occurrence as a binomial dataset) in S-PLUS 2000. As a tool for model selection, we used the AIC (see Burnham and Anderson, 2001) in GRASP (an extension to S-PLUS 2000) (Lehmann et al., 2003, 2004). The AIC method implies an untraditional way of thinking in ecological research (even though the method has existed for several decades and is implemented in several statistical packages), as it does not include traditional H_0 testing and interpretation of p-values. The candidate models were tested and ranked relative to each other. The best model according to AIC is the model receiving most support from the data, and which at the same time uses a low number of explanatory factors (the principle of parsimony, i.e. the trade-off between squared bias and variance versus the number of parameters in the model). We used the AICc calculations (as recommended by Burnham and Anderson, 2004), which is AIC adjusted to fit small sample sizes (AIC and AICc being equal at large sample sizes).

5.2.4 Spatial probability prediction and validation

A predictive model of the probability of channelled wrack occurrence based on the presence/absence data was developed in the S-PLUS 2000 extension GRASP (Lehmann et al., 2003). GRASP has solved one of the largest problems with predictive modelling (as described by Lehmann, 1998), as it has introduced a way of exporting the statistical models to GIS software (i.e. ArcView) for predictive GIS modelling. The predicted probability model was validated using a cross-validation method, a receiver-operating characteristic (ROC) test (Fielding and Bell, 1997), based on the presence/absence data. The cross-validation was made with five subsets (folds) of the entire dataset (fivefold cvROC).

5.3 Results and discussion

Our analyses show that the probability of finding channelled wrack is mainly determined by wave exposure (higher probability in sheltered than exposed areas) and slope (higher probability in gentle than steep slopes) (Model 5 in Table 5.1, Figure 5.1, fivefold cvROC= 0.73). These two factors were used for the predictive modelling (Figure 5.2). The effect of wave exposure is apparent in the probability map (Figure 5.2), where the probability of channelled wrack occurrence is higher in inner (i.e. sheltered) areas and on the east side of island compared to outer (i.e. more exposed) areas in the western part of the area. This is consistent with the findings of Hansen (1995) and Kvist and Lein (1999). Also, Subrahmanyan (1960) supports the results, as he showed that the density of fucales decreased with increasing exposure to prevailing winds. He found this to be particularly true for channelled wrack, which was completely absent at locations with high wave exposure levels. Channelled wrack depends on being periodically dried and will not survive if it is submerged in water for too long or exposed to too much water due to wave action (Schonbeck and Norton, 1978). Some studies have found channelled wrack to be able to stay dry up to 8 days (at low temperatures) (Pfetzing et al., 2000). Hence, high wave exposure levels will expose the algae to too much water and reduce the suitability of the site as a channelled wrack habitat.

Table 5.1 The results of the AICc model selection.

Model	AICc	Δ	W_i
5 WE + Slope	111.27	0.00	0.38
7 WE + Slope + SR	112.54	1.28	0.20
8 WE + Slope + (WE * Slope)	113.23	1.98	0.14
1 WE	113.70	2.44	0.11
4 WE + SR	113.92	2.66	0.10
9 WE + Slope + SR + (WE * Slope)	114.54	3.28	0.07
10 No effect from predictors	128.53	17.26	0.00
3 Slope	137.80	26.54	0.00
2 SR	138.58	27.32	0.00
6 SR + Slope	139.81	28.54	0.00

The response variable was occurrence. Predictor variables were wave exposure (WE), slope, solar radiation (SR) and the interaction WE * Slope. The models are listed from best to worst. Δ is the discrepancy from the best model, and W_i is the probability that the model is in fact the most appropriate (the Akaike weight).

Predictive Probability Modelling of Marine Habitats 61

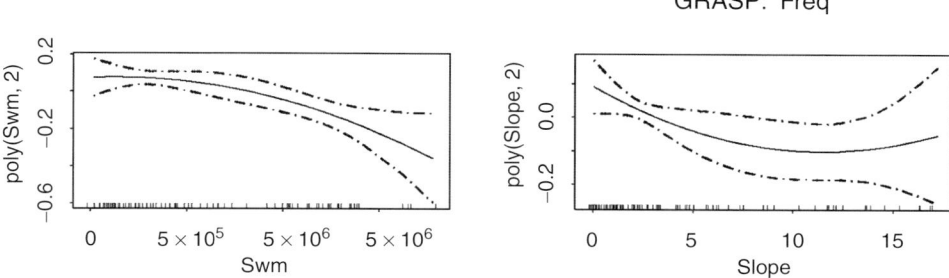

Figure 5.1 The figure shows the second-order polynomial regression response curve of the occurrence against wave exposure (WE) and slope. The WE x-axis represents the wave exposure index developed by Isæus (2004). The slope x-axis is presented in degrees. Note that the range of the y-axes differs between the two graphs. Dashed lines represent the upper and lower point-wise twice-standard-error curves.

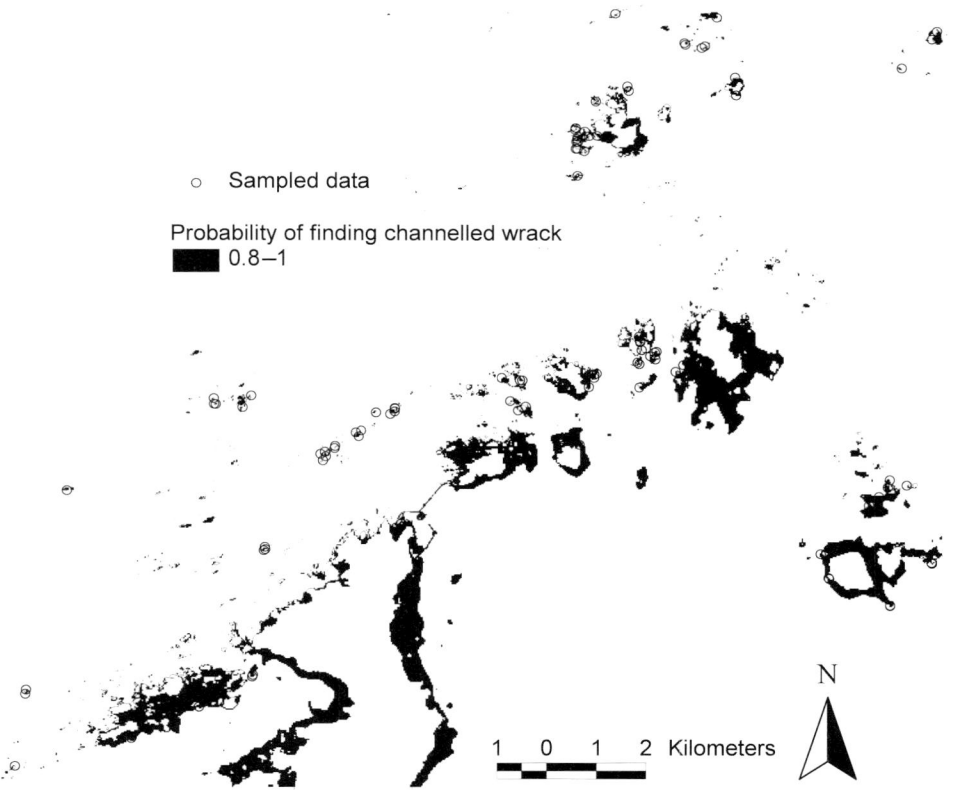

Figure 5.2 The figure shows the sampling locations (black circles) and areas with more than 80% predicted probability of finding channelled wrack (*Pelvetia canaliculata*). The grid resolution is 25 m.

A higher probability of finding channelled wrack in gentle than steep slopes was expected (Lalli and Parsons, 1997), simply due to the increase in available area within the littoral zone from the geometrical effect of a gentle compared to a steep slope. An alternative, or maybe additional explanation, may be that flat terrain in the littoral zone provides protection from waves and increases the width of the splash zone with an appropriate level of water exposure. We did not reveal any interaction between wave exposure and slope in this study (even though the models including both as single factors received good support, see below). This is most likely due to the effect of scale, as the influence of slope on wave exposure is probably at a much smaller scale than the modelled wave exposure, which has a grid resolution of 25 m.

Model 7 (including wave exposure, solar radiation and slope) and Model 8 (wave exposure, slope and the interaction) had a $\Delta < 2$ (Δ represents discrepancy from the best model, Table 5.1); i.e., both extended models had good support (Burnham and Anderson, 2001). However, these models are more complex than the model based on wave exposure and slope. As $\Delta < 2$ implies that the models are more or less equally good, the simplest model was selected. An effect of solar radiation was expected, as drought and heat stress damage algal species in the littoral zone (Schonbeck and Norton, 1978). However, the channelled wrack tolerance for desiccation is high (Schonbeck and Norton, 1978; Dring and Brown, 1982), and the species is not very sensitive to temperature fluctuations (White, 2003). Also, the solar radiation experienced by the algae in our study may not have been sufficient to result in an effect in our model. Also, desiccation effects may have been absent or small due to the relatively large tidal amplitude and the wave exposure levels at our sampling locations.

Even though Model 5 was the best models among the alternatives tested, the probability that this model in fact is the most appropriate (the Akaike weight, W_i, Table 5.1) was only 0.38. This implies that there are parameters of high relevance that have not been included in the analyses. These may be parameters that have been tested at a scale that has not revealed any influence (e.g. the interaction between wave exposure and slope), or these may be parameters not tested. Hence, even though we have found some of the factors of relevance, more information is needed in order to get a more accurate model.

All data available were used for the probability modelling, as we did not have enough data for an independent validation of the model. Hence, the fivefold cross-validation method was applied. However, for later studies, and as a general recommendation, independent dataset for model validation should be collected in the model area. Also, testing the model outside the model area is recommended, as this will test the general validity of the model.

5.4 Aplication in coastal zone management

Predictive modelling is a useful tool in nature management and can be very cost effective (Lehmann et al., 2003). Few studies use objective, comparable and reproducible datasets, and comparing results and finding trends over time and across regions using already sampled data are often difficult. This study shows that it is possible to develop area coverage models for important geophysical factors and by statistical analysis be able to develop probability maps of species and habitat distribution. Surveying and mapping all coastal areas with respect to all desired qualities is both time consuming and costly. In Norway, and most likely in most countries, it is at present not possible to achieve funding for total mapping of marine habitats and biodiversity. Hence, GIS modelling of the likely distribution of selected habitats have been developed as a tool to identify areas to be prioritised in order to ensure mapping of the most important and valuable habitats. In this chapter, we illustrate how GIS-based modelling may be used to integrate objective and reproducible information on environmental conditions to analyse and predict the distribution of species and habitats.

A high advantage of the predictive probability models is that it results in a probability grid that is easy to understand and disseminate to the various stakeholders. The mapped probability estimates include uncertainties in finding the habitat and species within an area. Often models are binomial, showing only the areas where the species/habitat is most likely found based on geophysical factors. These models do not include the uncertainties within the data and they do not present the actual probabilities. Probability maps provide users with the possibility of applying the precautionary principle approach or to reduce the focal area. Using the precautionary principle, you include all possible habitats for the given species (i.e. also including areas with low probabilities) in your mapping, monitoring or research activity. However, if you prefer guaranteed findings of the habitat, you will be interested in identifying areas with high probabilities and reduce the focal area.

The accuracy of the predictions is crucial for the value of models as a management decision tool. The model performance is set by the distributional variability of the species, our ecological knowledge, the quality and resolution of the predictor datasets and the modelling technique. The effect of predictor data scale differs from habitat to habitat, from species to species, but in general, more detailed data are needed for the purpose of developing good probability models for use by managers of the coastal zone.

Acknowledgements

The project was funded by the Research Council of Norway, the Norwegian institute for Nature Research and the Norwegian institute for Water Research.

References

Bekkby, T., Erikstad, L., Bakkestuen, V. and Bjørge, A. (2002) A landscape ecological approach to coastal zone applications. *Sarsia* **87**, 396–408.

Bekkby, T., Nilsson, H.C., Rygg, B., Isachsen, P.E., Olsgard, F. and Isæus, M. (2008a) Identifying soft sediments at sea using GIS-modelled predictor variables and Sediment Profile Image (SPI) measured response variables. *Estuarine, Coastal and Shelf Science* **79**, 631–636.

Bekkby, T., Rinde, E., Erikstad, L., Bakkestuen, V., Longva, O., Christensen, O., Isæus, M. and Isachsen, P.E. (2008b) Spatial probability modelling of eelgrass Zostera marina L. distribution on the West coast of Norway. *ICES Journal of Marine Science* **65**, 1093–1101.

Bekkby, T. and Rosenberg, R. (2006) Marine habitaters utbredelse – terrengmodellering i Gullmarsfjorden (Norwegian/Swedish, English abstract). Länsstyrelsen i Västra Götaland län, Vattenvårdsenheten, Rapport 2006:07, ISSN 1403–168X. 33 pp.

Burnham, K.P. and Anderson, D.R. (2001) Kullback-Leibler information as a basis for strong inference in ecological studies. *Wildlife Research* **28**, 111–119.

Burnham, K.P. and Anderson, D.R. (2004) Multimodel inference: understanding AIC and BIC in model selection. *Sociological Methods and Research* **33**, 261–304.

Bruntse, G., Lein, T.E., Nielsen, R. and Gunnarsson, K. (1999) Response to wave exposure by littoral species in the Faroe Islands. *Frødskaparrit* **47**, 181–198.

Christie, H., Jørgensen, N.M., Norderhaug, K.M. and Waage-Nielsen, E. (2003) Species distribution and habitat exploitation of fauna associated to kelp (*Laminaria hyperborea*) at the Norwegian coast. *Journal of the Marine Biological Association of the United Kingdom* **83**, 687–699.

Connel, J.H. (1961) Effects of competition, predation by *Thais lapillus* and other factors on natural populations of the barnacle *Balanus balanoides*. *Ecological Monographs* **31**, 61–104.

Crothers, J.H. (1976) On the distribution of some common animals and plants along the rocky shores of West Somerset. *Field Studies* **4**, 469–489.

Dring, M.J. and Brown, F.A. (1982) Photosynesis of intertidal brown algae during and after periods of emersion: a renewed search for physiological causes of zonation. *Marine Ecology Progress Series* **8**, 301–308.

Elith, J., Graham, C.G., Anderson, RP., Dudík, M., Ferrier, S., Guisan, A., Hijmans, R.J., Huettmann, F., Leathwick, J.R., Lehmann, A., Li, J., Lohmann, L.G., Loiselle, B.A., Manion, G., Moritz, C., Nakamura, M., Nakazawa, Y., Overton, J.M., Peterson, A.T., Phillips, S.J., Richardson, K., Scachetti-Pereira, R., Schapire, R.E., Sobero'n, J., Williams, S., Wisz, M.S. and Zimmermann, N.E. (2006) Novel methods improve prediction of species' distributions from occurrence data. *Ecography* **29**, 129–151.

Eriksson, B.K., Sandström, A., Isæus, M., Schreiber, H. and Karås, P. (2004) Effects of boating activities on aquatic vegetation in the Stockholm archipelago, Baltic Sea. *Estuarine, Coastal and Shelf Science* **64**, 339–349.

Fielding, A.H. and Bell, J.F. (1997) A review of methods for the assessment of prediction errors in conservation presence/absence models. *Environmental Conservation* **24**, 38–49.

Guisan, A. and Zimmermann, N.E. (2000) Predictive habitat distribution models in ecology. *Ecological Modelling* **135**, 147–186.

Hansen, E. (1995) Utvikling av biologisk eksponeringsskala på Nord-Vestlandet. Hovedfagsoppgave i marinbiologi til graden Cand. Scient, Institutt for fiskeri- og Marinbiologi, Universitetet i Bergen.

Isæus, M. (2004) *Factors Structuring Fucus Communities at Open and Complex Coastlines in the Baltic Sea*. Doctoral thesis, Department of Botany, Stockholm University, Sweden.

Kautsky, H. and Van Der Maarel, E. (1990) Multivariate approaches to the variation in phytobenthic communities and environmental vectors in the Baltic Sea. *Marine Ecology Progress Series* **60**, 169–184.

Kiirikki, M. (1996) Mechanisms affecting macroalgal zonation in the northern Baltic Sea. *European Journal of Phycology* **31**, 225–232.

Kvist, M. and Lein, T.E. (1999) Sammenheng mellom biologisk og fysiske mål på bølgeeksponering i Hordaland. IFM Report 2.

Lalli, C.M. and Parsons, T.R. (1997) *Biological Oceanography – An Introduction*. Butterworth-Heinemann, Oxford.

Lehmann, A. (1998) GIS modeling of submerged macrophyte distribution using generalized additive models. *Plant Ecology* **139**, 113–124.

Lehmann, A., Leathwick, J.R. and Overton, J.McC. (2004) *GRASP v.3.1. User's Manual*. Swiss Centre for Faunal Cartography, Switzerland.

Lehmann A., Overton J.C. and Leathwick, J.R. (2003) GRASP: generalized regression analysis and spatial predictions. *Ecological Modelling* **160**, 165–183.

Lein, T.E., Küfner, R. and Hansen, J.E. (1987) Konsekvensutredninger Barentshav syd. Artsammansetninger i fjæra i Finnmark. Del 1: hardbunn. Økoforsk program, Norges allmenvitenskapelige forskningsråd, 101 pp.

Norderhaug, K.M. (2004) Use of red algae as hosts by kelp-associated amphipods. *Marine Biology* **144**, 225–230.

Norton, T.A., Hiscock, K. and Kitching, J.A. (1977) The ecology of Lough Ine. The Laminaria forest at Carrigathorna. *Journal of Ecology* **65**, 919–941.

Økland, R.H. and Eilertsen, O. (1993) Vegetation–environment relationships of boreal Coniferous forests in the Solhomfjell area, Gjerstad, S Norway. *Sommerfeltia* **16**, 1–254.

Økland, T. (1990) Vegetational and ecological monitoring of boreal forests in Norway. I. Rausjømarka in Akershus county, SE Norway. *Sommerfeltia* **10**, 1–52.

Parker, K.C. (1988) Environmental relationships and vegetation associates of columnar cacti in the northern Sonoran desert. *Vegetatio* **78**, 125–140.

Paine, R.T. (1974) Intertidal community structure. Experimental studies on the relationship between a dominant competitor and its principal predator. *Oecologia* **15**, 93–120.

Pfetzing, J., Stengel, D.B., Cuffe, M.M., Savage, A.V. and Guiry, M.D. (2000) Effects of temperature and prolonged emersion on photosynthesis, carbohydrate content and growth of the brown intertidal alga *Pelvetia canaliculata*. *Botanica Marina* **43**, 399–407.

Rinde, E., Rygg, B., Bekkby, T., Isæus, M., Erikstad, L., Sloreid, S.-E. and Longva, O. (2006) Documentation of marine nature type models included in Directorate of Nature Management's database Naturbase (in Norwegian with English abstract). First generation models for the municipalities mapping of marine biodiversity 2007. NIVA Report LNR 5321–2006.

Ruuskanen, A., Bäck, S. and Reitalu, T. (1999) A comparison of two cartographic exposure models using *Fucus vesiculosus* as an indicator. *Marine Biology* **134**, 139–145.

Sandström, A., Eriksson, B.K., Karås, P., Isæus, M. and Schreiber, H. (2005) Boating and navigation activities influence the recruitment of fish in a Baltic Sea archipelago area. *Ambio* **34** (2), 125–130.

Schonbeck, M. and Norton, T. (1978) Factors controlling the upper limits of fucoid algae on the shore. *Journal of Experimental Marine Biology and Ecology* **31**, 303–313.

Schultze, K., Janke, K., Krüß, A. and Weidemann, W. (1990) The macrofauna and macroflora associated with *Laminaria digitata* and *L. hyperborea* at the island of Helgoland (German Bight, North Sea). *Helgolaender Meeresuntersuchungen* **44**, 39–51.

Sjøtun, K. and Fredriksen, S. (1995) Growth allocation in *Laminaria hyperborean* (Laminariales, Phaeophyceae) in relation to age and wave exposure. *Marine Ecology Progress Series* **126** (1–3), 213–222.

Sjøtun, K., Fredriksen, S. and Rueness, J. (1998) Effect of canopy biomass and wave exposure on growth in *Laminaria hyperborea* (Laminariaceae: Phaeophyta). *European Journal of Phycology* **33**, 337–343.

Subrahmanyan, R. (1960) Ecological studies on the Fucales. 1. *Pelvetia canaliculata* Dcne. et Thur. *Journal of Indian Botanical Society* **39**, 614–630.

Thomas, M.L.H. (1986) A physically derived exposure index for marine shorelines. *Ophelia* **25** (1), 1–13.

White, N. (2003) *Pelvetia canaliculata*. Channelled wrack. *Marine Life Information Network: Biology and Sensitivity Key Information Sub-Programme.* Marine Biological Association of the United Kingdom, Plymouth. Available at http://www.marlin.ac.uk.

Whittick, A. (1983) Spatial and temporal distributions of dominant epiphytes on the stipes of *Laminaria hyperborea* (Gunn.) Fosl. (Phaetophyta: Laminariales) in S.E. Scotland. *Journal of Experimental Marine Biology and Ecology* **73**, 1–10.

Section 2
Impacts on Coastal Systems

Chapter 6
Further Testing of the Approved EU Indicator to Measure the Progress in the Implementation of Integrated Coastal Zone Management in Europe

Alan H. Pickaver

Abstract

The current chapter reports on the results of tests using an indicator established to measure the progress of integrated coastal zone management (ICZM) implementation by EU Member States. This indicator will allow Member States to determine the extent of their national implementation of ICZM. Used together with indicators to measure sustainable development of the coastal zone, this progress indicator will determine whether there is any correlation, over time, between the ICZM decision-making process and improvement in the sustainability of coastal communities and coastal ecosystems and biodiversity. The indicator to show progress in the implementation of ICZM breaks down the cyclical, but stepwise, process into a series of successive actions which are needed to pass from a situation where no ICZM is being used to one where it is being fully implemented. This methodology allows the trend in implementation within any one country to be compared at regional and local levels. It can also be used to compare local, regional and national implementation. Measuring progress in ICZM implementation against improved sustainability of the coastal zone at all spatial levels is a necessary component to promote a collective and mutually supportive approach to tackling the challenges posed by coastal and marine issues. Results are presented from three different countries in Europe at local, regional and national levels.

6.1 Introduction

Growing concerns about the state of the coast in Europe prompted the European Commission and Member States to establish a 'demonstration programme' in 1996 to ascertain best practice in addressing coastal issues (European Commission, 1996). The outcome of six thematic studies together with the experience of 35 pan-European demonstration projects led to the presentation of two documents by the Commission in September 2000: a recommendation concerning the implementation of Integrated Coastal Zone Management (ICZM)

(European Commission, 2002a) and an ICZM Strategy for Europe (European Commission, 2002b). The recommendation was adopted by Council and Parliament on 30 May 2002.

An EU ICZM Expert Group was set up which agreed on a Working Group on Indicators and Data (WG-ID) to develop an indicator-based assessment. A document describing an indicator to measure the progress of ICZM implementation in the coastal zone was presented to the WG-ID at the second meeting of the Group of Experts in June 2003 by the present author (Pickaver et al., 2004). This original indicator had 26 actions divided into five phases. It allows Member States and acceding countries to determine the extent of their national implementation of ICZM and a means to assess whether that progress is leading to improved sustainability of the coastal resources. The Group of Experts agreed to adopt the progress indicator and begin its testing to determine whether the indicator would work in practice as well as in theory. It was the intention to modify the indicator, if necessary, according to feedback.

A series of preliminary, unpublished, tests were conducted in Spain, Germany, Italy, France, Poland and Lithuania for the Working Group. The results were collated and analysed and brought to the attention of the WG-ID, which following a lively but positive discussion agreed to modify the Indicator in line with the critique given. These changes were related primarily to language, i.e. comprehensibility although some additional actions were added which were felt to be important for a full assessment of ICZM implementation. The revised progress indicator of 31 actions divided into four phases with 5, 7, 12 and 7 action levels, respectively (see Table 6.1). It was adopted by the 5th Meeting of the EU ICZM Group of Experts in September 2005 with a recommendation that it now be used to measure the progress of ICZM in Member States.

Whilst the theory and advantages of the progress indicator have been reported in detail elsewhere (Pickaver et al., 2004), this report presents, for the first time, results of tests using the progress indicator. The work was conducted in the UK, France, Belgium and Ireland at local, regional and national levels as part of the Interreg IIIB Corepoint project.

6.2 Methodology

A series of one day workshops was organised in each of the regions where the progress indicator was tested, namely Wales, NE England, NW England, Belgium and NW France. The tests were conducted in English, French or Dutch (Flemish) accordingly using translations where appropriate. The participants were coastal and marine practitioners from different administrations, organisations, agencies and interest groups who were asked to complete the table together. They represented national, regional and/or local interests. Between 5 and 25 persons were involved in each test. Each person was asked to complete the indicator. Any questions relating to the indicator were answered by the author or someone with experience from the WG-ID. Normally, it took no more than 2 h to complete the work. Respondents answered each question with a simple affirmative, negative or 'don't know'. The results from all the participants were then collated and a simple majority used to determine whether or not a particular ICZM action is being conducted or not. These results are then colour coded with an affirmative response given as green, a negative response as red and an undecided as yellow. This is the standard methodology that has been adopted by the EU ICZM Group of Experts.

Table 6.1 The revised progress indicator as used in the tests (published for the first time).

Phase	Action	Description
Aspects of coastal planning and management are in place	1	Decisions about planning and managing the coast are governed by general legal instruments.
	2	Sectoral stakeholders meet on an ad hoc basis to discuss specific coastal and marine issues.
	3	There are spatial development plans which include the coastal zone but do not treat it as a distinct and separate entity.
	4	Aspects of the coastal zone, including marine areas, are regularly monitored.
	5	Planning on the coast includes the statutory protection of natural areas.
A framework exists for taking ICZM forward	6	Existing instruments are being adapted and combined to deal with coastal planning and management issues.
	7	Adequate funding is usually available for undertaking actions on the coast.
	8	A stocktake of the coast (identifying who does what, where and how) has been carried out.
	9	There is a formal mechanism whereby stakeholders meet regularly to discuss a range of coastal and marine issues.
	10	Ad hoc actions on the coast are being carried out that include recognisable elements of ICZM.
	11	A sustainable development strategy which includes specific references to coasts and seas is in place.
	12	Guidelines have been produced by national, regional or local governments which advise planning authorities on appropriate uses of the coastal zone.
Most aspects of an ICZM approach to planning and managing the coast are in place and functioning reasonably well	13	All relevant parties concerned in the ICZM decision-making process have been identified and are involved.
	14	A report on the state of the coast has been written with the intention of repeating the exercise every 5 or 10 years.
	15	There is a statutory coastal zone management plan.
	16	Strategic environmental assessments are used commonly to examine policies, strategies and plans for the coastal zone.
	17	A non-statutory coastal zone management strategy has been drawn up and an action plan is being implemented.
	18	There are open channels of communication between those responsible for the coast at all levels of government.

(Continued)

Table 6.1 (*Continued*)

Phase	Action	Description
An efficient, adaptive and integrative process is embedded at all levels of governance and is delivering greater sustainable use of the coast	19	Each administrative level has at least one member of staff whose sole responsibility is ICZM.
	20	Statutory development plans span the interface between land and sea.
	21	Spatial planning of sea areas is required by law.
	22	A properly staffed and properly funded partnership of coastal and marine stakeholders is in place.
	23	ICZM partnerships are consulted routinely about proposals to do with the coastal zone.
	24	Adequate mechanisms are in place to allow coastal communities to take a participative role in ICZM decisions.
	25	There is strong, constant and effective political support for the ICZM process.
	26	There is routine (rather than occasional) cooperation across coastal and marine boundaries.
	27	A comprehensive set of coastal and marine indicators is being used to assess progress towards a more sustainable situation.
	28	A long-term financial commitment is in place for the implementation of ICZM.
	29	End users have access to as much information of sufficient quality as they need to make timely, coherent and well-crafted decisions.
	30	Mechanisms for reviewing and evaluating progress in implementing ICZM are embedded in governance.
	31	Monitoring shows a demonstrable trend towards a more sustainable use of coastal and marine resources.

The test in Wales was done in July 2005 and was one of the first to be submitted and used the initial progress indicator as agreed by the second EU ICZM Group of Experts. The results from this test were used to complement all the other preliminary results which led to the improvement in the indicator. Thereafter, all workshops have used the revised progress indicator (Table 6.1).

6.3 Results

The specific results of the tests conducted are given in Table 6.2 for NW England (conducted in May 2006), NE England (conducted in October 2005) and Belgium (conducted in November 2005). The test conducted in France (October 2006) was deemed to be confidential, although the results were analysed and conclusions drawn from them.

Further Testing of the Approved EU Indicator 73

Table 6.2 Overview of test results conducted on the revised EU progress indicator. (To see this table in colour, please see colour plate 25.)

Action	NE England (national)	Belgium (national)	NE England (national)	Belgium (national)	NW England (regional)	NW England (local)	NE England (local)
1							
2							
3							
4							
5							
6							
7							
8							
9							
10							
11							
12							
13							
14							
15							
16							
17							
18							
19							
20							
21							
22							
23							
24							
25							
26							
27							
28							
29							
30							
31							

Green, affirmative response; red, negative response; yellow, undecided.

One advantage of the progress indicator is that it can be visualised in different ways. Table 6.2 compares across national, regional and local areas of different countries. However, it can also be used to compare the progress in ICZM for these same administrative entities within a country by laying the results directly side by side (Table 6.3, NE England). Figure 6.1 shows the pooled results from all the respondents in Belgium with respect to the 31 action levels. Nonetheless, it is not simply necessary to tabulate all the results individually. Figure 6.2 shows all the pooled results from Belgium, NE and NW England after grouping of all the actions within each phase. This gives a clear visual representation of the overall progress in ICZM with respect to each of the four phases.

6.4 Discussion

These first results confirm that the progress indicator is a legitimate methodology which reinforced similar comments made during preliminary testing outside the Corepoint project as reported to the WG-ID. The tables, again, confirm that the progress indicator is able to distinguish between different implementation levels at national, regional and local levels throughout the different areas of Europe where it is being used.

The use of workshops as a mechanism to transfer knowledge on ICZM is highly recommended. However, it was noted that these were costly with respect to both the respondent's time and travel. Other tests conducted elsewhere by the author have been done by electronic exchange, telephone conversations and one-on-one meetings. In terms of gaining meaningful results, there is little difference between the methodologies. However, the workshop allows face-to-face communication with persons having different ICZM expertise. There was a consistent feeling that this process, itself, is of enormous benefit in bringing together different, but relevant, stakeholders to discuss the ICZM process. Persons dealing with ICZM on a daily basis are not often given the opportunity to discuss ICZM issues with persons from other departments or fields of work.

It was also considered an easy methodology to use. Although the results can only be considered semi-quantitative at best, there was often real agreement (or disagreement) about the actions under discussion. However, there was also general agreement that the binary scale was not sensitive enough and there should be some sort of semi-quantitative breakdown, e.g. replacing the yes/no/undecided with a scale of 0–5. In fact, the original methodology of the progress indicator deliberately chose a binary response in order to commit people to answering as honestly as possible. It was felt that most respondents would be reluctant to say either yes or no if given a middle-of-the-road choice. At the very least, the Corepoint testers felt that a 'don't know' was very useful. Nonetheless, some of the Corepoint testers still felt the choice of replies offered to be an inadequate response since it did not allow the multi-dimensional realities of ICZM to be fully expressed.

The respondents also felt that there needed to be some breakdown of the actions in particular at local level of assessment. Again, a choice was made at the onset to keep the indicator as simple as possible, although the EU Group of Experts is now looking at the potential to subdivide some of the actions. Again, the persons doing the test felt that support notes and an explanation on how to run the test could also improve the effectiveness of the test, and this recommendation was also taken on board for the revised progress indicator. Again, one of the groups observed that the new explanatory notes were not detailed or

Table 6.3 Overview of test results in three different administrative units of NE England. (To see this table in colour, please see colour plate 26.)

Action	National	Regional	Local
1			
2			
3			
4			
5			
6			
7			
8			
9			
10			
11			
12			
13			
14			
15			
16			
17			
18			
19			
20			
21			
22			
23			
24			
25			
26			
27			
28			
29			
30			
31			

Green, affirmative response; red, negative response; yellow, undecided.

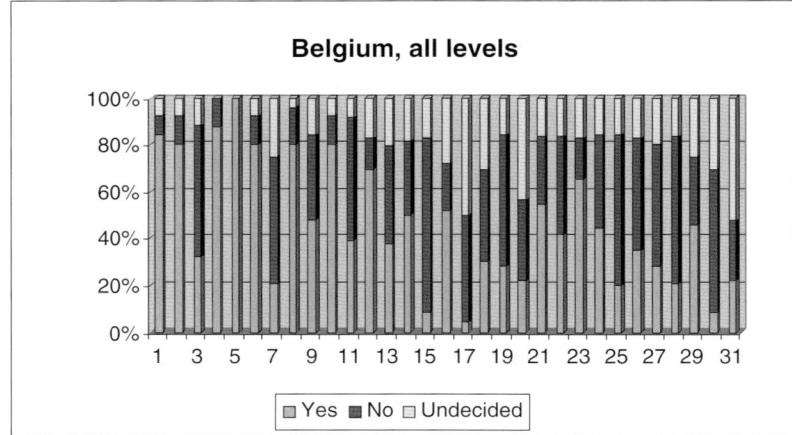

Figure 6.1 The results from Belgium showing the all the responses for the 31 action levels. (To see this figure in colour, please see colour plate 5.)

precise enough to be useful and that a response to a question could still be 'interpreted' according to their knowledge and background which may not give a consistent answer.

A final comment was that a new action was needed which would look at including EU Directives/plans in the description, e.g. Water Framework Directive and Marine Spatial Planning.

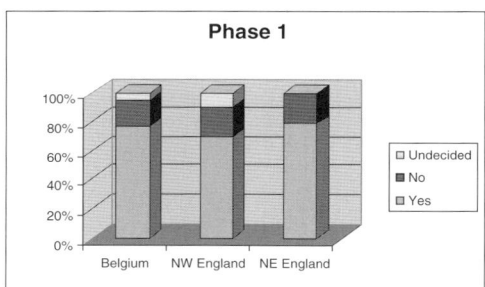
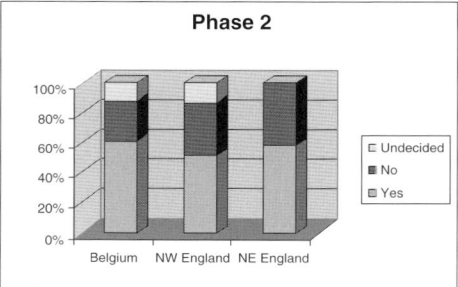
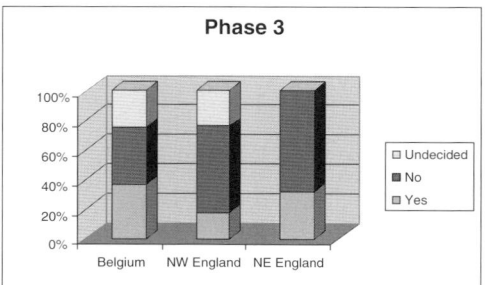
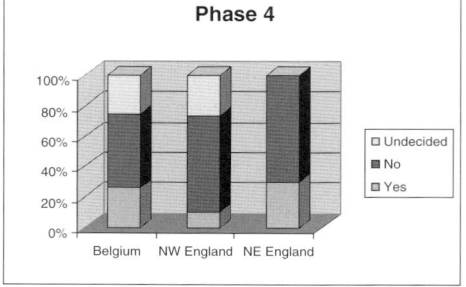

Figure 6.2 An overview of all results from Belgium, NW and NE England, Phases 1–4. (To see this figure in colour, please see colour plate 6.)

One other key point was the apparent sensitivity to the publication of the answers. Even with the confines of the Corepoint project, one of the regions tested only did so provided that the results would not be made available outside the group. As in tests conducted outside of Corepoint, some participants are reluctant to have their answers published even if they, themselves, remain anonymous. This was not totally unexpected since, during discussions within the WG-ID, it had been clear that some Member States were fearful that a comprehensive testing of coastal Member States would eventually lead to the publication of 'league tables' which would show some countries in a bad light, even if this was not the objective. This was always considered by the Member States to be the weakest point of developing a progress indicator. This concern is also linked to the seeming 'competence' of the person answering the questions and to their placement in the hierarchy of the authority they are representing. In other words, who is actually responsible within local, regional and/or national government for determining the state of progress of ICZM?

There were also some, rather country-specific concerns. The main concern in France was that, although the progress indicator allows national, regional and local responses to be compared, these particular administrative boundaries are not really applicable to France. The regional bodies which are independent from the national administration do not have legal competency for the coastal zone. This is done by the prefectures but these are state representatives of the region. In fact, there are two administrative levels between national and local, namely the Regional Councils ('Régions') and the County Councils ('Départements'). Neither does the local level have any link with the classical, administrative boundaries. There are also still some concerns about the language used in some of the questions. These relate to an understanding of specific English words used in the descriptions of the action levels by those having English as a second language. This led to difficulties in translating those terms into other European languages. However, it was also felt that some of the questions were open to interpretation in the English language.

Despite such criticisms, the tables, again, confirm that the progress indicator is able to distinguish between different implementation levels at national, regional and local levels throughout the different areas of Europe where it is being used

The results from the Corepoint project have proven very useful and will be carried forward to next meeting of the EU ICZM Group of Experts.

6.5 Conclusion

The results collated from the tests in France, Belgium and the UK, in this study, using the progress indicator show that Phase 1 has evolved acceptably. This phase, covering aspects of coastal planning and management that should be in place, has been completed in practically all the countries involved in this study, even though a sectoral approach is still pre-dominant. In Phase 2 where a framework for ICZM is in existence, actions 11 and 12 present the greatest problems. However, in general, other actions are being implemented which, although still largely sectoral, now has a greater tendency towards integration. It is this phase which has shown most progress in the last few years. Some countries have even begun clearly to work in the direction of integration, e.g. France and Belgium. But the positive trends were also evident in the UK. With respect to Phases 3 and 4, relating to having an ICZM planning and management approach in place and functioning well as well as having an efficient, adaptive and integrative process embedded in all levels of governance, some progress has been made but it is largely ad hoc, i.e. no trends are present or are very small. Any improvements have been largely determined by priorities set by each country. Quite clearly, further progress in ICZM needs to be seen in these two action levels in particular.

Although these results will come as no surprise to those working in ICZM, the progress indicator now provides the means by which such conclusions can be drawn and a methodology for determining whether progress in implementation really is being achieved.

Acknowledgements

Thanks are due to those Corepoint partners who organised workshops for the testing of the progress indicator, namely Jeremy Gault (CMRC, Ireland), Jeremy Hills (Envision, UK), Graham Lymbery (Sefron Council, UK), Kathy Belpaeme, (Coordination Centre for ICZM, Belgium), and Manuelle Philippe (University of Brittany, France).

Thanks are also due to the Interreg IIIB Programme for funding this work within the Corepoint project.

References

Commission of the European Communities (2002a) Council Recommendation of the European Parliament and of the Council of 30 May 2002 concerning the implementation of Integrated Coastal Zone Management in Europe. *Official Journal of the European Union* L 148/24.

Commission of the European Communities (2002b) Communication from the Commission to the Council and the European Parliament on integrated coastal zone management: a strategy for Europe. Brussels. COM 547 final.

European Commission (1996) Demonstration Programme on Integrated Management of Coastal Zones. Document XI/102/96.

Pickaver, A.H., Gilbert, C. and Breton, F. (2004) An indicator set to measure the progress in the implementation of integrated coastal zone management in Europe. *Ocean & Coastal Management* **47**, 449–462.

Chapter 7
The Response of Hyperbenthos and Infauna to Hypoxia in Fjords along the Skagerrak: Estimating Loss of Biodiversity Due to Eutrophication

Lene Buhl-Mortensen, Eivind Oug and Jan Aure

Abstract

At present there is a need for ecological quality criteria in environmental management of the marine bottom environment. Unique long-time series of oxygen measurements exists for many fjord basins along the Norwegian Skagerrak coast where concentrations in bottom water have been measured since 1950 or earlier. Recently, several of these have experienced oxygen levels below 2 mL L^{-1}, the level at which most marine invertebrates are significantly affected. These long-time measurements provide an opportunity to study the fauna in fjords with a known oxygen history to detect effects of hypoxia. This study investigates the response of hyperbenthos and macro-infauna to eutrophication-related hypoxia in 11 fjord basins, which are part of the ongoing oxygen-monitoring programme. These fjords represent three categories of hypoxia defined by the historic oxygen concentration in bottom water: (1) <2 mL^{-1}, (2) 2–3 mL^{-1} and (3) >3 mL^{-1}, and with 3, 3 and 5 basins within the respective categories. A very strong linear correlation was observed between species richness of hyperbenthos and the near-bottom oxygen minimum during the last 5 years ($R^2 = 0.91$). The number of species decreased from 42–55 in well-oxygenated basins to 27–30 in the intermediate hypoxia situation, and 0–19 in the most hypoxic basins. For infauna, the correlation was also clear ($R^2 = 0.81$) and the corresponding number of species in the three hypoxia categories was 67–123, 28–63 and 2–25. Amongst the infauna organisms, molluscs seemed to be the most susceptible group. Of the crustacean groups belonging to the hyperbenthos fauna amphipods, tanaids and ostracods showed the strongest response to hypoxia and thus would be good candidates for indicators that can provide early warning of changes in oxygen conditions. The loss of biodiversity due to increased eutrophication after 1980 was estimated. The 'category 1' fjords have experienced a marked worsening of oxygen condition and have lost 50–90% of bottom species. The 'category 2' fjords have lost 50–35% of the species, whereas 'category 3' fjords appear not to have been affected.

7.1 Introduction

Fjord localities along the Norwegian Skagerrak coast have experienced a clear decrease in oxygen concentration since 1975 (Johannessen and Dahl, 1996; Aure et al., 1997).

This decrease in oxygen concentration is caused by an increased flux of organic material to the basins. This is suggested to be due to regional increase in organic matter, but additionally some fjords are also affected by locally increased input of organic matter. In several of the fjord localities, oxygen concentrations have been measured regularly back at least to 1950 as part of a coastal environmental monitoring programme (Johannessen and Dahl, 1996). Recently, several of the fjord basins have experienced oxygen levels below 2 mL L^{-1}, which is the level at which most marine invertebrates are significantly affected (Diaz and Rosenberg, 1995). Today, 2 mL L^{-1} is used as a rule of thumb for when the benthic fauna is significantly affected. This value is based on studies carried out on infauna that provide a good picture mainly of molluscs and polychaetes. Consequently, species indicative of pollution has been sought in these fauna groups (Gray and Pearson, 1982; Pearson et al., 1983). However, the mobile fauna living at the sediment–water interface, the 'hyperbenthos', is assumed to be particularly sensitive to low oxygen levels. This fauna comprises a number of crustacean species which in general appear to be the most sensitive group while polychaetes and molluscs are more tolerant (Theede, 1973; Dries and Theede, 1974; Josefson and Widbom, 1988; and studies cited in Diaz and Rosenberg, 1995). Unfortunately, little information is available on the response of the more mobile and sensitive crustacean fauna living at or close to the bottom.

At present, there is a need for ecological quality criteria in environmental management, especially for the marine bottom environment (Anon, 1999; Lanters et al., 1999). Early warnings of environmental degradation are essential for a precautionary management of marine systems and thus the identification of sensitive indicator species is crucial. The purpose of this chapter was to investigate differences in species diversity of 'hyperbenthos' and macro-infauna in fjord localities influenced by decreasing oxygen levels. Comparing the responses to hypoxia of different fauna groups will provide a test of the hypothesis that the more mobile crustacean fauna is less tolerant to hypoxia than other benthic fauna groups.

Several fjord and coastal basins with different oxygen regimes on the Norwegian Skagerrak coast were selected for study, together forming a gradient from benign conditions to severe hypoxia. Altogether, 11 fjord localities were included, of which 8 has been part of the oxygen monitoring programme. The monitoring data provide an opportunity to study the effects in localities with a known oxygen history. The ongoing eutrophication can be considered a large-scale experiment that offers possibilities to make valuable observations on how increased carbon flux and low oxygen concentrations affect different parts of the marine bottom fauna. Based on the oxygen history, an estimate of the loss of species due to gradually evolving hypoxia has been made.

7.2 Materials and methods

7.2.1 Study area

The 11 fjord basins investigated in August 2003 are situated along the Norwegian Skagerrak coast (Figure 7.1 and Table 7.1). The basins were selected to represent a range in oxygen history. Based on oxygen minimum measured in bottom water during the last 5 years

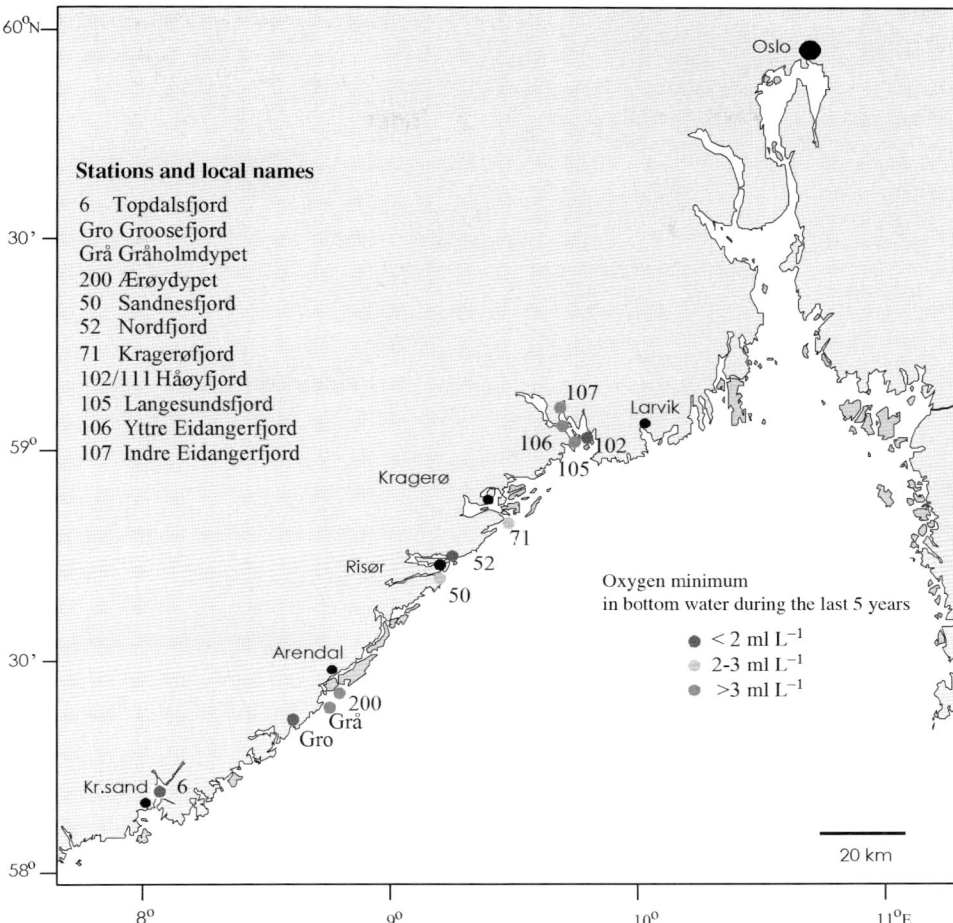

Figure 7.1 The 11 basins studied are adjacent to each other and the main differences between them are the local input of organic carbon and topography. Sill depths range from 25 to 90 m and basin depth from 64 to 200 m. (To see this figure in colour, please see colour plate 7.)

(O_2 min 5Y), the 11 fjord basins were divided into three categories: **(1)** oxygen minimum above 3.0 mL L^{-1} (stn 105, 106, 107, 200, Grå), **(2)** minimum values between 3.0 and 2.0 mL L^{-1} (stn 50, 71) and **(3)** minimum below 2.0 mL L^{-1} (stn 6, 52, 102, Gro). The fjords differ in topography (Table 7.1). The most topographically enclosed basins, stn 52, 102 and 71, have shallow sills (18–35 m) and deep basins (140–200 m) with a sill- to basin-depth ratio of ∼1/6. Fjord basins stn Gro, 6 and 50 have shallow sills (21–30 m) and shallow basins (64–75 m) and a ratio of 1/3. The basins 106, 107, 105, Grå and 200 have relatively deep sills (50–90 m) and deep basins (100–190 m) and ratio of ∼1/2.

Table 7.1 Position of sampling stations and topographic information of the investigated fjord basins.

Fjord basin	Station	Position N	Position E	Af (km²)	Ht (m)	H_{max} (m)	H_m (m)
Nordfjorden	52	58°44′	09°15′	20.00	28	180	57
Håøyfjorden	102/111	59°02′	09°48′	10.00	35	200	48
Groosefjorden	Gro	58°19′	08°36′	4.10	21	64	25
Topdalsfjorden	6	58°10′	08°04′	8.70	25	75	22
Kragerøfjorden	71	58°50′	09°28′	14.00	18	140	37
Sandnesfjorden	50	58°41′	09°10′	4.40	30	70	15
Inner Eidangerfjorden	107	59°06′	09°43′	6.20	50	100	40
Outer Eidangerfjorden	106	59°04′	09°42′	6.20	50	100	40
Langesundsfjorden	105	59°02′	09°36′	6.50	50	120	40
Ærøydypet	200	58°24′	08°46′	4.70	60	110	28
Gråholmdypet	Grå	58°22′	08°43′	3.40	90	198	46

Af, surface area; Ht, sill dept; H_{max}, maximum depth of basin; H_m, mean basin depth below sill. The fjords are ordered in three categories based on measured oxygen minimum concentration in bottom water during the last 5 years. The lower group of fjords has only experienced values higher that 3 mL L^{-1}, for the middle group minimum has been 2–3 mL L^{-1} and for the uppermost group the minimum values are bellow 2 mL L^{-1}.

7.2.2 Oxygen monitoring

Every year since 1927, oxygen, salinity and temperature were measured in Topdalsfjorden (stn 6), Nordfjorden (stn 52), Sandnesfjorden (stn 50) and Kragerøfjorden (stn 71) at fixed stations and depths (Figure 7.1). The measurements have been carried out in September/October, which is the time of the year when the oxygen concentrations are usually close to a minimum in the fjord basins (Dahl and Danielssen, 1992; Johannessen and Dahl, 1996). In Håøyfjorden (stns 102/111), Langesund/Eidangerfjorden (stns 105–107) and Ærøydypet (stn 200) regular observations in September/October started in 1952–1953. In Gråholmdypet (Grå), Sandsnesfjorden (stn 50) and Nordfjorden (stn 52) monthly observations were executed in 1990–1991. After 2000 regular observations, about 6–8 times a year, started in Håøyfjorden (stns 102/111), Nordfjorden (stn 52) and Langesund/Eidangerfjorden (stns 105–107). In Groosefjorden (Gro) oxygen was observed in 1966, 1981–1985, 1995–1998 and in 2003. In fjord basins with observations only in September/October each year, the O$_2$ min values are reduced to December – values to get more realistic O$_2$ min values before the ordinary inflow of new water to the fjord basins in the winter period, using the oxygen consumption rate given in Equation 7.1 (Aure et al., 1996):

$$\frac{dO_2}{dt} = \frac{m\left[(a-b)Ht\right]}{Hb} \text{ (mL L}^{-1}\text{ month}^{-1}\text{)} \tag{7.1}$$

where Hb is mean basin depth (m), Ht is sill depth (m) and $m = 2.43$ (mL O$_2$ gC^{-1}) where C is carbon.

Before 1980 in Skagerrak fjords: $a = 5.4$ (gC m^{-2} month^{-1}) and $b = 0.07$ (gC m^{-2} month^{-1} m^{-1}). After 1980 in Skagerrak fjords: $a = 8.5$ (gC m^{-2} month^{-1}) and $b = 0.11$ (gC m^{-2} month^{-1} m^{-1}).

7.2.3 Sampling and analysis of environmental factors

Sediment samples were collected at each station in July/August 2003 using a 0.1-m² box corer. Two plastic-core sub-samples of surface sediment (0–2 cm) were taken for analysis of grain size, water content, organic carbon (%C), and total nitrogen (%N) and one sub-sample (0–1 cm) was analysed for content of Chlorophyll a (Chl *a*) and pheopigments (Pheo). Replicate redox measurements were taken in two plastic-core samples from each box core. Redox measurements were taken every centimetre from sediment surface down to 10 cm depth using a platinum electrode introduced from the sediment surface. Water samples for oxygen measurements were siphoned from just above the sediment surface of the box core samples and analysed using the Winkler method (Winkler, 1888). Grain-size fraction <0.063 mm was determined using the pipette method (Buchanan, 1984), and the fraction >0.063 mm was dry sieved. The weight percentage of organic carbon (%C) and total nitrogen (%N) was determined with a Carlo Erba model 1106 CHN analyser. To obtain carbonate-free sediment, samples were treated with 0.25 M HCl until no visible reaction occurred and then dried at 50°C overnight. Pigments were extracted with 90% aceton and centrifuged and measured using a fluorometer (Turner Design model 10 AU).

7.2.4 Sampling and analysis of bottom fauna

Hyperbenthos was sampled with a modified Rothlisberg–Percy (RP) epibenthic sledge described in Fosså et al. (1988). The mean speed during the 15-min-long hauls was 1 knot (1.85 km h^{-1}), which gives a maximum sampled area of about 450 m². Samples were sieved through a 0.5-mm screen and preserved in 4% neutralised formalin. Before identification, the samples were transferred to 96% alcohol. Sampling took place in July–August 2003. Infauna was sampled with a 0.1-m² van Veen grab. In each fjord, one station consisting of five replicate samples was taken at the deepest point of the basin and close to the course of the RP-sledge samples. The samples were sieved on 1 mm screens and preserved in 4–6% formaldehyde in seawater. Collected specimens were identified to species level as far as possible. The samples were collected about 1 week after the RP-sledge samples.

7.2.5 Numerical methods

The linear correlation between environmental factors and the abundance (N) and diversity of hyperbenthos and infauna was investigated. Three different measures of diversity were used: number of species (S), expected number of species in a random sample, ES (sample size) (Hurlbert, 1971; Simberloff, 1978), and the diversity index H' calculated for each sample using the formula:

$$H' = -\sum_{i=1}^{S} \left(\frac{N_i}{N}\right) \log_2 \left(\frac{N_i}{N}\right)$$

where S is the total number of species, N is the total number of individuals and N_i is the number of individuals of the ith species (Shannon and Weaver, 1949).

7.3 Results

7.3.1 Environmental factors

Sediment composition and oxygen values are given in Table 7.2. All fjord basins had fine muddy bottom with a water content of 58–76% and a clay-silt fraction of 100–91%. Several of the environmental factors, i.e. O_2 in bottom water, %C, %N, %water and pigments in the sediment, were strongly inter-correlated ($r > 0.80$) (Table 7.3). The fjords with low historical oxygen levels ($O_2 < 2.4$ mL L^{-1}) had in general a high carbon content in surface sediments ($C > 5.5\%$). The exception to this pattern was stn 102/111. This is a basin with very restricted bottom-water exchange, often experiencing 5 years between bottom-water renewals. Here minimum oxygen concentration in bottom water was 0.1 mL L^{-1} even with carbon content in the sediment of 4.3%. The two fjords with permanent low oxygen concentrations, stn 52 and stn Gro, had carbon content $>7.3\%$. The fjords with high amount of organic matter in the sediment also had relatively high values for plant pigments; again stn 102/111 was an exception with 12.8–14.5 µg g^{-1} dry sediment.

7.3.2 Environmental factors and diversity of hyperbenthos and infauna

In general, the fjord basins with oxygen levels above 3 mL L^{-1} had a species-rich and diverse fauna, whereas the basins with hypoxic conditions (<2 mL L^{-1}) had an impoverished fauna (Table 7.4). The number of hyperbenthos species per sledge sample decreased from 42–55 per sample in well-oxygenated basins to 27–30 in the intermediate hypoxia situation and to 0–19 in the most hypoxic basins. The number of infauna species in five grab samples (0.5 m^2) decreased from 67–123 per station in well-oxygenated basins to 28–63 in the intermediate hypoxia situation and to 2–28 in the most hypoxic environment. The total abundances (N) generally decreased from well-oxygenated to hypoxic conditions, but the patterns were less regular for infauna where some of the highest densities were found in the basins with intermediate oxygen levels.

The correlation coefficients between environmental variables and the faunal parameters: abundance (N), number of species (S), ES and H' are shown for hyperbenthos and infauna in Table 7.5 and for dominating organism groups in Table 7.6. In general, the faunal parameters correlated best with the oxygen minimum concentration during the last 5 years (O_2 min 5Y) with exception for the abundance of infauna which correlated poorly with most environmental variables. The linear relation between faunal parameters (N, S, ES_{300}/ES_{100}, H') and (O_2 min 5Y) for all basins are illustrated in Figure 7.2. Linear regressions fitted the data well in most cases and the regression line has been calculated for each parameter (for $r > 0.65$, $p < 0.05$).

There were some clear differences in response to hypoxia between the two fauna groups. Hyperbenthos S showed the strongest correlation ($r = 0.96$) with a corresponding high fit of the linear regression with an explained variance (R^2) of 0.91. The linear correlation of ES_{300} was also high with an R^2 of 0.81. Abundance (N) also showed a clear and significant response to O_2 min 5Y ($r = 0.88$), but the response of H' was weak. Amongst the hyperbenthos fauna groups the amphipods, tanaids and ostracods appears to be most susceptible to hypoxia. For infauna, the correlation with O_2 min 5Y was equally strong for S

Table 7.2 Environmental information based on two replicate samples except for oxygen measurements.

Station number	Organic matter			Grain size and water content					O_2 water and sediment			Pigments	
	Nitrogen (%)	Carbon (%)	C/N	Clay (%)	Silt (%)	Sand (%)	Gravel (%)	Water (%)	O_2 min monitoring (mL L^{-1})	O_2 2003 (mL L^{-1})	RDL (cm)	Chl a (µg g^{-1})	Pheo (µg g^{-1})
52A	0.81	7.59	9.37	29	70	1	0	76	0.00	0.00	0.0	29.36	94.03
52B	0.90	7.25	8.06	40	59	1	0	76	0.00	0.00	0.0	25.39	94.51
102/111A	0.48	4.23	8.81	44	55	1	0	66	0.10	3.80	6.0	14.54	37.25
102/111B	0.45	4.25	9.44	42	57	1	0	70	0.10	3.80	6.0	12.75	33.22
GroA	1.05	8.58	8.17	40	58	2	0	71	0.50	0.72	0.0	80.32	146.63
GroB	1.00	8.17	8.17	42	56	1	0	70	0.50	0.72	0.0	67.19	117.08
6A	0.58	6.00	10.34	32	67	1	0	68	0.75	2.35	0.0	44.29	90.54
6B	0.56	5.99	10.70	31	68	1	0	69	0.75	2.35	0.0	45.69	91.45
71A	0.70	5.72	8.17	34	66	1	0	69	2.09	4.85	1.5	13.15	49.60
71B	0.70	5.88	8.40	31	68	1	0	69	2.09	4.85	1.5	9.21	44.91
50A	0.62	5.50	8.87	48	51	0	0	71	2.37	3.00	1.5	14.88	67.28
50B	0.60	5.45	9.08	46	54	1	0	69	2.37	3.00	1.5	16.42	68.34
106A	0.29	3.36	11.59	32	64	3	0	63	3.13	4.02	7.0	8.35	17.57
106B	0.24	3.12	13.00	29	68	3	0	64	3.13	4.02	7.0	6.46	21.73
107A	0.30	3.41	11.37	35	58	6	0	66	3.50	4.16	3.0	6.38	28.07
107B	0.22	3.01	13.68	32	60	9	0	59	3.50	4.16	3.0	5.00	30.36
105A	0.28	3.04	10.86	42	51	7	0	59	3.86	3.86	3.0	8.01	41.81
105B	0.28	3.23	11.54	42	51	7	0	60	3.86	3.86	3.0	8.85	43.14
200A	0.36	3.28	9.11	45	52	4	0	62	3.59	5.42	4.0	7.30	42.91
200B	0.37	3.34	9.03	46	50	3	0	62	3.59	5.42	4.0	13.60	25.67
GråA	0.41	3.38	8.24	50	48	1	0	64	4.25	5.25	4.0	7.05	31.18
GråB	0.39	3.33	8.54	51	47	2	0	58	4.25	5.25	4.0	9.94	28.86

RDL, redox discontinuity layer. Organic matter and grain size composition is analysed in homogenised samples from the upper 2 cm of sediment. Oxygen concentration in bottom water is given as lowest value measured by the IMR monitoring programme in the period 1999–2003 (O_2 min 5Y) and field measurements in August 2003. Depth of RDL is based on the information from two replicate redox profiles indicating the transition between oxic and anoxic sediment. Pigments are measured in homogenised samples of the upper 1 cm of the sediment and are given in µg per gram dry sediment.

Table 7.3 Linear correlation between environmental factors that faunal diversity showed the strongest response to.

Station number	Nitrogen (%)	Carbon (%)	C/N	Water (%)	O_2 min monitoring (mL L^{-1})	O_2 2003 (mL L^{-1})	RDL (cm)	Chl a (μg g^{-1})	Pheo (μg g^{-1})
%N	1.00								
%C	0.97	1.00							
C/N	−0.69	−0.54	1.00						
%Water	0.80	0.84		1.00					
O_2 min last 5Y	−0.75	−0.81		−0.83	1.00				
O_2 2003	−0.75	−0.83		−0.75	0.78	1.00			
RDL	−0.74	−0.79		−0.58		0.65	1.00		
Chl a	0.78	0.83		0.51	−0.64	−0.73	−0.64	1.00	
Pheo	0.88	0.92		0.68	−0.70	−0.85	−0.82	0.92	1.00

RDL, redox discontinuity layer.
The values are correlation coefficients, only $r > 0.50$ are listed.

and H' ($r = 0.91$) with a high liner fit ($R^2 = 0.81$). The correlation for ES$_{100}$ was lower ($r = 0.84$). The correlation between N and hypoxia was not significant ($r = 0.55$). Amongst the infauna organism groups, the molluscs appear to be most sensitive to hypoxia. The response in species richness for polychaetes was equally high to %C in the sediment and O_2 min 5Y, $r = -0.81$ and 0.82, respectively. Comparison of the hypoxia response of hyperbenthos and infauna indicates that there was a stronger relationship between the presence of species

Table 7.4 Abundance (N), number of species (S), Shannon–Wiener index (H') and Hulbert's expected number of species, ES (sample size is given in parenthesis) for samples from eleven fjord basins along the Norwegian Skagerrak coast. Numbers for infauna refers to five grab samples (0.5 m^2) and for hyperbenthos one sledge sample.

Fjordbasin, stn id	O^2 min 5Y	Infauna				Hyperbenthos			
		N	S	ES (100)	H'	N	S	ES (300)	H'
Nordfjorden, 52	0	2	2	2	1	1	0.5	0	0
Håøyfjorden, 111/102	0.1	1045	25	11.1	2.1	166	4	3.35	0.35
Groosefjorden, Gro	0.5	21	4	1.1	1.1	550	4.5	3.5	0.3
Topdalsfjorden, 6	0.75	287	28	19	3.3	341	18.5	18.35	2.45
Kragerøfjorden, 71	2.09	3742	63	17.4	3.6	4096	26.5	17.95	2.15
Sandnesfjorden, 50	2.37	2760	74	21.7	3.8	1310	30	20.2	1.95
Y. Eidangerfjorden, 106	3.5	2033	104	31	4.7	4375	44.5	26.7	2
I. Eidangerfjorden, 107	3.5	2297	123	38	5.3	4874	48	30.75	3
Ærøydypet, 200	3.59	1225	103	36.6	5.2	4698	49.5	34.35	3
Langesundsfjorden, 105	3.86	3939	116	28.4	4.4	4181	55	33.15	2.8
Gråholmdypet, Grå	4.25	897	67	28.1	4.3	7397	42	24.6	2

Table 7.5 Correlation between abundance and diversity of hyperbenthos and infauna, and environmental factors.

Environment	Hyperbenthos				Infauna			
	N	S	H'	ES (300)	N	S	H'	ES (100)
%Carbon	−0.70	−0.82	−0.67	−0.78		−0.84	−0.85	−0.71
Carbon/Nitrogen		0.52		0.52		0.61	0.52	
%Water	−0.79	−0.84	−0.68	−0.79		−0.77	−0.77	−0.65
O_2 min 5Y	**0.88**	**0.96**	**0.78**	**0.91**	0.55	**0.90**	**0.90**	**0.84**
O_2 in 2003	0.76	0.73	0.69	0.73	0.54	0.72	0.82	
RDL (cm)						0.58	0.56	
Chl a (μg g^{-1})	−0.58	−0.63		−0.58	−0.61	−0.71	−0.69	
Pheo (μg g^{-1})	−0.69	−0.70	−0.57	−0.65	−0.52	−0.76	−0.77	−0.51
Sill depth (m)	0.77	0.65		0.58		0.53	0.60	0.60

RDL, redox discontinuity layer; N, abundance; S, number of species; H', Shannon–Wiener index; ES, Hulbert's expected number of species, sample size is given in parenthesis.
Linear correlation $r > 0.50$ is shown and for $r > 0.65$ and $p < 0.05$.
"Strong correlations are indicated using bold numbers."

of hyperbenthos and oxygen levels in the bottom water than for infauna. The infauna, however, showed a clearer response in H' suggesting that changes in dominance of species (evenness) as response to hypoxia is stronger in this group rather than species richness.

The correlation of faunal parameters with %C and %water in the sediment was also strong, and this is most likely explained by the inter-correlation between the environmental factors (Table 7.3). It may be noted that the bottom oxygen measured at the same time as the samples were taken (O_2 in 2003) consistently correlated less well than oxygen minimum concentration during the last 5 years (O_2 min 5Y). This suggests that the fauna reflects the near history of oxygen conditions in the fjord basins rather than the present oxygen situation. The number of species for all organism groups are strongest correlated to O_2 min 5Y (Table 7.6).

7.3.3 Biodiversity loss

Historical mean oxygen minimum values in the fjord basins before and after 1980 are summarised in Figure 7.3. All basins have been subjected to increased oxygen consumption due to increased sedimentation of organic material after about 1980. A main source to the problem has been nutrient and material coming from the Southern North Sea and Kattegat. This resulted in 50–100% increase in the oxygen consumption and consequently decreased oxygen minimum concentrations in the sill basins along the Skagerrak coast (Aure et al., 1996, 1998).

The loss of biodiversity due to increased eutrophication after 1980 has been estimated on the basis of the observed linear relationships between O_2 min 5Y and species richness (S) for hyperbenthos and infauna. The regression equations for hyperbenthos (Sh = 11.9 × O_2 min 5Y + 2.9) and infauna (Si = 24.5 × O_2 min 5Y + 10), for the interval 0–3.5 mL L^{-1} (Figure 7.2), were used together with historic minimum oxygen concentration for the fjord basin before and after 1980 (Table 7.7) to calculate number of species present. The upper limit of 3.5 mL L^{-1} for the linear relation was based on the observations

Table 7.6 Correlation between abundance and diversity (number of species, S) of the main organism groups of hyperbenthos and infauna, and environmental factors.

Environmental factors	Hyperbenthos							Infauna			
	Amphipoda	Cumacea	Isopoda	Decapoda	Tanaidacea	Mysidacea	Ostracoda	Polychaeta	Mollusca	Crustacea	Echinodermata
% Carbon	−0.77	−0.77	−0.79	−0.58	−0.80	−0.84	−0.75	**−0.81**	−0.77	−0.73	−0.62
Carbon/Nitrogen	0.60	0.65	0.56		0.53	0.75		0.60		0.73	
% Water	−0.81	−0.76	−0.91	−0.66	−0.88	−0.87	−0.82	−0.77	−0.77	−0.68	**−0.76**
O_2 min 5Y	**0.91**	**0.87**	**0.88**	**0.77**	**0.91**	**0.89**	**0.90**	**0.82**	**0.93**	**0.79**	**0.75**
O_2 2003	0.67	0.64	0.65	**0.75**	0.70	0.62	0.68	0.70	0.74	0.55	0.61
RDL (cm)					0.51	0.50				0.58	
Chl a μg g^{-1}	−0.61	−0.65	−0.52	−0.44	−0.64	−0.64	−0.60	−0.68	−0.71	−0.61	−0.53
Pheo μg g^{-1}	−0.63	−0.69	−0.65	−0.47	−0.72	−0.70	−0.60	−0.71	−0.71	−0.71	−0.53
Sill depth	0.54		0.80	0.55	0.66	0.59	0.78		0.80	0.55	

RDL, redox discontinuity layer.
Linear correlation $r > 0.50$ is shown and for $r > 0.65$ and $p < 0.05$
"Strong correlations are indicated using bold numbers."

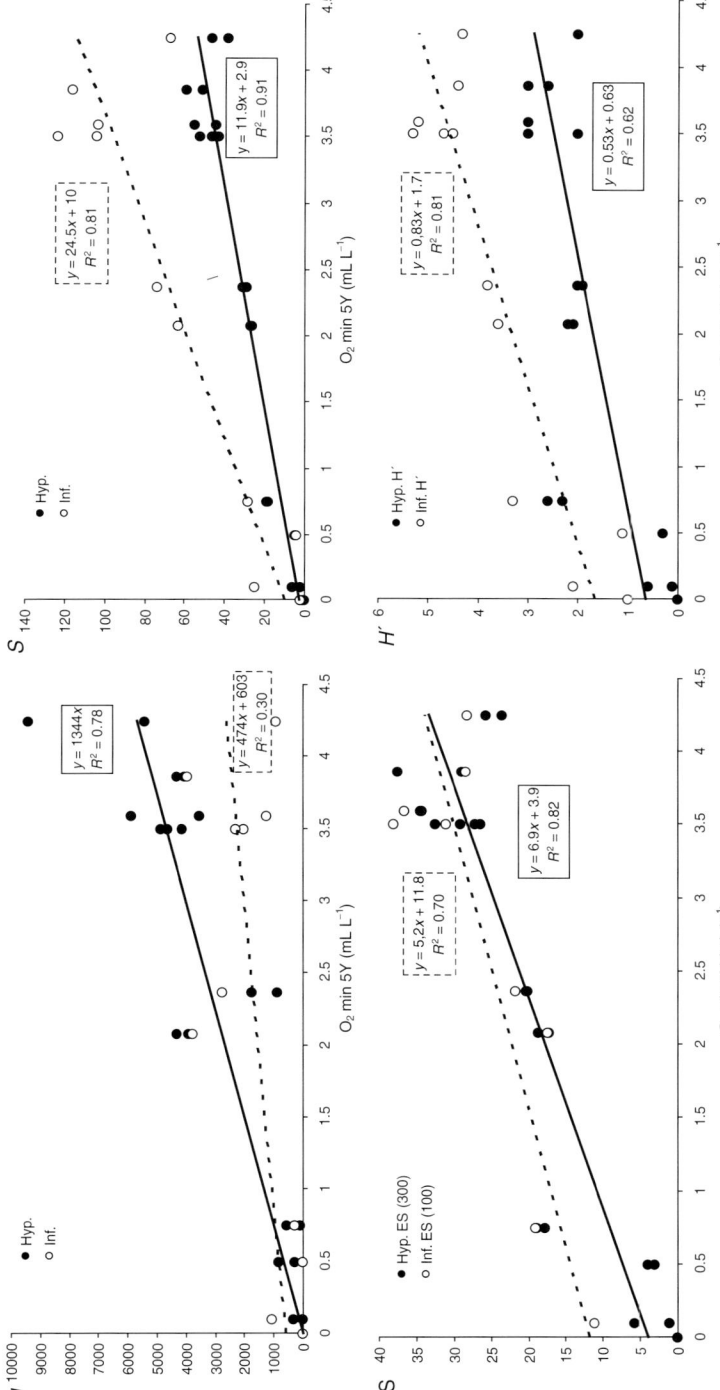

Figure 7.2 Plot of linear relation between oxygen minimum during the last 5 years (O_2 min 5Y) and the abundance and diversity of hyperbenthos (Hyp) and infauna (Inf). N, abundance; S, number of species; ES, expected number of species (Hurlbert's index) and H', Shannon–Wiener's diversity index. Hyperbenthos is represented by two sledge samples per station and infauna (open circles) is summed value from five grab samples. Regression line is shown together with R^2 and the equation of the line.

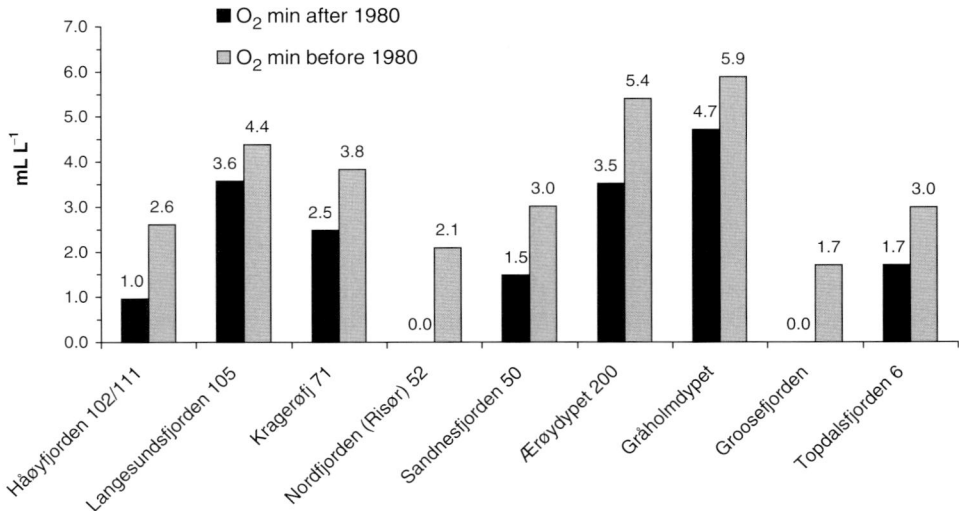

Figure 7.3 Mean oxygen minimum (mL L^{-1}) in the 11 investigated basins before and after 1980.

that at higher O$_2$ min 5Y there seemed to be no further systematic change in fauna composition (Figure 7.2). This is also to be expected because increased oxygen concentration above what represents a healthy bottom environment should have little effect on the bottom fauna and thus not change the composition. Thus a maximum number for S is estimated as a presumed reference value for oxygen levels higher than 3.5 mL L^{-1}, this number is 50 for hyperbenthos and 110 for infauna. The species loss since 1980 is the difference in species number between the two periods before and after 1980. The estimated loss of biodiversity in each fjord is illustrated in Figure 7.4. The 'group 1' fjords, which have the lowest oxygen levels and have experienced a marked worsening of the oxygen condition, have lost 50–90% of the bottom species. The 'group 2' fjords have lost 50–35% of the species, whereas 'group 3' fjords appear not to have been affected. The results suggest that species richness of infauna and in particular of hyperbenthos are a useful assessment parameter for eutrophication.

7.4 Discussion

At present there is a need for ecological quality criteria in environmental management, especially for the marine bottom environment as was highlighted at a workshop on ecological quality objectives (EcoQOs) for the North Sea (Anon., 1999; Lanters et al., 1999). Early warnings of environmental degradation are essential for a precautionary management of marine systems and thus the identification of sensitive indicator species is crucial. It is an established method to use infauna to monitor changes in marine bottom environment. However, it is also known that organisms living in the sediment that are not capable of avoiding bad conditions are less susceptible to environmental changes compared to mobile fauna.

Table 7.7 Species loss in the soft bottom fauna of fjord basins along the Skagerrak coast due to increased hypoxia after 1980.

Fjord	O_2 min (mL L^{-1})			Number of species before 1980		Number of species after 1980		% species loss	
	Before 1980	After 1980	Difference O_2	Hyperbenthos	Infauna	Hyperbenthos	Infauna	Hyperbenthos	Infauna
Nordfjorden	2.1	0	2.1	27.9	61.5	2.9	10.0	90	84
Groosefjorden	1.7	0	1.7	23.1	51.7	2.9	10.0	87	81
Håøyfjorden	2.6	1	1.6	33.8	73.7	14.8	34.5	56	53
Topdalsfjorden	3	1.7	1.3	38.6	83.5	23.1	51.7	40	38
Sandnesfjorden	3	1.5	1.5	38.6	83.5	20.8	46.8	46	44
Kragerøfj	3.8	2.5	1.3	50.0	110.0	32.7	71.3	35	35
Langesundsfjorden/Eidangerfjorden	4.4	3.6	0.8	50.0	110.0	50.0	110.0	0	0
Ærøydypet	5.4	3.5	1.9	50.0	110.0	50.0	110.0	0	0
Gråholmdypet	5.9	4.7	1.2	50.0	110.0	50.0	110.0	0	0

Species loss is estimated on the basis of the observed relation between species number and oxygen minimum (number of species of hyperbenthos = $11.9 \times O_2$ min + 2.9 and for infauna = $24.5 \times O_2$ min + 10.0 for the interval 0–3.5 mL L^{-1}). Expected number of species when O_2 min > 3.5 mL L^{-1} is 50 for hyperbenthos and 110 for infauna. Oxygen minimum values are from the monitoring programme conducted by the Institute of Marine Research (IMR), Norway.

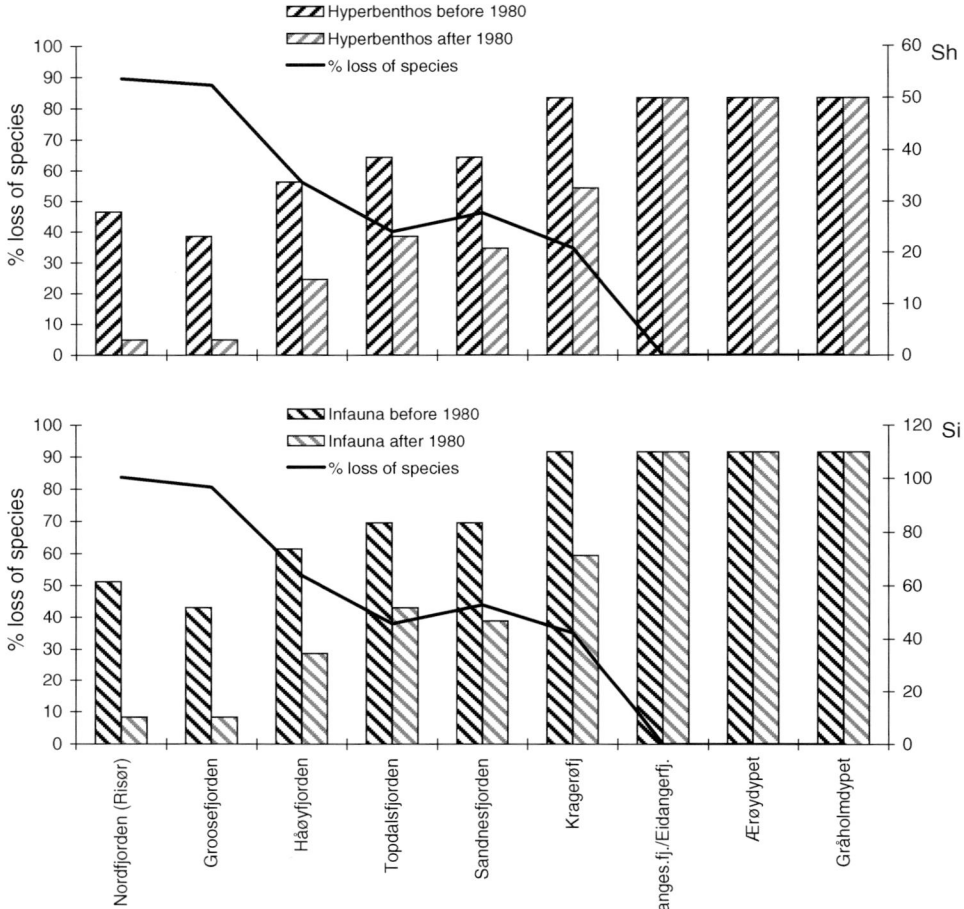

Figure 7.4 Species loss in the soft bottom fauna of fjord basins along the Skagerrak coast due to increased hypoxia after 1980. Species loss is estimated on the basis of the observed relation between species number and oxygen minimum. Sh: number of species of hyperbenthos = 11.9 × O$_2$ min + 2.9 and Si: number of infauna species = 24.5 × O$_2$ min + 10 for the interval 0–3.5 mL L^{-1}. Expected number of species when O$_2$ min > 3.5 mL L^{-1} is 50 for hyperbenthos and 110 for infauna. Oxygen minimum values are from a monitoring programme conducted by the Institute of Marine Research (IMR) (Johannessen and Dahl, 1996).

This study of infauna and hyperbenthos in fjord basin that represents different states of hypoxia has confirmed the general relationships between bottom fauna diversity and oxygen concentration of a correlated decrease (Pearson and Rosenberg, 1978). The relationships between oxygen concentration and species richness were generally stronger for hyperbenthos than for infauna. This supports earlier studies indicating that crustaceans are a particularly hypoxia sensitive group (Theede, 1973; Dries and Theede, 1974; Josefson and Widbom, 1988; and studies cited in Diaz and Rosenberg, 1995). Probably the infauna is more dependent on other factors such as the composition of the sediment. The dominating group in numbers of species of this fauna is polychaetes and most likely many of these species are adapted to sustain periods of low oxygen concentration in the sediment. Interestingly species richness of polychaetes correlates equally well with oxygen concentration and %C in the sediment. Amongst the infauna organisms, molluscs seemed to be the most susceptible group. Of the crustacean groups belonging to the hyperbenthos fauna amphipods, tanaids and ostracods showed the strongest response to hypoxia and thus would be good candidates for indicators that can provide an early warning of changes in oxygen conditions. Studies of amphipods (Buhl-Jensen, 1986; Buhl-Jensen and Fosså, 1991; Buhl-Mortensen, 1996) and molluscs and foraminifera (Buhl-Mortensen and Høisæter, 1993; Klitgaard-Kristensen and Buhl-Mortensen, 1999) in fjords have shown that in particular not only the amphipod fauna but also molluscs decrease in species number in response to increased content of organic carbon in the sediment. Unfortunately, the detailed oxygen history of the fjords in these studies was not known. Our results show that to be able to monitor early changes in the bottom fauna as a response to eutrophication the standard method of infauna monitoring should be complimented with sledge samples of 'hyperbenthos'. The much stronger correlation between fauna patterns and the oxygen minimum during the last 5 years of the fjord basins than with oxygen concentrations measured when sampling the fauna in 2003 suggests that an earlier situation of hypoxia may affect the fauna for several years. This implies that detailed knowledge of the oxygen history of fjord basins is crucial to be able to document and relate faunal changes to hypoxia. It is important to recognise that the response to a period of restricted bottom-water exchange in fjords will affect the bottom-fauna composition for several years. Thus a mismatch of low diversity and high oxygen concentration may occur after such a period. Such a 'mismatch' is hardly understood if oxygen is not monitored frequently and reflects the near-bottom minimum concentration during longer periods.

In a recent study, Tett et al. (2003) aimed to understand the processes that determine the trophic status of fjords through the development of models to define, predict and assess eutrophication. They found that occasional extremes could control situation and highlighted the importance of choice of parameters; for example, 'the level of extreme to enter into the model: should it be the summer maximum in a typical year, or the greatest value observed, for example, during 1 year in 10?' Because there is a clear link between fjord topography and bottom-water exchange and between effects of organic enrichment on near-bottom oxygen concentration and bottom-water exchange (Aure and Stigebrant, 1989, 1990), the same load of organic matter (level of eutrophication) will affect the oxygen level differently in fjord basins depending on topography. These are features of fjords that are well known; however, they implies that to be able to document the effects of eutrophication on the bottom environment, it is crucial to undertake frequent, preferable monthly, monitoring of near-bottom oxygen concentration to capture the resent history that has affected the fauna present in the fjord.

In this study, we have been able to show a very strong connection between hypoxia and species richness of bottom fauna in fjords because regular measurements of bottom-water oxygen concentrations were available from most of the investigated fjords going back to the 1950s (Johannessen and Dahl, 1996). Based on the recorded oxygen history and the observed linear relation between species richness and hypoxia, we calculated what the species number (S) was before 1980 and after to get an estimate of species loss. This would not have been possible without the historic oxygen information. The estimated loss of species should be taken as an indication of the magnitude of loss due to eutrophication not as an exact number. However, it is our hope that the model we have developed will prove useful not only to estimate losses due to eutrophication, but also to indicate what is the historic 'good' and 'high' ecological status the fjords should be restored to by 2015 to fulfil the goals of the Water Framework Directive (WFD) (C.E.C., 2000).

7.5 Conclusions

- The species richness of bottom fauna was strongly correlated ($r > 0.90$) to bottom-water oxygen concentration. The correlation was clearly stronger for hyperbenthos than for infauna.
- Hyperbenthos are accordingly good candidates as sensitive indicators of eutrophication. The most sensitive groups were amphipods, tanaids and ostracods.
- The relation between minimum oxygen concentration and species richness can be used to estimate loss of species on the basis of historical records of changes in near-bottom oxygen. The general relationship between hyperbenthos species-richness Sh and bottom oxygen conditions of fjord basins in the study area is Sh = $12 \times O_2$ minimum during the last 5 years (mL L^{-1}) + 2.9. It is important to recognise that the response to a period of restricted bottom water of fjords will affect the fauna composition for several years after improved oxygen conditions. If the 5-year oxygen minimum value is not used, a mismatch of low diversity and high oxygen concentration may occur due to effects of a recent hypoxia situation.
- Estimated species loss in the 11 fjords along the Skagerrak coast indicates that many of the fjords have lost 50% or more of their bottom-fauna species due to eutrophication since 1980.

References

Anon. (1999) Workshop on the Ecological Quality Objectives (EcoQOs) for the North Sea. *TemaNord* **591**, 72.

Aure, J., Dahl, F.E., Golmen, L.G., Johannessen, T. and Molvær, J. (1997) Vurdering av oxygenutvikling og organisk belastning på kyststrekningen Jomfruland-Stavanger, report no. 3555–96. Norwegian Institute for Water Research, Oslo.

Aure, J., Danielsen, D. and Sætre, R. (1996) Assessment of eutrophication in Skagerrak coastal waters using oxygen consumption in fjordic basins. *ICES Journal of Marine Science* **53**, 589–595.

Aure, J., Danielsen, D.S. and Svendsen, E. (1998) The origin of Skagerrak coastal water off Arendal in relation to variations in nutrient concentrations. *ICES Journal of Marine Science* **55**, 610–619.

Aure, J. and Stigebrant, A. (1989) On the influence of topographic factors upon the oxygen consumption rate in sill basins of fjords. *Estuarine, Coastal and Shelf Science* **28**, 59–69.

Aure, J. and Stigebrant, A. (1990) Quantitative estimates of the eutrophication effects of fish farming on fjords. *Aquaculture* **90**, 135–156.

Buchanan, J.B. (1984) Sediment analysis. In: Holme, N.A. and McIntyre, A.D. (eds). *Methods for the Study of Marine Benthos*. IBP Handbook 16, 2nd edn. Blackwell, Oxford, pp. 41–65.

Buhl-Jensen, L. (1986) The benthic amphipod fauna of the west-Norwegian continental shelf compared with the fauna of five adjacent fjords. *Sarsia* **71**, 193–208.

Buhl-Jensen, L. and Fosså, J.H. (1991) Hyperbenthic crustacean fauna of the Gullmarfjord area (western Sweden): species richness, seasonal variation and long-term changes. *Marine Biology* **109**, 245–258.

Buhl-Mortensen, L. (1996) Amphipod fauna along an offshore-fjord gradient. *Journal of Natural History* **30**, 23–49.

Buhl-Mortensen, L. and Høisæter, T. (1993) Mollusc fauna along an offshore-fjord gradient. *Marine Ecology Progress Series* **97**, 209–224.

C.E.C. (2000) Directive 2000/60/EC of the European Parliament and of the Council of 23 October 2000 establishing a framework for Community action in the field of water policy. *Official Journal of the European Communities* **L 327**, 1–73.

Dahl, E. and Danielssen, D.S. (1992) Long-term observations of oxygen in the Skagerrak. *ICES Marine Science Symposia* **195**, 455–461.

Diaz, R.J. and Rosenberg, R. (1995) Marine benthic hypoxia: a review of its ecological effects and the behavioural responses on benthic macrofauna. *Oceanography and Marine biology* **33**, 245–303.

Dries, R.R. and Theede, H. (1974) Saurstoffmangelresistenze mariner Bodenvertebraten aus der Westlichen Ostsee. *Marine Biology* **25**, 327–333.

Fosså, J.H., Larsson, J. and Buhl-Jensen, L. (1988) A pneumatic, bottom activated, opening and closing device for epibenthic sledges. *Sarsia* **73**, 299–302.

Gray, J.S. and Pearson, T.H. (1982) Objective selection of sensitive species indicative of pollution-induced changes in benthic communities. I. Comparative methodology. *Marine Ecology Progress Series* **9**, 111–119.

Hurlbert, S.H. (1971) The nonconcept of species diversity: a critique and alternative parameters. *Ecology* **52** 577–586.

Johannessen, T. and Dahl, E. (1996) Declines in oxygen concentrations along the Norwegian Skagerrak coast, 1927–1993: a signal of ecosystem changes due to eutrophication? *Limnology and Oceanography* **41**, 766–778.

Josefson, A.B. and Widbom, B. (1988) Differential response of benthic macrofauna and meiofauna to hypoxia in the Gullmar fjord-basin. *Marine Biology* **100**, 31–40.

Klitgaard-Kristensen, D. and Buhl-Mortensen, L. (1999) Benthic foraminifera along an offshore-fjord gradient: a comparison with amphipods and molluscs. *Journal of Natural History* **33**, 317–350.

Lanters, R.L.P., Skjoldal, H.R. and Noji, T.T. (1999) *Ecological Quality Objectives for the North Sea*. Fisken og Havet **10**, 57 pp.

Pearson, T.H., Gray, J.S. and Johanessen, P.J. (1983) Objective selection of sensitive species indicative of pollution-induced changes in benthic communities. 2. Data analysis. *Marine Ecology Progress Series* **12**:237–255.

Pearson, T.H. and Rosenberg, R. (1978). Macrobenthic succession in relation to organic enrichment and pollution of the marine environment. *Oceanography and Marine Biology Annual Review* **16**, 229–311.

Shannon, C.E. and Weaver, W. (1949) *The Mathematical Theory of Communication.* University of Illinois Press, Urbana, 117 pp.

Simberloff, D. (1978) Use of rarefaction and related methods in ecology. In: Dickson, K.L., Carins, J. and Livingstion R.J. (eds). *Biological Data in Water Pollution Assessment: Quantitative and Statistical Analyses, ASTM STP 652*. American Society for Testing and Materials, Philadelphia, pp. 150–165.

Tett, P., Gilpin, L., Svendsen, H., Erlandsson, C.P., Larsson, U., Kratzer, S., Fouilland, E., Janzen, C., Lee, J-Y., Grenz, C., Newton, A., Ferreira, J.G., Fernandesa, T. and Scory, S. (2003)

Eutrophication and some European waters of restricted exchange. *Continental Shelf Research* **23**, 1635–1671.

Theede, H. (1973) Comparative studies on the influence of oxygen deficiency and hydrogen sulphide on marine bottom invertebrates. *Netherlands Journal of Sea Research* **7**, 244–252.

Winkler, L.W. (1888) Die Bestimmung des in Wasser gelösten Sauerstoffen. *Berichte der Deutschen Chemischen Gesellschaft* **21**, 2843–2855.

Chapter 8
The Effects of Human Impact on Benthic Foraminifera in the Augusta Harbour (Sicily, Italy)

Elena Romano, Luisa Bergamin, Maria Grazia Finoia, Maria Celia Magno, Isabel Mercatali, Antonella Ausili and Massimo Gabellini

Abstract

The Augusta harbour (Sicily, Italy) is a natural bay enclosed by dams, characterised by low hydro-dynamism and connected with the open sea by two narrow inlets. Currently, it is one of the most important Italian harbours and shelters a big petrochemical pole. The research is focused on two critical areas of the harbour, located in front of industrial plants, close to wharfs. A total of 20 samples were analysed for benthic foraminifera, heavy metals (As, Cd, Cr, Cu, Fe, Hg, Ni, Pb, Zn), polycyclic aromatic hydrocarbons (PAHs), polychlorinated biphenyls (PCBs) and total organic carbon (TOC). Very strong pollution, mainly due to Hg, PAHs and PCBs, was recorded in the southern sector of the harbour. The distributions of pollutants and foraminifera were compared by means of the coinertia analysis and the results showed a set of 'pollution-tolerant' species correlated to the main pollutants. The abundance of pollution-tolerant species associated with the most polluted samples, the presence of stunted assemblages, and significant percentages of abnormal specimens contributed to recognise an environmental stress due to high pollution in the benthic environment of the southern sector of the harbour.

8.1 Introduction

The Augusta harbour is a complex area affected by a high degree of contamination mainly due to harbour and industrial activities.

The use of foraminifera as environmental indicators in marine and transitional environments has been successfully implemented since the end of the 1950s (see Nigam et al., 2006, for a review of the topic). Due to their short life cycle, abundance in sediments and high taxonomic diversity, benthic foraminifera offer a considerable potential response to environmental changes and, particularly, to pollution, which determines modifications on the whole assemblage and on single specimens. Diversity and density generally decrease, while the assemblage composition and structure may change. The increase in abundance of few pollution-tolerant species and the decrease of pollution sensitive species determine a faunal shift and the lowering of the assemblage evenness. In addition, single specimens may be affected by abnormalities and changes in chemistry and microstructure of the test. Finally, the size of foraminifera may be reduced (Yanko et al., 1999). The response of foraminifera to the anthropogenic impact in harbours has been already studied by several authors such as Vilela et al. (2004), Burone et al. (2006) and Ferraro et al. (2006). Signals of environmental stress such as lowered foraminiferal density and diversity, increased percentages of abnormal tests and reduced size were recognised in the above studies.

The research aimed at the environmental characterisation of the Augusta harbour, including the study of benthic foraminifera, started in 2005. The results from two critical areas near the wharfs of the industrial plants are presented and discussed in this study. Aim of this work is to characterise marine sediments with regards to grain size, chemical contaminants and foraminiferal content, with focus on the response of benthic foraminifera to environmental stress determined by pollution. The characterisation was carried out by identifying foraminiferal features that indicate environmental stress, and by correlating them with chemical and textural characteristics of sediments.

8.2 Materials and method

8.2.1 The study area

The Augusta harbour (Figure 8.1) is comprised in a natural bay 8 km long and 4 km wide (Azzaro et al., 2001). It is delimited in the northern sector by the Augusta town and closed to the south and east by artificial dams, built in the early 1960s. Two main inlets allow connection with the open sea. The north-western part of the harbour is influenced by the outflows of the streams Mulinello, Marcellino and Cantera. Their contributions, with regards to freshwater and sediments, are conditioned by seasonality and are generally poor (Anon., 1995). In the southernmost part of the harbour patchy areas of emerging rocky substrate are present nearshore and near the dams. The whole harbour is a micro-tidal area characterised by poor water circulation with a general clockwise motion and an average speed of 0.4 km h^{-1} (Anon., 1992). Such low hydrodynamics is unfavourable to significant sediment transport, while a limited re-mobilisation may occur due to shipping activity.

The Augusta harbour may be considered one of the most important Italian harbours for bunker operations, ships repairs and maintenance, and goods loading and unloading. Besides, it is also one of the most important petrochemical poles worldwide, with a commercial port devoted mainly to oil refineries and, particularly, to the production of 'green' (lead-free) petrol. Although the first industries were built along the coast in the last years of the nineteenth century, it was in the 1950s that industrialisation increased dramatically. Since then, the strong industrial activity has determined heavy consequences on the terrestrial and marine environment. Particularly, marine sediments have been polluted by heavy metals and polycyclic aromatic hydrocarbons (PAHs). In addition, untreated urban and agricultural wastes seem responsible for eutrophication phenomena due to the low water exchange in the enclosed environment of the bay. However, such eutrophication does not lead to severe hypoxia in the benthic environment (Magazzù et al., 1995).

The Effects of Human Impact on Benthic Foraminifera in the Augusta Harbour   99

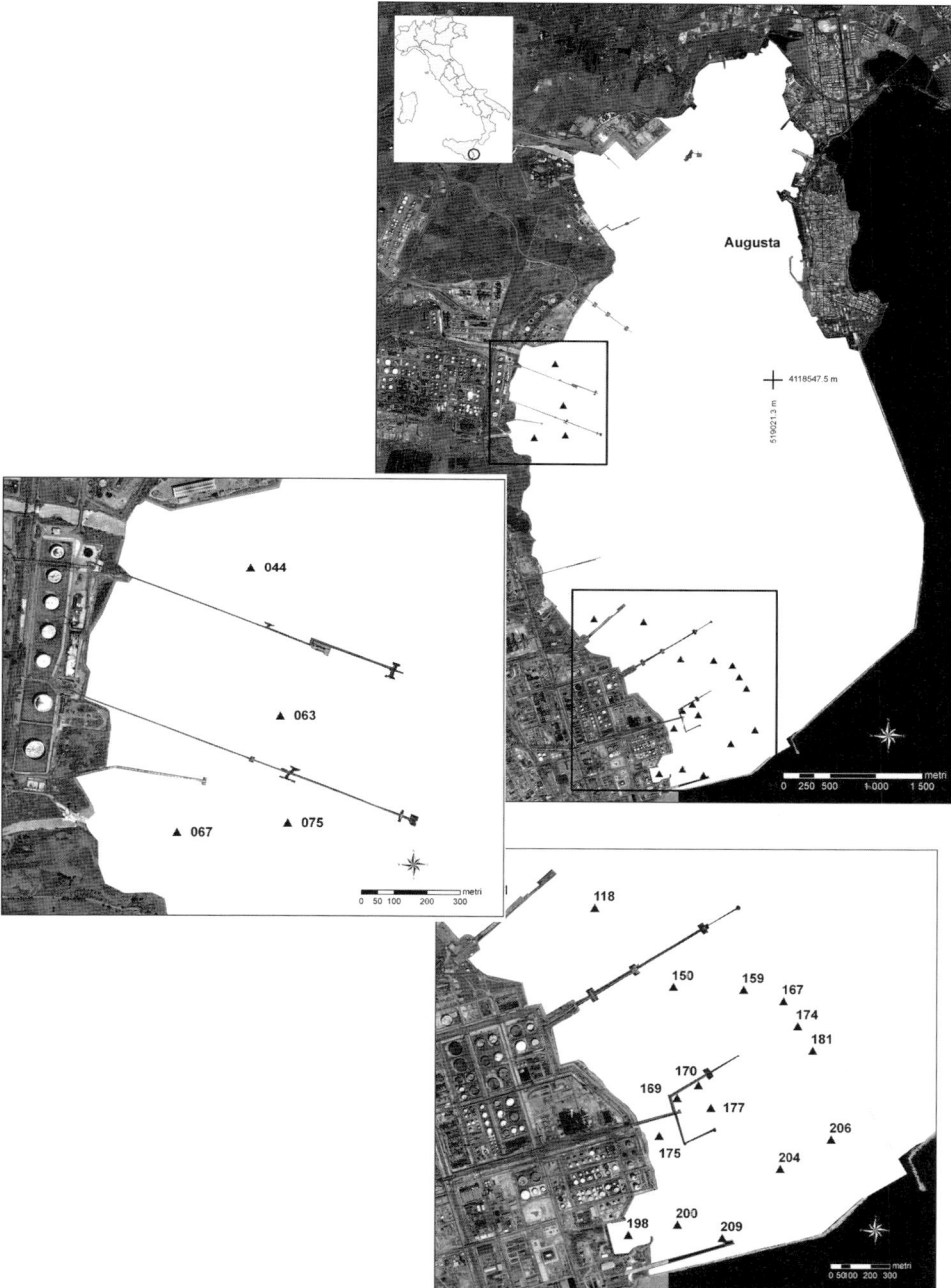

Figure 8.1   Location map of the study area and sampling stations.

Due to this environmental deterioration, The Augusta harbour was included among the 'contaminated sites of national relevance' in 1999 by the Italian legislation.

8.2.2 *Sampling*

A total of 20 samples were collected using a Van Veen grab during a cruise carried out in February 2005. Five distinct sub-samples were taken for the analyses on foraminifera and for the determination of grain size, metals and trace elements (As, Cd, Cr, Cu, Fe, Hg, Ni, Pb, Zn), PAHs, polychlorinated biphenyls (PCBs) and total organic carbon (TOC).

8.2.3 *Analyses*

8.2.3.1 *Grain-size and chemical analyses*

Samples for grain-size analysis were treated with a 30% H_2O_2 solution and separated by wet sieving. The >63 µm fraction was dried and fractionated by ASTM series sieves, while the fine fraction was analysed by the x-ray sedigraph in order to determine silt and clay fractions separately (Romano et al., 1998).

Metals and trace elements were determined using homogenised sample aliquots and analysed by atomic absorption spectrometry, according to Giani et al. (1994).

The analysed PAHs include 15 of the 16 ones indicated by Environmental Protection Agency (EPA) as important toxicological contaminants. Analyses were performed by a preliminary extraction and a successive purification on silica gel treatment; successively, the determination by HPLC with spectrofluorimetric detector was carried out (Ausili et al., 1998).

PCBs were analysed by gas chromatography with electron capture detector, following the procedures described by Cicero et al. (2003).

For TOC analyses, the dried samples were weighted in silver cups and treated with hydrochloric acid (1 N, 6 N, 8 N) to oxidise inorganic carbonate. To determine the TOC, a CHNS-O EA 1110 Thermo Electron was used. The sample, held in a lightweight silver cup, was energetically oxidised yielding a gas mixture, which was swept into a chromatographic column from which any eluted pure combustion gas passed through a thermo-conductivity detector. It generated an electrical output signal proportional to the amount of eluted gas. The resulting components were detected as CO_2. Marine sediment reference material (MESS-2) was used for the calibration curve. The value of total carbon (2.14 ± 0.03) is certified by the National Research Council Canada. The results are expressed in percentage of dry weight.

8.2.3.2 *Foraminifera*

Samples were collected from the grab opening the windows in the upper part of the device. The top 2 cm of the central, most undisturbed part of the retained sediment were taken. Although some fine sediment may be washed away during the recovery operations, the Van Veen grab is considered suitable for studies on the foraminiferal total assemblage (Schafer, 2000; Scott et al., 2001). Samples were washed over a 63-µm sieve in order to eliminate mud particles and then oven dried at 40°C. The quantitative analysis was based on the count of at least 300 specimens (Blanc-Vernet, 1969). Foraminifera are Protozoa (Phylum

Granuloreticulosa, Class Foraminifera) bearing a mineralised shell called 'test'. Only well-preserved tests without breakages or abrasion signs were picked, counted and classified.

The use of the total or living assemblage, as describers of the mean environmental conditions, is extensively debated by researchers and several significant contributions have improved the discussion. Scott and Medioli (1980) assessed the validity of the use of the total assemblage in ecological studies. They found that the high seasonal variability of the living assemblage may be attributed to seasonal meteorological changes and does not represent changes in the prevailing marine environment. Also, Debenay et al. (2001) and Arminot du Châtelet et al. (2004) preferred to study the total assemblage as indicator of average environmental conditions for the study of the foraminiferal response to pollution in harbour areas. Morvan et al. (2006) found that isolated samplings of living assemblages may provide different or even contradictory results, while only the total assemblage is useful to obtain integrated information over a given period of time, taking into account the possible influence of taphonomic processes.

On the other hand, Murray (1982) stated that ecological studies must be based only on the living assemblage, carried over a period of time, in order to determine the relationships between living and dead assemblages. Alve and Murray (1994) found that, due to post-mortem processes such as dissolution of calcareous tests or transport influencing the dead assemblage, only the study on the living assemblage is reliable. Murray (2000) evidenced the main problem concerning the total assemblages: they combine data on living assemblages (not influenced by taphonomic changes) together with those on dead assemblages (modified by taphonomic processes). In addition, the proportion of live and dead tests is influenced by several factors such as the thickness of sampling, temporal variations of standing crop and the sedimentary rate.

From this discussion emerges that the living assemblage is certainly autochthonous, but it is affected by wide temporal changes due to the high irregularity of the foraminifera life cycles and to their patchy distribution. Consequently, it may be considered representative of the overall environmental conditions only if several samplings are carried out over the year. Conversely, the total assemblage is affected by post-mortem processes, but it may be considered as representative of the average environmental conditions during the time span corresponding to the deposition of the sediment sample. It may be preferred in environments in which taphonomic processes are limited and autochthonous/allochthonous specimens may be recognised, because it is simpler, more practical and low cost.

The total assemblage was considered suitable for the ecological study in this survey because the main taphonomic processes that may affect it, such as post-mortem transport, dissolution of calcareous tests and desegregation of agglutinated tests have a limited influence in this area. A large-scale sediment transport that modifies significantly the foraminiferal assemblage may be considered absent in the Augusta harbour due to the low hydro-dynamism and the micro-tidal conditions. Only a small-scale sediment re-mobilization due to the shipping activity may be hypothesised. In addition, the abundance of thick porcellaneous taxa without evident signs of calcite dissolution suggests that water conditions are favourable to the deposition of calcite, while post-mortem dissolution, in general, should not be a common process. Finally, the low agglutinated/calcareous ratio that generally characterises the Mediterranean coastal assemblages gives scarce importance to a potential loss of agglutinated tests.

The generic classification was made according to Loeblich and Tappan (1987), and most species were determined by comparison with those figured by Jorissen (1988),

Cimerman and Langer (1991) and Sgarrella and Moncharmont-Zei (1993). The foraminiferal number (FN) (Schott, 1935) was calculated as number of specimens per gram of dry sediment (specimens/g), while the relative abundance of abnormal specimens was reported as the foraminiferal abnormality index (FAI) (Coccioni et al., 2005). Species with attached lifestyle were not included in the count of abnormal specimens because their abnormality could be due to the substrate irregularity (Geslin et al., 2000). The species diversity was given by the α-index (Fisher et al., 1943) which was calculated by using the statistical package PAST (PAlaeontological STatistics).

8.2.3.3 Data processing

Two distinct matrices, the first one containing the results of chemical and grain-size analyses and the second one with the results of the quantitative analysis on benthic foraminifera, were prepared for the statistical analysis. As regards foraminifera, the relative abundance of the 22 commonly occurring species, i.e. species more abundant than 5% in at least one sample, was used (Kovach, 1987, 1989). Two samples PR067 and PR075 were not included in the matrix due to the very low foraminiferal abundance that did not allow the count of a statistically significant number of specimens. The coinertia analysis (CA) was applied in order to measure the concordance between two datasets, plotting two distinct multivariate analyses on the same factorial plain (Dolédec and Chessel, 1994; Dray et al., 2003). The main function of this data analysis is the matching of two data tables: a 'sites × environmental variables' table and a 'sites × species' table, in order to study the relationships between species distribution and their environment. The Monte Carlo test (rv), a non-parametric test based on the results of 999 random permutations processed on tables, was applied to the CA in order to verify the goodness of the matching of the two data tables (Heo and Gabriel, 1997). In addition, a correlation matrix was calculated using the Spearman's coefficient (rho) for evidencing the non-parametric correlation between pair of variables (Best and Roberts, 1975). Such coefficient was preferred to the most used Pearson's one because it does not require the assumption that the correlation between variables is linear and because it evidences more easily correlations when the number of cases is low.

8.3 Results

8.3.1 Grain-size and chemical analyses

The grain-size analysis (Table 8.1) showed heterogeneous sediments, ranging from sand to silty clay. The Shepard diagram (Figure 8.2) exhibits samples aligned along a rather constant silt/clay ratio, with highly variable sand percentage. Although a general trend with the decreasing percentages of sand from nearshore to offshore may be recognised, sediment distribution on sea-bottom appeared to be rather patchy. From the examination of the sandy fraction under stereomicroscope, the dominance of the bioclastic fraction was observed in most samples.

The results of chemical analyses are given in Table 8.2. Among the analysed elements and compounds, Cd, Ni and Pb (up to 1.25 mg kg^{-1} d.w., 49 mg kg^{-1} d.w. and 42 mg kg^{-1} d.w., respectively) were below or only occasionally slightly exceeded the limits according to the Decree of the Italian Ministry of Environment (D.M. 367/2003) (2004)

Table 8.1 Results of grain-size analyses. Percentages of grain-size fractions and Shepard (1954) classification.

	Gravel (%)	Sand (%)	Silt (%)	Clay (%)	Shepard (1954) classification
PR044	0.0	88.0	6.4	5.6	Sand
PR063	0.0	20.0	49.9	30.1	Loam
PR067	0.0	82.7	10.7	6.6	Sand
PR075	0.0	15.7	45.3	39.0	Clayey silt
PR101	0.0	20.0	37.1	42.9	Loam
PR118	0.0	10.1	43.9	46.0	Silty clay
PR150	0.0	23.1	36.7	40.2	Loam
PR159	0.0	14.2	39.4	46.4	Silty clay
PR167	1.6	50.0	21.4	27.0	Loam
PR169	0.0	10.4	36.1	53.5	Silty clay
PR170	0.0	23.7	41.2	35.1	Loam
PR174	0.0	60.3	21.3	18.4	Sand
PR175	0.0	47.5	25.1	27.4	Loam
PR177	4.2	18.6	31.8	45.4	Loam
PR181	0.0	21.9	39.6	38.5	Loam
PR198	0.0	57.2	30.9	11.9	Silty sand
PR200	0.0	82.0	9.6	8.4	Sand
PR204	5.6	32.5	34.2	27.7	Loam
PR206	0.0	36.8	37.5	25.7	Loam
PR209	29.4	34.8	21.9	13.9	Silty sand

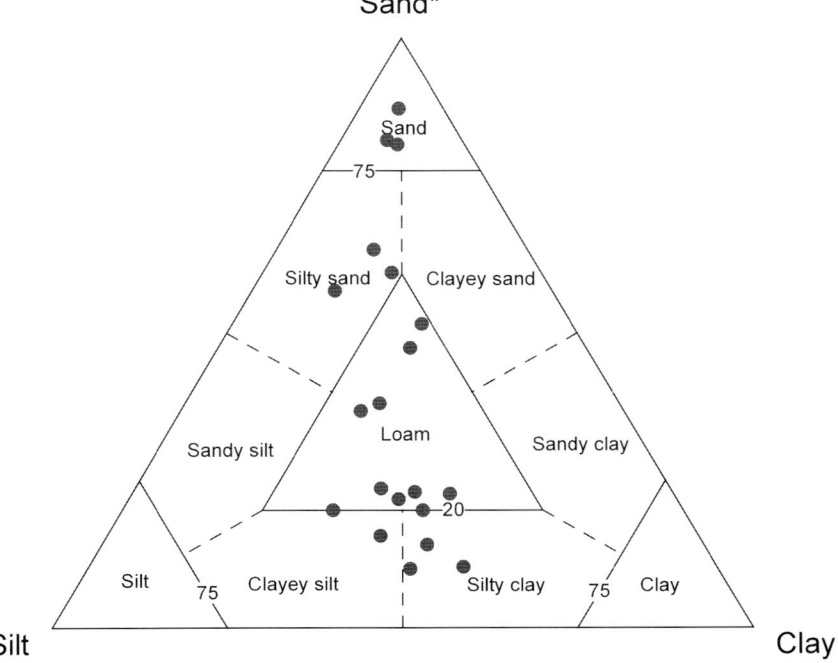

Figure 8.2 Shepard (1954) diagram with results of grain-size analyses. *, including gravel fraction.

Table 8.2 Results of chemical analyses with average values and standard deviation. Reference values from literature are reported for comparison.

	As (mg kg^{-1} d.w.)	Cd (mg kg^{-1} d.w.)	Cr (mg kg^{-1} d.w.)	Cu (mg kg^{-1} d.w.)	Fe (mg kg^{-1} d.w.)	Hg (mg kg^{-1} d.w.)	Ni (mg kg^{-1} d.w.)	Pb (mg kg^{-1} d.w.)	Zn (mg kg^{-1} d.w.)	PAHs (ng g^{-1} d.w.)	PCBs (ng g^{-1} d.w.)
PR044	18	0.08	28	17	15 777	1.1	18	6	44	175	8
PR063	14	0.23	59	34	33 137	7.2	34	19	95	536	65
PR067	13	0.18	31	14	19 686	0.2	31	3	40	2491	5
PR075	23	0.22	59	29	32 912	1.2	49	11	68	773	28
PR101	19	0.23	59	52	28 875	37.0	27	24	107	639	421
PR118	22	0.30	58	54	31 414	57.2	28	25	104	633	546
PR150	16	0.92	61	119	24 736	177.2	24	25	132	1358	3146
PR159	19	0.33	54	43	26 898	40.1	24	23	95	1265	121
PR167	16	0.15	30	20	16 926	33.6	13	11	48	819	885
PR169	7	0.93	49	86	18 192	98.6	19	25	180	19 498	1606
PR170	12	0.79	52	115	22 630	198.3	27	29	192	631	3070
PR174	22	0.04	26	18	13 512	11.8	12	6	28	638	174
PR175	19	1.00	40	140	20 379	215.7	29	20	169	2370	3754
PR177	21	1.02	52	206	23 703	321.6	17	25	166	4468	2768
PR181	20	0.28	45	37	24 731	36.8	19	13	65	1293	14
PR0198	27	1.25	31	392	193 500	59.1	29	42	325	1718	1670
PR200	16	0.63	29	58	29 385	21.8	12	18	100	1550	266
PR204	9	0.46	35	45	21 765	56.6	15	27	79	707	1126
PR206	13	0.14	25	19	18 053	10.6	15	9	42	519	76
PR209	17	0.61	34	84	37 621	80.9	17	26	146	881	1108
Mean	17	0.49	43	79	32 692	73.3	23	19	111	2148	1043
SD	4.9	0.4	13.2	88.8	38 402.7	87.5	9.2	9.7	71.1	4198.4	1226.4
USEPA (1993) 'moderately polluted sediments'	3–8	1–6	25–75	25–50	17 000–25 000	0.3–1.0	20–50	40–60	90–200		
UNEP (1996)	40–1400	0.02–64		0.6–1890		0.05–1.1		3–3300	1.7–6200	0.2–797	<0.1–16 000
D.M. 367/2003			50			0.3	30	30		200	4

PAHs, polycyclic aromatic hydrocarbons; PCBs, polychlorinated biphenyls.

and/or the limits for unpolluted/moderately polluted sediments by USEPA (1993). Based on the same references, the analysed sediments may be considered moderately polluted as regards Cr and Zn (up to 61 mg kg^{-1} d.w. and 325 mg kg^{-1} d.w., respectively), and between moderately and heavily polluted as regards Cu and Fe (up to 392 mg kg^{-1} d.w. and 19 3500 mg kg^{-1} d.w., respectively). Sediments should be considered heavily polluted by As (up to 27 mg kg^{-1} d.w.), according to USEPA (1993) that gives a lower limit of 8 mg kg^{-1} d.w. for heavily polluted sediments. Nevertheless, the As concentration was hardly interpretable because it may cover very wide ranges. Values from 40 to 1400 mg kg^{-1} d.w. were recorded in the Athens harbour and in the Gulf of Saronikos (UNEP, 1996), while values ranging from <0.15 mg kg^{-1} d.w. to 480 mg kg^{-1} d.w. are summarised by Garnaga et al. (2006) for the Baltic Sea.

Heavy pollution was due to the very high concentrations of Hg, PAHs and PCBs (up to 322 mg kg^{-1} d.w., 4468 ng g^{-1} d.w. and 3754 ng g^{-1} d.w., respectively). Hg exceeded the 0.3 mg kg^{-1} d.w. limit (D.M. 367/2003) more than 100 times in most stations of the southern harbour and PCBs exceeded the 4 ng g^{-1} d.w. limit (D.M. 367/2003) more than 100 times in the majority of the stations of the same area. Finally, PAHs exceeded the 200 ng g^{-1} d.w. limit (D.M. 367/2003) in all the analysed samples, except for station PR044. Considering these last contaminants, the most polluted stations were PR150, PR169, PR170, PR175 and PR177.

The TOC values ranged between 1.1 and 3.9%, with the mean of 2.8% (Table 8.3). These values were high when compared with those reported by Di Leonardo et al. (2006),

Table 8.3 Results of analyses on TOC with average values and standard deviation.

	TOC (% d.w.)
PR044	1.13
PR063	2.43
PR067	2.36
PR075	3.87
PR101	3.35
PR118	3.31
PR150	3.64
PR159	2.94
PR167	1.61
PR169	3.22
PR170	3.73
PR174	1.08
PR175	3.37
PR177	3.25
PR181	2.51
PR198	2.63
PR200	3.07
PR204	2.51
PR206	2.26
PR209	2.79
Mean	2.75
SD	0.80

TOC, total organic carbon.

who found TOC comprised between 0.1 and 0.8% in samples collected offshore the south-western coast of Sicily, from 14 to 488 m water depth. Conversely, a much wider range, from 1.3 to 12.0% was recorded by Debenay et al. (2001) in an enclosed harbour area.

8.3.2 Foraminifera

The relative abundance of species higher than 5% in at least one sample, together with α-index, FN and FAI, is given in Table 8.4. From the quantitative analysis on benthic foraminifera, a total of 152 species were identified. Among them, 6 belong to Textulariina (3.9%), 60 to Miliolina (39.5%) and 86 to Rotaliina (56.6%). On the whole, as regards the absolute abundance, *Lobatula lobatula, Quinqueloculina lata, Rosalina bradyi* and *Miliolinella subrotunda* were the most abundant species with 304, 289, 279 and 274 specimens, respectively. *L. lobatula* and *R. bradyi* are typical epiphytic species that may live also on detritic sands, attached to bioclastic fragments (Murray, 2006); *Q. lata* is a typical infralittoral species which was recognised by Bergamin et al. (2003) and Romano et al. (2008) as the most abundant species in the heavy-metal-polluted sands from the Bagnoli Bay (Naples, Italy). *M. subrotunda* is a ubiquitous species occurring from infralittoral to bathyal mud; it may have epiphytic lifestyle in infralittoral environment (Langer, 1993), but it may be abundant also in pelitic sands (Albani et al., 1998). Bergamin et al. (2003) found it associated with the highest pelitic content in Pb- and Zn-polluted sands and considered it a pollution-tolerant species. Also, Yanko et al. (1999) reported *M. subrotunda* as a pollution-tolerant species in the coastal zone.

The assemblages were well diversified, with α-index which may be considered as mean/high values for the normal marine shelf environment (Murray, 1991). As regards the faunal abundance, the FN covered a wide range, from the very low values of stations PR67 and PR75 (8 and 19, respectively) to the considerably high value of 2300 recorded at station 230. The FAI recorded the highest values at stations PR150, PR174, PR175 and PR198, with 2.0, 2.3, 2.3 and 2.4, respectively.

Although a biometric study was not performed, it was possible to recognise assemblages smaller than normal. In Figure 8.3, foraminiferal assemblages picked into a micropaleontology slide were photographed under stereomicroscope at the same magnification. Most foraminifera from sample PR170 were smaller than 125 µm and nearly all were smaller than 250 µm. In sample PR175, most foraminifera were smaller than 250 µm. The assemblages from samples PR044 and PR167, with most specimens larger than 250 µm, are shown for comparison in Figure 8.3.

8.3.3 Statistical analysis

By mean of CA (Figure 8.4) two distinct multivariate analyses (factors loading) applied distinctly on the results of grain-size and chemical analyses (Figure 8.4d) and on the results of foraminiferal analysis were plotted on the same factorial plane (Figure 8.4e). Therefore, the response of foraminifera to the selected environmental features (pollutants and grain size) was highlighted. The factors scores common to the two factorial analyses permits referencing the variables from both the multivariate analyses and the samples (Figure 8.4f). The Monte Carlo test, applied on the CA, showed a significant association between the two datasets (rv = 0.52, $p < 0.003$) (Figure 8.5). The correlation matrix based on the Spearman's coefficient (rho) reported in Table 8.5 showed correlations by numerical

Table 8.4 Results of quantitative analyses on benthic foraminifera. Relative abundance of species more abundant than 5% in at least one sample, α-index, FN and FAI.

	PR044	PR063	PR067	PR075	PR101	PR118	PR150	PR159	PR167	PR169	PR170	PR174	PR175	PR177	PR181	PR198	PR200	PR204	PR206	PR209
Ammonia parkinsoniana	20.9	6.3	11.1	0.0	1.3	1.0	0.7	1.0	1.0	0.0	0.0	0.0	2.9	0.3	0.0	1.6	2.0	1.0	0.0	4.0
Ammonia tepida	2.6	14.0	0.0	27.8	5.2	5.8	0.3	1.6	0.3	5.3	5.9	0.3	8.8	1.0	1.0	4.5	1.5	0.0	1.7	0.7
Asterigerinata mamilla	5.0	3.0	0.0	2.8	3.6	1.0	5.2	1.0	7.2	1.0	0.0	1.7	2.0	2.6	0.3	2.5	6.3	2.9	3.0	5.9
Bolivina aenariensis	1.0	1.0	0.0	2.8	1.3	1.3	0.0	1.6	2.3	1.7	0.7	0.7	0.3	1.7	1.9	3.8	3.9	5.4	2.0	5.3
Bolivina catanensis	6.0	0.7	0.0	2.8	1.0	0.6	0.3	0.0	0.0	0.3	0.7	1.0	1.0	0.0	1.0	0.3	0.0	0.3	0.3	0.0
Bolivina seminuda	0.0	5.6	0.0	11.1	5.6	3.9	0.0	0.0	0.0	0.0	5.3	0.0	7.5	0.0	0.0	0.0	0.0	0.0	4.7	5.0
Bolivina variabilis	0.0	0.0	0.0	0.0	0.3	0.0	2.3	9.5	3.0	8.0	3.9	2.3	2.6	2.6	2.9	5.4	2.9	4.5	0.0	0.0
Cornuspira involvens	0.0	0.0	0.0	0.0	1.6	1.3	0.7	3.9	0.7	8.0	7.9	0.0	3.6	4.6	3.2	1.9	1.0	8.6	1.7	3.6
Elphidium crispum	1.7	0.7	0.0	0.0	0.3	0.0	0.3	0.0	0.7	1.0	0.0	0.3	0.0	0.3	0.0	0.0	7.3	0.0	0.0	0.0
Elphidium granosum	0.7	6.0	0.0	5.6	1.6	3.9	3.6	1.0	0.7	2.7	0.0	0.3	2.0	0.3	0.0	3.5	0.5	0.3	0.3	2.0
Elphidium pulvereum	0.0	0.0	0.0	0.0	0.0	0.0	0.0	0.0	0.0	0.0	0.0	0.0	9.8	0.0	0.0	9.6	0.0	1.9	0.0	0.0
Haynesina depressula	2.0	2.3	0.0	2.8	1.3	0.6	2.0	1.0	0.0	0.0	0.0	1.3	0.3	1.7	1.9	5.4	0.0	1.6	0.3	0.0
Haynesina germanica	0.0	6.6	0.0	5.6	4.9	3.2	0.3	0.0	0.0	0.0	2.6	0.0	0.7	0.3	1.0	1.6	2.0	1.6	0.3	3.6
Lobatula lobatula	9.6	4.0	22.2	0.0	7.2	6.5	6.5	4.6	10.2	2.7	2.6	7.0	2.6	5.0	3.9	5.4	13.7	2.9	5.3	4.0
Miliolinella subrotunda	0.7	0.0	0.0	0.0	5.2	6.5	4.6	6.2	6.9	5.7	2.6	3.0	2.9	7.6	9.4	1.0	2.0	9.3	15.6	1.3
Neoconorbina posidonicola	0.3	0.0	0.0	0.0	0.0	2.6	2.0	1.0	1.3	1.7	0.0	5.3	0.0	0.0	3.6	0.0	1.0	1.6	2.0	0.0
Planorbulina mediterranensis	0.0	0.0	0.0	0.0	0.0	0.0	2.3	1.0	2.0	0.0	0.0	5.0	0.3	0.0	0.0	0.0	0.5	0.6	0.0	0.0
Quinqueloculina lata	0.0	0.0	0.0	0.0	0.3	0.3	3.9	7.8	2.3	12.3	8.9	1.3	5.5	12.5	6.2	6.4	3.4	8.6	7.3	8.9
Quinqueloculina pygmaea	0.7	0.0	0.0	0.0	1.6	1.6	0.7	1.0	0.7	1.0	1.6	1.0	1.0	4.0	1.0	2.9	2.0	0.3	1.0	3.0
Quinqueloculina stelligera	0.7	0.0	0.0	0.0	1.6	1.9	5.5	5.9	2.0	1.0	1.6	7.3	1.0	2.0	8.1	2.5	1.0	3.5	6.3	2.0
Rosalina bradyi	11.6	4.3	0.0	0.0	11.7	4.2	13.0	6.5	11.8	7.7	2.9	15.3	6.9	8.0	8.4	14.6	18.6	5.4	3.7	10.9
Sigmoilinta costata	0.7	1.7	0.0	2.8	2.6	5.8	5.5	7.5	0.7	1.7	0.0	6.0	1.0	1.7	3.6	0.3	2.0	3.8	4.0	3.0
Spirillina vivipara	0.0	0.7	0.0	0.0	2.0	1.9	2.9	2.0	3.0	5.7	8.9	1.0	1.6	3.0	1.9	0.3	0.5	7.3	2.0	0.7
α-Index	17.1	19.7	7.9	16.3	27.1	30.9	25.2	21.8	27.2	15.6	12.3	23.1	17.0	23.6	24.0	16.3	17.8	17.3	18.7	17.6
FN	192	519	8	19	1113	915	1612	1317	917	281	208	1734	177	183	1106	182	99	1680	893	464
FAI	1.0	1.3	0.0	0.0	1.3	0.3	2.0	1.3	1.0	1.0	0.3	2.3	2.3	1.0	1.3	0.6	2.4	1.6	1.0	1.3

FN, foraminiferal number; FAI, foraminiferal abnormality index.

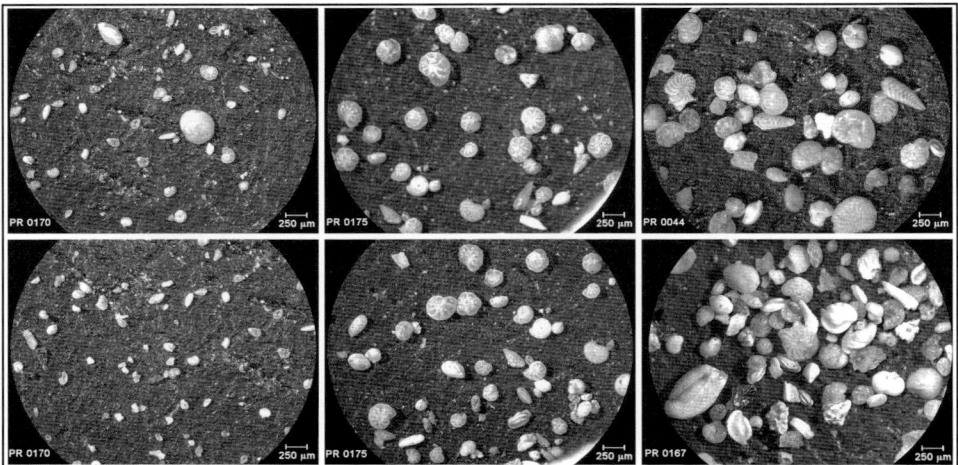

Figure 8.3 Foraminiferal assemblages photographed under stereomicroscope at the same magnification. Samples PR170 and PR175 are constituted by assemblages considerably smaller than normal. Samples PR044 and PR167 are figured for comparison.

viewpoint and supported the main results found by the CA. The comparison between CA and correlation matrix was made taking into account that two different statistical approaches were used. The correlation matrix was obtained by a bivariate analysis that considers only pairs of variables, while the CA was based on two multivariate analyses that process the variables on the whole. For this feature, the multivariate analysis may be considered a more powerful tool for showing correlations.

The factorial analysis on pollutants, grain-size fractions and TOC showed that all pollutants plotted positively on the first factor together with TOC and that they were scarcely correlated with the grain-size fractions. The high number of correlations found in the correlation matrix confirmed the numerous relationships existing among most pollutants. Only Cr showed a clear correlation with grain size, because it had a significant rho with sand, silt and clay (-0.805, 0.684 and 0.781, respectively).

Also, the negative correlation of sand with silt and clay was shown by both the statistical analyses. The CA provided evidence that the main pollutants (Hg, PAHs and PCBs) had a very similar distribution and were associated with species such as *Ammonia tepida*, *Q. lata*, *Cornuspira involvens*, *Haynesina germanica*, *Bolivina aenariensis*, *Bolivina seminuda* and *Bolivina variabilis*. These two groups of variables were both referable mainly to samples PR169, PR170, PR175 and PR177. The correlation matrix confirmed the correlation of *Q. lata* and *C. involvens* with most pollutants among which Hg, PAHs and PCBs for the first species and Hg and PCBs for the second one.

Because all the pollutants plotted positively on the first axis, species with negative values such as *Planorbulina mediterranensis*, *Bolivina catanensis*, *L. lobatula*, *Elphidium crispum*, *Asterigerinata mamilla* were negatively correlated to pollutants; they were referable to less polluted samples (PR044, PR167, PR174, PR200). Among these species, *L. lobatula* had a negative correlation coefficient with Hg in the correlation matrix (-0.536). Two other elements that determine heavy pollution, Fe and Cu, were negatively correlated to the second axis, together with *Haynesina depressula* and *Elphidium pulvereum* and

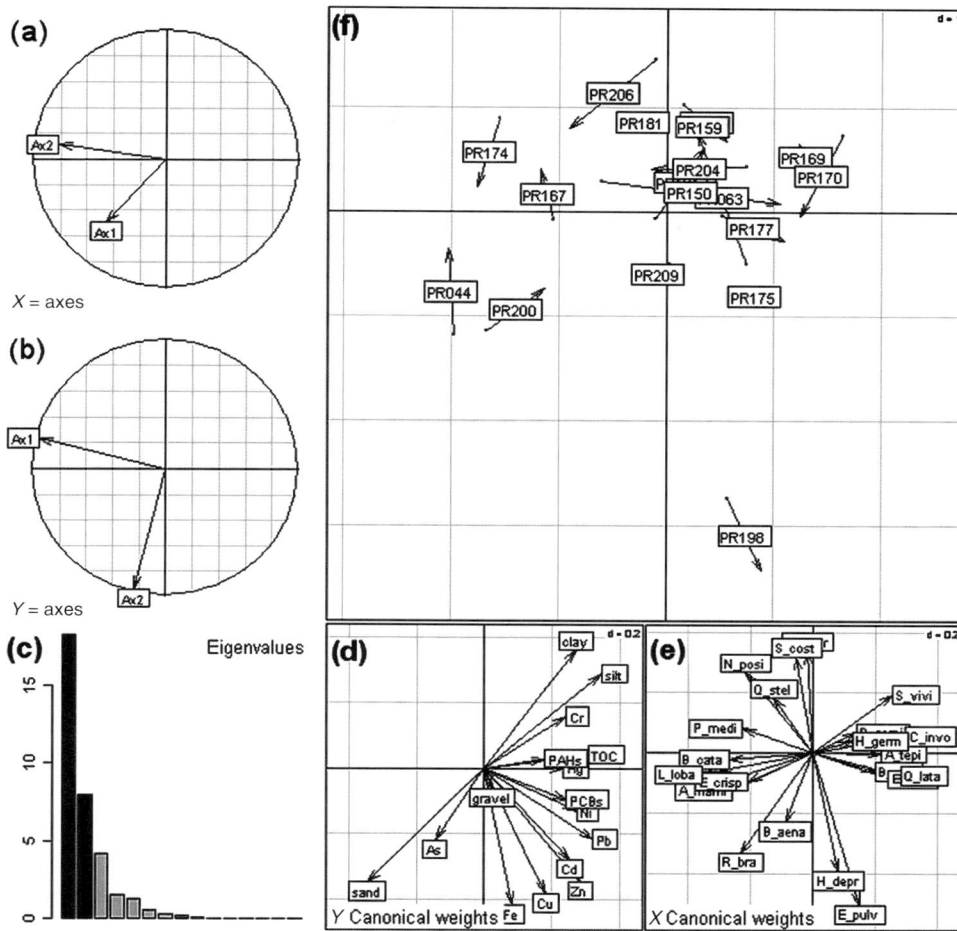

Figure 8.4 Output of coinertia analysis (CA). (a) Principal component analysis (PCA) applied to foraminiferal data: plot of factors on the coinertia plane; (b) PCA applied to grain-size and chemical data: plot of factors on the coinertia plane; (c) histogram of eigenvalues; (d) PCA applied to grain-size and chemical data: plot of factors loading on the coinertia plane; (e) PCA applied to foraminiferal data: plot of factors loading on the coinertia plane; (f) plot of factor scores on the coinertia plane.

characterised strongly sample PR198. Such correlation, which was very clear in the CA, was not highlighted by the Spearman index. On the other hand, the Spearman correlation pointed out the significant correlation of Hg and PCBs with Cd, Cu, Pb and Zn. Similarly, Romano et al. (2004) highlighted the correlation of Hg with Cd, Pb and Zn in the heavy metal polluted coastal zone of Bagnoli (Naples, Italy).

8.4 Discussion

Very strong and diffused pollution mainly due to Hg, PAHs and PCBs (with mean concentrations of 73.3 mg kg^{-1} d.w., 2148 ng g^{-1} d.w. and 1043 ng g^{-1} d.w., respectively) was

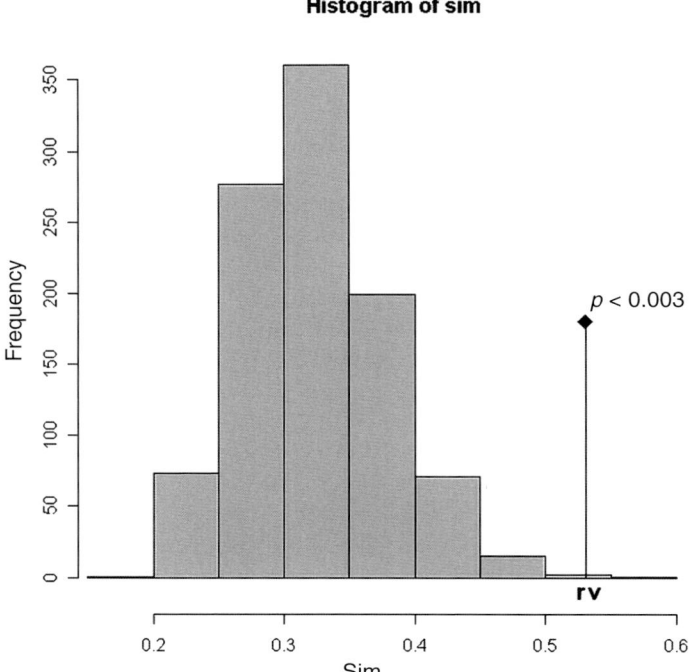

Figure 8.5 Monte Carlo test applied to the coinertia analysis (CA), evidences the significant association between the two datasets (rv = 0.52, $p < 0.003$).

observed in samples from the southern sector of the study area, with the highest concentrations in samples PR150, PR169, PR170, PR175 and PR177. They were localised around the pronged pier, where oil products loading and unloading are carried out. The distribution of such pollutants did not appear significantly influenced by sediment grain size. In order to evaluate the extent of the Hg pollution, the background concentration of 0.038 mg kg^{-1} d.w., calculated by Di Leonardo et al. (2006) for Sicilian marine sediments, must be considered. Also Cu and Fe contribute to pollution in the same area, with the highest concentration in sample PR198, located in the southernmost harbour. The effects of the very high concentrations of these pollutants on the ecological health of benthic environment may be recognised by the study of benthic foraminifera.

Q. lata, *A. tepida*, *C. involvens*, *H. germanica*, *B. aenariensis*, *B. seminuda* and *B. variabilis* have been recognised as 'pollution-tolerant species' by the comparison of pollutants and species distribution. In fact, the statistical analysis showed the positive correlation of such species with the main pollutants in the most polluted samples. In addition to *Q. lata*, other species of this group had been already recognised as 'pollution-tolerant species'. *A. tepida* was considered as a bioindicator of pollution in sediments from the Naples harbour (Ferraro et al., 2006). *H. germanica* was recognised as 'pioneer-tolerant' species in moderately polluted sediments from Atlantic French harbours (Debenay et al., 2001; Arminot du Châtelet et al., 2004) and in heavily polluted sediments from the industrial site of Bagnoli (Naples, Italy) (Romano et al., 2008). Finally, *B. variabilis* was found to be a tolerant species for PAHs in the Baia harbour (unpublished data from the authors). The

Table 8.5 Correlation matrix using Spearman's rho. Correlations between pairs of species and between pairs of grain-size fractions are not showed.

	As	Cd	Cr	Cu	Fe	Hg	Ni	Pb	Zn	PAHs	PCBs	TOC	Gravel	Sand	Silt	Clay
Ammonia parkinsoniana	0.114	−0.011	−0.005	−0.009	0.397	−0.224	0.299	−0.057	0.055	−0.091	−0.135	−0.097	0.102	0.321	−0.339	−0.454
Ammonia tepida	0.007	0.152	0.288	0.189	0.218	0.071	**0.699**	0.098	0.408	−0.141	0.009	0.331	−0.561	−0.264	0.416	0.187
Asterigerinata mamilla	−0.235	−0.184	−0.247	−0.156	0.042	−0.316	−0.356	−0.249	−0.233	−0.060	−0.072	−0.242	0.376	0.472	−0.549	−0.515
Bolivina aenariensis	−0.187	0.063	−0.408	−0.028	0.243	−0.146	−0.422	0.156	−0.082	0.199	−0.146	−0.344	0.560	0.131	−0.224	−0.259
Bolivina catanensis	0.168	−0.429	0.018	−0.349	−0.351	−0.285	0.295	−0.349	−0.228	−0.441	−0.281	−0.095	−0.561	0.088	0.122	−0.036
Bolivina seminuda	−0.104	−0.039	0.276	0.097	0.248	0.100	0.522	0.123	0.234	−0.337	0.104	0.339	−0.135	−0.160	0.396	0.039
Bolivina variabilis	−0.147	0.470	−0.001	0.331	−0.068	0.343	−0.074	0.324	0.335	0.564	0.337	0.155	0.042	−0.086	−0.045	0.249
Cornuspira involvens	−0.260	0.580	0.135	0.502	0.055	**0.657**	0.032	**0.599**	0.524	0.440	0.471	0.419	0.371	−0.407	0.214	0.417
Elphidium crispum	−0.200	−0.224	−0.116	−0.268	−0.259	−0.405	−0.325	−0.493	−0.226	0.017	−0.259	−0.264	−0.106	0.220	−0.438	−0.169
Elphidium granosum	0.107	0.183	0.420	0.232	0.503	0.066	**0.593**	0.232	0.317	0.094	0.121	0.195	−0.235	−0.254	0.149	0.136
Elphidium pulvereum	0.164	0.462	−0.154	0.411	0.068	0.322	0.307	0.361	0.329	0.326	0.462	0.100	0.096	0.281	−0.195	−0.246
Haynesina depressula	0.412	−0.062	0.212	−0.082	0.174	−0.216	0.354	−0.041	−0.157	−0.120	−0.237	−0.262	−0.182	0.008	0.132	−0.055
Haynesina germanica	−0.030	0.114	0.345	0.233	**0.712**	0.055	0.449	0.443	0.289	−0.220	0.032	0.322	0.063	−0.190	0.416	−0.030
Lobatula lobatula	0.263	−0.451	−0.255	−0.362	−0.028	−0.536	−0.358	−0.439	−0.434	−0.224	−0.364	−0.321	−0.105	0.395	−0.432	−0.360
Miliolinella subrotunda	−0.073	−0.127	0.005	−0.148	−0.331	0.096	−0.362	−0.088	−0.294	0.104	−0.026	−0.026	0.239	−0.386	0.280	0.476
Neoconorbina posidonicola	0.007	−0.428	−0.207	−0.471	−0.405	−0.314	−0.431	−0.350	−0.584	−0.127	−0.333	−0.337	−0.203	−0.065	0.088	0.154
Planorbulina mediterranensis	−0.011	−0.121	−0.128	−0.163	−0.323	−0.065	−0.412	−0.217	−0.332	0.149	0.137	−0.119	0.081	0.295	−0.360	−0.090
Quinqueloculina lata	−0.253	**0.611**	−0.030	0.510	−0.001	**0.642**	−0.189	0.499	0.471	0.504	0.485	0.272	0.446	−0.256	0.053	0.255
Quinqueloculina stelligera	0.267	−0.117	−0.156	−0.120	−0.119	0.043	−0.318	0.015	−0.287	0.066	−0.057	−0.224	0.138	−0.053	0.128	0.112
Rosalina bradyi	0.385	−0.003	−0.294	0.038	0.030	−0.158	−0.321	−0.224	−0.069	0.358	−0.005	−0.203	0.007	0.531	**−0.709**	−0.437
Sigmoilinita costata	0.135	−0.329	0.131	−0.317	0.042	−0.183	−0.230	−0.125	−0.434	−0.115	−0.311	−0.086	−0.084	−0.320	0.243	0.311
Spirillina vivipara	−0.500	0.184	0.324	0.175	−0.336	0.452	−0.137	0.311	0.158	0.114	0.435	0.370	0.267	−0.500	0.387	**0.634**
As	1.000	0.040	0.005	0.150	0.234	0.086	0.207	−0.057	0.049	0.183	−0.038	−0.020	−0.159	0.062	−0.146	−0.059
Cd	0.040	1.000	0.321	**0.959**	0.401	**0.845**	0.352	**0.709**	**0.889**	**0.781**	**0.796**	**0.676**	0.122	−0.219	0.020	0.234
Cr	0.005	0.321	1.000	0.389	0.484	0.430	**0.664**	0.455	0.437	0.154	0.271	**0.651**	−0.126	**−0.805**	**0.684**	**0.781**
Cu	0.150	**0.959**	0.389	1.000	0.461	**0.889**	0.416	**0.754**	**0.936**	**0.713**	**0.851**	**0.781**	0.098	−0.239	0.051	0.251
Fe	0.234	0.401	0.484	0.461	1.000	0.195	0.544	0.567	0.476	0.187	0.084	0.360	0.006	−0.316	0.352	0.104
Hg	0.086	**0.845**	0.430	**0.889**	0.195	1.000	0.307	**0.725**	**0.835**	**0.639**	**0.899**	**0.789**	0.248	−0.387	0.119	0.451
Ni	0.207	0.352	0.430	0.416	0.544	0.307	1.000	0.447	0.550	0.028	0.222	0.482	−0.418	−0.416	0.575	0.337
Pb	−0.057	**0.709**	0.455	**0.754**	0.567	**0.725**	0.447	1.000	**0.781**	0.311	**0.684**	**0.626**	0.258	−0.386	0.311	0.311
Zn	0.049	**0.889**	0.437	**0.936**	0.476	**0.835**	0.550	**0.781**	1.000	0.587	**0.775**	**0.781**	0.001	−0.307	0.139	0.294
PAHs	0.183	**0.781**	0.154	**0.713**	0.187	**0.639**	0.028	0.311	0.587	1.000	0.569	0.406	0.145	−0.162	−0.232	0.259
PCBs	−0.038	**0.796**	0.271	**0.851**	0.084	**0.899**	0.222	**0.684**	**0.775**	0.569	1.000	**0.703**	0.255	−0.112	−0.071	0.214
TOC	−0.020	**0.676**	**0.651**	**0.781**	0.360	**0.789**	0.482	**0.626**	**0.781**	0.406	**0.703**	1.000	−0.136	**−0.472**	0.337	0.529

Grey bold: correlation is significant at the 0.05 level (2-tailed); black bold: correlation is significant at the 0.01 level (2-tailed). PAHs, polycyclic aromatic hydrocarbons; PCBs, polychlorinated biphenyls; TOC, total organic carbon.

existence of correlations among variables, highlighted by the statistical analysis, indicates the co-occurrence of events, but this does not necessarily imply a direct cause-and-effect relationship. However, the comparable results obtained by distinct statistical analyses and the confirmation from literature of several correlations recognised in this research, concur in indicating the relevant effect of pollutants on the foraminifera distribution. Thus, the adaptation of the foraminiferal assemblage to the environmental stress determined by pollution, with the increasing of resistant species in the most polluted samples, may be deduced.

The presence of stunted assemblages, particularly in samples PR170 and PR175, which are among the most polluted ones, may be considered due to the strong pollution. Assemblages with reduced size are a rather common feature in heavy-metal-polluted sediments. Yanko et al. (1994) found about 70% of specimens smaller than 250 μm in the Haifa Bay. Foraminifera with reduced size from heavy-metal-polluted sediments are reported also by Samir and El-Din (2001) and Bergamin et al. (2005).

The foraminiferal abundance is a highly variable parameter and it is strongly influenced by biotic and abiotic factors such as the reproductive rate of species and the sedimentary rate, which may determine different degrees of dilution of tests in sediment. Cearreta et al. (2002) found in the Ashua core from the Bilbao estuary values of FN ranging between 180 and 1515 in the lower unpolluted part, and from 0.2 to 1.2 in the upper heavily polluted part. Samir and El-Din (2001) reported values of FN ranging from 12 to 259 in the heavy-metal-polluted El-Mex Bay, and values from 55 to 1636 in the Miami Bay, an area which received urban wastes but was not affected by heavy metal pollution. Taking into account these references, the values of FN recorded at station PR067 and PR075 (8 and 19, respectively) appear very low, suggesting a high degree of environmental stress. However, this feature is not apparently due to the high degree of pollution, because these stations show low concentrations of the analysed pollutants, except for PAHs, that are over the limit of the Italian regulatory (2491 ng g^{-1} d.w. at station PR067 and 773 ng g^{-1} d.w at PR075). More detailed analyses of environmental parameters should be performed to explain the very scarce presence of foraminifera.

In order to estimate the FAI values, experiments under controlled conditions may give reference for background values. Stouff et al. (1999) recorded about 1% of abnormalities under normal saline conditions. *A. tepida* living in unpolluted microcosm with 35‰ salinity showed the 1.75% of abnormal tests (Morvan et al., 2004). Values exceeding the above-reported ones, recorded at stations PR150, PR174, PR175 and PR198 (2.0%, 2.3%, 2.3% and 2.4%, respectively), may be considered as evidence of environmental stress. Some of these are not among the most polluted samples. However, the generalised high pollution degree affecting the whole southern sector of the harbour could be above a critical threshold and thus it could determine the significantly high FAI values. However, the complexity of factors determining vital processes in natural environment makes it difficult to find linear correlations between concentration of single pollutants and percentages of abnormal tests.

In conclusion, some evidences of environmental stress attributable to industrial pollution have been recognised in the foraminifera assemblages. The abundance of pollution-tolerant species associated with the most polluted samples, the presence of stunted assemblages and significant percentages of abnormal specimens concur in indicating the existence of a considerable environmental stress in the benthic environment of the southern sector of the study area.

Acknowledgements

We are indebted to Giancarlo Pierfranceschi, Chiara Maggi, Jessica Bianchi and Giulio Sesta for physical chemical analyses and to Francesco Venti and Andrea Salmeri for images editing. We are also grateful to two anonymous reviewers, who gave us precious suggestions for the improvement of the article and to the volume editors for the final revision of the manuscript.

References

Albani, A.D., Favero, V.M. and Serandrei Barbero, R. (1998) Distribution of sediment and benthic foraminifera in the Gulf of Venice, Italy. *Estuarine, Coastal and Shelf Science* **46**, 251–265.

Alve, E. and Murray, J.W. (1994) Ecology and taphonomy of benthic foraminifera in a temperate mesotidal inlet. *Journal of Foraminiferal Research* **24**, 18–27.

Anon. (1992) Sistema integrato per il monitoraggio automatico della rada di Augusta. I - Studi preliminari per il posizionamento di boe oceanografiche. Istituto Sperimentale Talassografico CNR Messina. *Rapporti* **6**, 1–120.

Anon. (1995) Studio sedimentologico e bionomico dei fondi mobili della Rada di Augusta. Istituto Sperimentale Talassografico CNR Messina. *Rapporti* **9**, 1–72.

Arminot du Châtelet, E., Debenay, J.P. and Soulard, R. (2004) Foraminiferal proxies for pollution monitoring in moderately polluted harbours. *Environmental Pollution* **127**, 27–40.

Ausili, A., Pellegrini, D., Onorati, F. and De Ranieri, S. (1998) Valutazione delle qualità di sedimenti del Porto di Viareggio da sottoporre ad escavo. *Acqua Aria* **1**, 67–71.

Azzaro, F., Azzaro, M., Bergamasco, A. and Giacobbe, S. (2001) Rada di Augusta Lagoon, Italy. Available at http://www.dsa.unipr.it/lagunet/infosheet/11-augusta.pdf.

Bergamin, L., Romano, E., Celia Magno, M., Ausili, A. and Gabellini, M. (2005) Pollution monitoring of Bagnoli Bay (Tyrrhenian Sea, Naples, Italy), a sedimentological, chemical and ecological approach. *Aquatic Ecosystem Health and Management* **8** (3), 293–302.

Bergamin, L., Romano, E., Gabellini, M., Ausili, A. and Carboni, M.G. (2003) Chemical-physical and ecological characterisation in the environmental project of a polluted coastal area: the Bagnoli case study. *Mediterranean Marine Science* **4** (1), 5–20.

Best, D.J. and Roberts, D.E. (1975) Algorithm AS 89: the upper tail probabilities of spearman's rho. *Applied Statistics* **24**, 377–379.

Blanc-Vernet, L. (1969) Contribution à l'étude des foraminifères da Méditerranée. These de Doctorat Etat. Travaux de la Station marine d'Endoume, Marseille.

Burone, L., Venturini, N., Sprechmann, P., Valente, P. and Muniz, P. (2006) Foraminiferal responses to polluted sediments in the Montevideo coastal zone, Uruguay. *Marine Pollution Bulletin* **52**, 61–73.

Cearreta, A., Irabien, M.J., Leorri, E., Yusta, I., Quantanilla, A. and Zabaleta, A. (2002) Environmental transformation of the Bilbao estuary, N. Spain: microfaunal and geochemical proxies in the recent sedimentary record. *Marine Pollution Bulletin* **44**, 487–503.

Cicero, A.M., Finoia, M.G., Gabellini, M., Pietrantonio, E., Romanelli, G. and Romano, E. (2003) Assessment of chlorinated organic pollutants in sediments of a coastal area of the Tyrrhenian Sea (Ombrone River - Italy): a case study of multivariate approach for marine sediments characterisation. In: Munawar, M. (ed.). *Sediment Quality Assessment and Management: Insight and Progress*. Ecovision World Monograph Series, Burlington, Canada, pp. 125–138.

Cimerman, F. and Langer, M. (1991) Mediterranean Foraminifera. Slovenska Akademija Znanosti in Umetnosti, Academia Scientiarum Artium Slovenica, Classis IV, Historia Naturalia 30, Ljubliana.

Coccioni, R., Frontalini, F., Marsili, A. and Troiani, F. (2005) Foraminiferi bentonici e metalli in traccia: implicazioni ambientali. *Quaderni del Centro di Geobiologia Università degli Studi di Urbino 'Carlo Bo'* **3**, 57–92.

Debenay, J.P., Tsakiridis, E., Soulard, R. and Grossel, H. (2001) Factors determining the distribution of foraminiferal assemblages in Port Joinville Harbor (Ile d'Yeu, France): the influence of pollution. *Marine Micropaleontology* **43**, 75–118.

Decree of the Italian Ministry of Environment (D.M. 367/2003) (2004) Regolamento concernente la fissazione di standard di qualità nell'ambiente acquatico per le sostanze pericolose, ai sensi dell'articolo 3, comma 4, del decreto legislativo 11 maggio 1999, n. 152. Gazzetta Ufficiale della Repubblica Italiana, 8 Gennaio 2004, 5, 17–29.

Di Leonardo, R., Tranchida, G., Bellanca, A., Neri, R., Angelone, M. and Mazzola, S. (2006) Mercury levels in sediments of central Mediterranean Sea: a 150+ year record from box-cores recovered in the Strait of Sicily. *Chemosphere* **65**, 2366–2376.

Dolédec, S. and Chessel, D. (1994) Co-inertia analysis: an alternative method for studying species–environment relationships. *Freshwater Biology* **31**, 277–294.

Dray, S., Chessel, D. and Thioulouse, J. (2003) Co-inerta analysis and the linking of ecological data tables. *Ecology* **84** (11), 3078–3089.

Ferraro, L., Sprovieri, M., Alberico, I., Lirer, F., Prevedello, L. and Marsella, E. (2006) Benthic foraminifera and heavy metals distribution: a case study from the Naples Harbour (Tyrrhenian Sea, Southern Italy). *Environmental Pollution* **142**, 274–287.

Fisher, R.A., Corbet, A.S. and Williams, C.B. (1943) The relationship between the number of species and the number of individuals in a random sample of an animal population. *Journal of Animal Ecology* **12**, 42–58.

Garnaga, G., Wyse, E., Sabine, S., Stankevičius, A. and de Mora, S. (2006) Arsenic in sediments from the southeastern Baltic Sea. *Environmental Pollution* **144**, 855–861.

Geslin, E., Stouff, V., Debenay, J.P. and Lesourd, M. (2000) Environmental variation and foraminiferal test abnormalities. In: Martin, R. (ed.). *Environmental Micropaleontology*. Kluwer Academic/Plenum Publishers, New York, pp. 91–215.

Giani, M., Gabellini, M., Pellegrini, D., Costantini, S., Boccaloni, E. and Giordano, R. (1994) Concentration and partitioning of Cr, Hg and Pb in sediments of dredge and disposal sites of the northern Adriatic sea. *Science of the Total Environment* **158**, 97–112.

Heo, M. and Gabriel, K.R. (1997) A permutation test of association between configurations by means of the RV coefficient. *Communications in Statistics – Simulation and Computation* **27**, 843–856.

Jorissen, F.J. (1988) Benthic foraminifera from the Adriatic Sea: principles of phenotypic variations. *Utrecht Micropaleontological Bulletin* **37**, 1–174.

Kovach, W.L. (1987) Multivariate methods of analyzing paleoecological data. *Paleontological Society* **3** (Special publication), 72–104.

Kovach, W.L. (1989) Comparisons of multivariate analytical techniques for use in Pre-Quaternary plant paleoecology. *Review of Palaeobotany and Palynology* **60**, 255–282.

Langer, M.R. (1993) Epiphytic foraminifera. *Marine Micropaleontology* **20**, 235–265.

Loeblich, A.R. and Tappan, H. (1987) *Foraminiferal Genera and Their Classification*. Van Nostrand Reinhold Company, New York.

Magazzù, G., Romeo, G., Azzaro, F., Dicembrini, F., Oliva, F. and Piperno, A. (1995) Chemical pollution from urban and industrial sewages in Augusta Bay (Sicily). *Water Science & Technology* **32** (9–10), 221–229.

Morvan, J., Debenay, J.P., Jorissen, F., Redois, F., Bénéteau, E., Delplancke, M. and Amato, A.S. (2006) Patchiness and life cycle of intertidal foraminifera: implication for environmental and paleoenvironmetal interpretation. *Marine Micropaleontology* **61**, 131–154.

Morvan, J., Le Cadre, V., Jorissen, F. and Debenay, J.P. (2004) Foraminifera as potential bio-indicators of the 'Erika' oil spill in the Bay of Bourgneuf: field and experimental studies. *Aquatic Living Resources* **17**, 317–322.

Murray, J.W. (1982) Benthic foraminifera: the validity of living, dead or total assemblages for the interpretation of palaeoecology. *Journal of Micropalaeontology* **1**, 137–140.

Murray, J.W. (1991) *Ecology and Paleoecology of Benthic Foraminifera*. Longman Scientific & Technical, New York.

Murray, J.W. (2000) The enigma of the continued use of total assemblages in ecological studies of benthic foraminifera. *Journal of Foraminiferal Research* **30** (3), 244–245.

Murray, J.W. (2006) *Ecology and Applications of Benthic Foraminifera*. Cambridge University Press, Cambridge.

Nigam, R., Saraswat, R. and Panchang, R. (2006) Application of foraminifers in ecotoxicology: retrospect, perspect and prospect. *Environment International* **32**, 273–283.

Romano, E., Ausili, A., Zharova, N., Celia Magno, M., Pavoni, B. and Gabellini, M. (2004) Marine sediment contamination of an industrial site at Port of Bagnoli, Gulf of Naples, Southern Italy. *Marine Pollution Bulletin* **49**, 487–495.

Romano, E., Bergamin, L., Finoia, M.G., Carboni, M.G., Ausili, A. and Gabellini, M. (2008) Industrial pollution at Bagnoli (Naples, Italy): benthic foraminifera as tool in integrated programs of environmental monitoring. *Marine Pollution Bulletin* **56**, 439–457.

Romano, E., Gabellini, M., Pellegrini, D., Ausili, A. and Mellara, F. (1998) Metalli in tracce e contaminanti organici provenienti da differenti aree marine costiere in relazione alla movimentazione dei fondali. *Proceedings of the 12th A.I.O.L (Associazione Italiana Oceanologia e Limnologia) Conference 2*, Eolian Islands, 473–486.

Samir, A.M. and El Din, A.B. (2001) Benthic foraminiferal assemblages and morphological abnormalities as pollution proxies in two Egyptian bays. *Marine Micropaleontology* **41**, 193–227.

Schafer, C.T. (2000) Monitoring nearshore marine environments using benthic foraminifera: some protocols and pitfalls. *Micropaleontology* **46** (1), 161–169.

Schott, W. (1935) Die Foraminiferen in den Äquatorialen Teil des Atlantischen Ozeans. *Deutsche Atlantische Expedition* **6**, 411–616.

Scott, D.B. and Medioli, F.S. (1980) Living vs total foraminiferal populations: their relative usefulness in paleoecology. *Journal of Paleontology* **54** (4), 814–831.

Scott, D.B., Medioli, F.S. and Schafer, C.T. (2001) *Monitoring of Coastal Environments Using Foraminifera and Thecamoebian Indicators*. Cambridge University Press, Cambridge.

Sgarrella, F. and Moncharmont-Zei, M. (1993) Benthic Foraminifera of the Gulf of Naples (Italy): systematics and autoecology. *Bollettino della Società Paleontologica Italiana* **32** (2), 145–264.

Shepard, F.P. (1954) Nomenclature based on sand-silt-clay ratios. *Journal of Sedimentary Petrology* **24**, 151–158.

Stouff, V., Geslin, E., Debenay, J.P. and Lesourd, M. (1999) Origin of morphological abnormalities in Ammonia (Foraminifera): studies in laboratory and natural environments. *Journal of Foraminiferal Research* **29** (2), 152–170.

UNEP (1996) State of marine and coastal environment in the Mediterranean region. MAP Technical Reports Series 100, UNEP, Athens.

USEPA (1993) Selecting remediation techniques for contaminated sediment. EPA-823-B93–001. U.S. Environmental Protection Agency, Office of Water, Washington, DC.

Vilela, C.G., Batista, D.S., Baptista-Neto, J.A., Crapez, M. and McAllister, J.J. (2004) Benthic foraminifera distribution in high polluted sediments from Niterói Harbor (Guanabara Bay), Rio de Janeiro, Brazil. *Anais da Academia Brasileira de Ciências* **76** (1), 161–171.

Yanko, V., Arnold, A. and Parker, W. (1999) Effect of marine pollution on benthic foraminifera. In: Sen Gupta, B. (ed). *Modern Foraminifera*. Kluver Academic, Dordrecht, the Netherlands, pp. 217–235.

Yanko, V., Kronfeld, J. and Flexer, A. (1994) Response of benthic foraminifera to various pollution sources: implications for pollution monitoring. *Journal of Foraminiferal Research* **24** (1), 1–17.

Chapter 9
Eco-Friendly Sustainable Shrimp Aquaculture in Bangladesh: Minimizing Coastal Degradation

Mohammad A.L. Siddique and John P. Volpe

Abstract

Coastal shrimp aquaculture contributes significantly to the world seafood supply despite generating potentially adverse environmental effects on coastal ecosystems. Coastal shrimp farming in Bangladesh is undergoing a shift from traditional extensive methods to semi-intensive methods so as to increase the expected economic benefit of increased production. However, if long-term sustainability of the industry is to be realized managers must consider precautionary management policies, in particular trade-offs between economic and ecological performance and longevity. The lack of an integrated regulatory framework governing shrimp farms has led to improper sitting, overcrowding, inappropriate changes in land use pattern, habitat conversion/loss, nutrient and organic release resulting in water quality issues, and a suite of secondary biodiversity related issues such as collection of wild seed. All of these challenges ultimately manifest as social issues of equity and multi-stakeholder conflict. Such problems emphasize the importance of identification and promotion of best management practices that explicitly seek to address ecological, social and economic trade-offs. A sustainable shrimp-farming system has to be biotechnically as well as socioeconomically viable in Bangladesh. This will entail institutional reform in order for environmental security and social equity to play roles of equal importance to economic performance.

9.1 Introduction

A major challenge of the next millennium will be to secure enough culturally appropriate, nutritious food without jeopardizing the ability of future generations to do the same. In the case of seafood, commercial scale aquaculture is rapidly substituting traditional capture and culture practices; however, there is doubt as to whether such production models can be relied upon to meet the demands of future generations. For instance, the global shrimp-farming industry experienced rapid growth in the 1980s mainly due to technological advances in hatcheries and feeds, high market demand for shrimp resulting in high price and high profit and public support. However, its rate of growth has slowed since 1991 (Shang et al., 1998). Serious disease outbreaks have been reported in most of the major producing countries and in particular, viral diseases have reduced shrimp production dramatically. Global farmed shrimp production in 1995 was 712 000 mt, or 27% of total shrimp production from both wild-caught and farm-raised sources. Asia produced about 78% of farmed shrimp and Western countries 22% (ADB/NACA, 1996).

The coast of Bangladesh is seen as a zone of vulnerabilities as well as opportunities where the culture of shrimp is a 100% export-oriented activity. Mirroring other shrimp-farming nations, commercial scale farming of shrimp grew to dominance here in the 1980s (MoWR, 2005) and today is the second largest national export product just after garments accounting for over US$300 million in trade and directly employing over 600 000 people (Fleming, 2004). This sector alone contributes more than 70% to the total export earnings from all agro-based products, including tea, raw jute, vegetables, and fruit (Karim, 2003). Bangladesh's 9000 shrimp farms – 18% of all shrimp farms globally – cover 170 000 ha (EC, 2002; MoWR, 2006). Bangladesh grew to be the 8th largest producer of farmed shrimp in 2000 from the 17th position in 1998 (EC, 2002).

The potential for adverse effects resulting from poorly managed shrimp farms on coastal ecosystems has long been identified (Hopkins et al., 1995). A key concern is the effect of the periodic discharge of shrimp pond water containing high concentrations of suspended solids and nutrients, particularly nitrogen (Jackson et al., 2003). The discharge of high loads of nutrients and suspended solids has the potential to generate algal blooms and creating anoxic conditions in the receiving environment (Funge-Smith et al., 1998; Naylor et al., 1998). Until recently nutrient loading from Bangladesh shrimp farms has on average been relatively low, reflecting the dominance of the *improved traditional* production model. This model depends on extensive inputs and is characterized by lower production intensity but balanced by greater ecological resilience. The extensive *improved traditional* system is rapidly being replaced by *semi-intensive* production models which rely to a much greater degree on external inputs, and in so doing realize increased production and thus increased economic return. The integrated coastal zone management (ICZM) plan, a government initiative, seeks to facilitate an industry-wide shift from extensive to semi-intensive production and has set a target for production increases of at least 5–10% each year for next 5 years (MoWR, 2006). This recent trend in increasing production intensity forces consideration of precautionary management policies to ensure the long-term sustainability of the industry.

Both significant positive and negative impacts have been associated with this policy change and the populations of affected communities are polarized regarding the present and probable future trend of shrimp farming in Bangladesh. While its supporters see it as a valuable way of generating foreign exchange, those opposed point to environmental damage, social disruption and rising domestic inequalities that result from trying to meet luxury demands of Western consumers. Both arguments are equally important in the context of true sustainability of this sector, which is required to meet a growing world demand for high-quality shrimp produced in an environmentally and socially sound way. So, this study is an attempt to answer two questions: (a) can shrimp be farmed commercially to minimize socio-ecological costs? and (b) what are the trade-offs between profit margin and socio-ecological subsidies in an eco-friendly sustainable shrimp-farming industry? Although shrimp farming in Bangladesh encompasses both marine and freshwater environments, this chapter focuses only on coastal/marine production, the dominant sector.

9.2 Coastal shrimp farming in Bangladesh

9.2.1 Area and production

Coastal area of Bangladesh (defined as the area inundated by tidal water and storm surges) comprises 3.22 million hectares, representing 25% of the total land area and the country has been quick to exploit these natural characteristics-vast tracts of low-lying flats and a tidal range that ensure a ready supply of saline water (Barkat and Roy, 2001). During the early stages of development in the 1970s, shrimp farming was largely restricted to peripheral land between flood embankments and the main river systems, but agricultural land used for domestic food production was soon targeted for conversion to more lucrative shrimp farming aimed at satisfying foreign markets (DoF, 2002). Continued increases in market demand triggered expansion in the coastal belt of the country from the Cox's Bazar district in the east to the Satkhira, Khulna and Bagerhat districts in the west (Figure 9.1). Exponential increases in the land area under culture occurred and by 1986 accounted for 115 000 ha from 52 000 ha in 1982–83 (Karim, 2003). Embankments around *polders* (reclaimed land) were easily breached and farmlands flooded with saltwater (Barkat and Roy, 2001). Elsewhere salt pans, abandoned and marginal lands, and wetlands including mangroves and marshes, were also appropriated for conversion (Barkat and Roy, 2001). Of this total, 170 000 ha are used for bagda (marine shrimp, *Penaeus monodon*) production, and 30 000 ha are under golda (giant freshwater prawn, *Macrobrachium rosenbergii* de Man) production (DoF, 2002).

9.2.2 Culture systems

The culture system of both marine and freshwater species is conducted in a gher (local term; large pond or number of ponds together in a farm). The average marine aquaculture farm is 4.5 ha, while the inland freshwater ponds are usually no greater than 0.3 ha (Banks, 2002). There are approximately 38 000 *P. monodon* and 105 000 *M. rosenbergii* ghers in Bangladesh (WARPO, 2003), though the *P. monodon* is the dominant product by weight due to the larger scale of production. In 1984, the percent of global shrimp supply from wild capture fisheries was approximately 20% higher than it is today and is continuing to decline, although it has been steady in recent years (Fleming, 2004). On the other hand, government subsidies for shrimp farming through bank loans, tax breaks, technical support, etc., together with foreign aid have allowed aquaculture to grow rapidly. The declining wild production combined with various forms of financial assistance triggered the booming expansion of this sector in Bangladesh (Table 9.1).

9.2.3 Existing institutional arrangements

The coastal areas including the shrimp-farming areas are under the jurisdiction of Director of Inland Fisheries, although the coastal area consists of both inland and marine waters, and all types of fishing activities are conducted here. The Director of Marine Fisheries is responsible for protection, preservation and development of marine resources. Out of a total of 4425 officers and staff, only 181 are tasked with 'marine' duties, in spite of the coastal zone representing one third of the country by area. In addition, significant

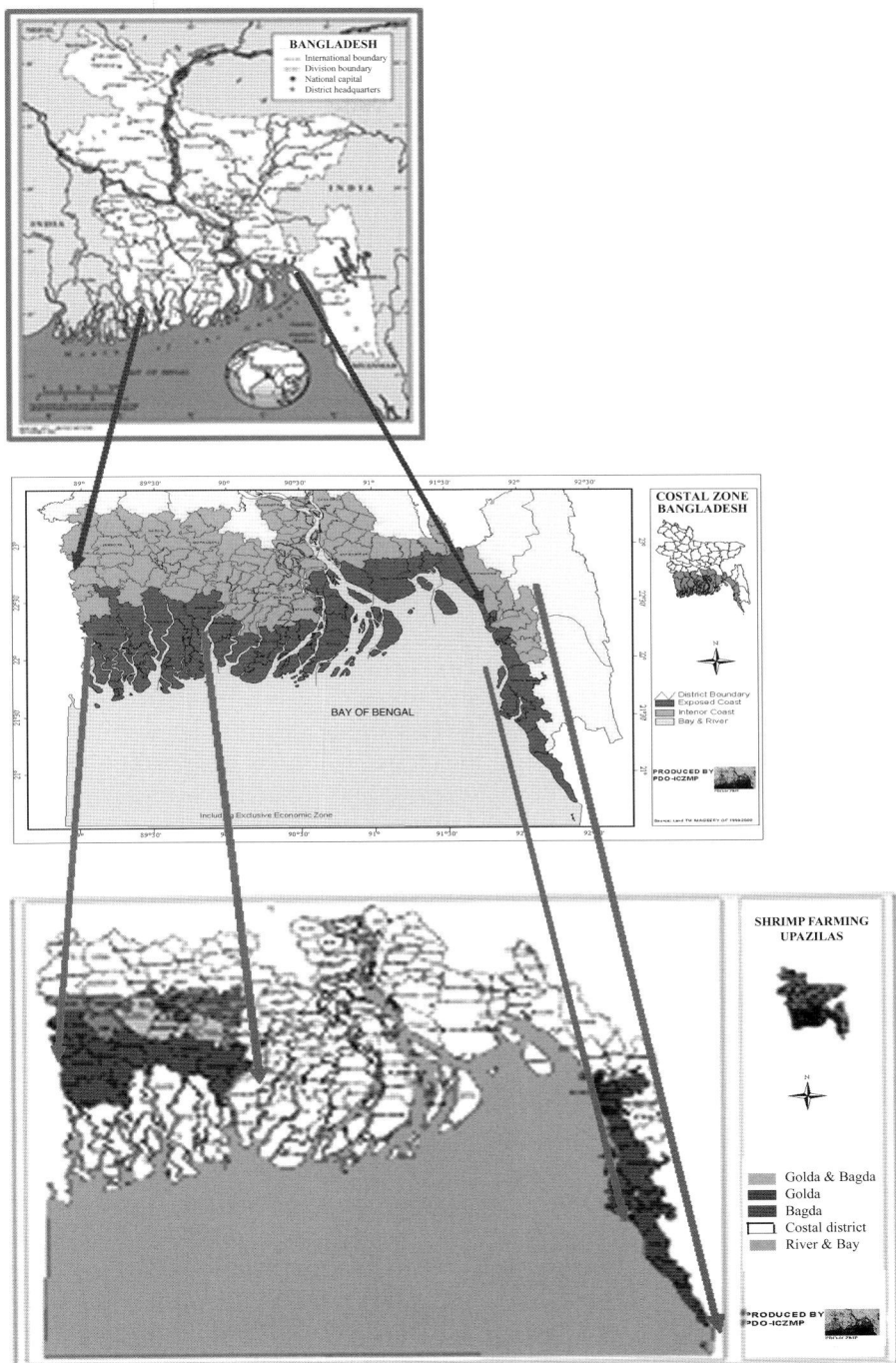

Figure 9.1 Coastal zone and main shrimp-farming *upazilas* (administrative areas; the country is divided into 463 *upazilas*) in Bangladesh. (To see this figure in color, please see color plate 8.)

Table 9.1 Profile of shrimp farm production models in the major farm-shrimp producing Asian countries.

Countries	Extensive (%)	Semi-intensive (%)	Intensive (%)
Bangladesh	90	10	00
China	10	85	05
India	70	25	05
Indonesia	45	45	10
Philippines	35	50	15
Thailand	05	10	85
Vietnam	80	15	05

Source: Rosenbery (1995), cited by Shang et al. (1998) data for 1994.

interministerial conflicts further reduce efficacy (much of this may be minimized through successful implementation of ICZM (Khatun, 2004)).

9.2.4 Legal regime

Bangladesh lacks legislation specifically dealing with the cultivation, farming, management, marketing, processing, etc., of shrimp. The only law that has direct bearing on shrimp is the Shrimp Cultivation Tax Act, 1992, which is targeted on revenue collection (PDO-ICZMP, 2003). The law of the land on fish and fisheries in general applies to shrimp also. The Protection and Conservation of Fish Act, 1950, did not include shrimp or prawn in its definition of fish; however, the law was amended in 1982 to include both (PDO-ICZMP, 2003). This Act does not cover broader biodiversity issues (e.g. intra- and interspecific impacts resulting from wild shrimp fry collection). Similarly, the Fish and Fish Products Ordinance (from 1983) is concerned only with revenue earnings and is silent on the livelihood of other stakeholders. Finally, the Marine Fisheries Ordinance (from 1983) addresses the licensing systems and the delimitation of fishing areas for artisanal and industrial shrimp fleets and has no jurisdiction within the aquaculture sector. Obviously, issues of ecological and social sustainability pertaining to shrimp aquaculture in Bangladesh have no legislative voice. A resilient, profitable and socially just future for this sector in Bangladesh depends on harmonizing the relevant legislature and while addressing the broader ecological and social deficiencies.

9.3 Social–ecological–economic interactions in industrial aquaculture

Industrial scale commercial aquaculture tends to focus on species of high ecological and economic value such as shrimp, salmon and tuna (ranching). Industrial scale culture of such species universally generates significant socio–ecological–economic friction. The primary objective of any commercial enterprise is to realize maximum profit. Profit increases as production cost per unit decreases and/or market value increases. In aquaculture, economic profits peak through the production of high trophic level species, which are in, naturally short supply and therefore demand the highest market price. The core of the success of these operations is that production volume is not limited by resources within the system, as is the case in commercial fisheries. The high degree of ecological capital required to produce a single carnivore ensures that nature does not produce many. In aquaculture, resources can be collected from afar and brought into the system, ad libitum. Therefore, the basic economic model is one that produces high volumes of high-value product that is produced through the conversion from low-value raw materials (forage fish). Production costs are shaved down as the intensity of the operation increases allowing ever-greater economies of scale. This seemingly sound business plan has won support from communities and governments around the world. For a period after establishment, such initiatives are often heralded as successes – creating job opportunities, earning foreign currency and increasing food security, though this last one remains contentious. Eventually, though booming expansion of these industries generates new challenges, declining unit price reflecting increased supply instigates further expansion to make up the lost earning power, while socio-ecological subsidies are sought more vigorously in order to further reduce the cost of production. For instance, intensification in salmon farming increases the rate of ecological degradation by increasing the spread of farm-derived sea lice to wild salmonids (Krkošek et al., 2005) to the point of now threatening widespread extinction of British Columbian wild salmon populations (Krkošek et al., 2007). While intensification drives up the costs of ecological effects, the potential socioeconomic benefits diminish. For instance, according to the report of WWF (2004), farmed salmon production in British Columbia had increased 895 000 times from 1998 to 2002, while the unit value had declined dramatically (55%). The value decline was more than compensated for by the six orders of magnitude production increase. Similarly, Mediterranean ranches are flooding world markets, resulting in a decline in value of ranch tuna by 50% from 2003 to 2004 (WWF, 2004).

The rapid expansion of the farm shrimp industry in Bangladesh has similarly demanded increasing socio-ecological subsidies. These include the reduction of indigenous fish stocks and the loss of economically important mangrove flora and fauna species in addition to the direct destruction of mangrove forests and the discharging of untreated wastewater to the coast (EJF, 2004). Wastewater is responsible or eutrophication of the coastal water through unused feed, and contamination of aquatic food products with residual harmful chemicals and antibiotics are among the identified issues regarding both aquatic resource conservation and human health (EJF, 2004). Since the mid-1990s, other shrimp-farming countries such as India, Thailand and Philippines have experienced similar types of serious challenges and even occasional shrimp production collapses (Alongi et al., 2000). Indeed, disease outbreaks, overexploitation of seed stock, environmental degradation and poor management continue to challenge shrimp aquaculture sustainability throughout Southeast Asia (Primavera, 1998). The consistency of experience across shrimp-producing countries suggests common underlying drivers. We argue here that not only is this the case but due to the inverse relationship of economic to socio-ecological benefit, these scenarios are predictable and unavoidable if maximum profit generation is the sole objective of the enterprise. If we plot the relationship of three subsystems (ecological, economic and social), we see a nonlinear inverse relationship between economic or production and socio-ecological factors.

As intensification of production increases, socio-ecological degradation is initially modest but increases with production. With further production increases, the relationship reaches a tipping point and socio-ecological benefits are eroded at an increasingly rapid pace

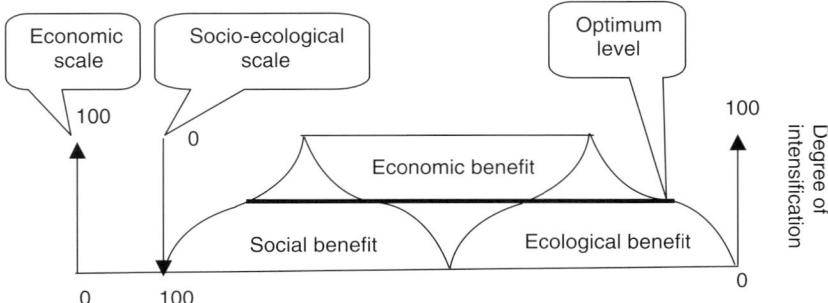

Figure 9.2 Socio-ecological-economical domes in industrial aquaculture where socio-ecological benefits go down along with the increasing trend of economic benefit or production due to intensification more and more.

(Figure 9.2). If the production goes to the asymptote level by virtue of super intensification, literally all socio-ecological benefits are lost, or in other words, social and ecological subsidies to the production system are 100%. The salient point here is that maximum socio-ecological-economic benefit cannot be simultaneously attained. The profit that lays beyond the optimum level in Figure 9.2 is derived from liquidating social and ecological capital – an inherently unsustainable position. Therefore, it would be a wise for the industrial aquaculture industry not to cross this threshold and to seek, in their own interest, optimal production strategies. The scale of production at which the threshold sits will vary across different species, production systems, governance models and biogeographic regions. From a purely economic perspective, Bangladesh remains in an advantageous position with regard to shrimp farming due to the availability of low-cost labor, low environmental compliance enabled by lack of enforcement and low environmental awareness. Destruction of mangrove and some coastal fish species during collection of wild shrimp post-larvae were the visible environmental degradation. The government has already enacted regulations to stop conversion of mangroves to shrimp farms and to prohibit the collection of wild shrimp post-larvae, though enforcement suffers due to lack of institutional capacity. On the other hand, due to lack of comprehensive research, awareness is low regarding issues associated with discharging untreated effluent, escaping cultured species and the spread of potentially harmful disease pathogens and parasites.

9.4 Sustainability model

Sustainable development (in the agriculture, forestry and fisheries sectors) conserves land, water, plant and animal genetic resources, and is environmentally nondegrading, technically appropriate, economically viable and socially acceptable. It is important now to develop policies and practices that ensure environmental sustainability, including environmentally sound impact assessments, appropriate technologies, resource-efficient farming systems and integration of aqua farms into coastal area management plans Bangladesh. The recently developed ICZM would be an effective basis for this. A schematic model of the eco-friendly sustainable aquaculture in the field of coastal shrimp farming is shown in Figure 9.3.

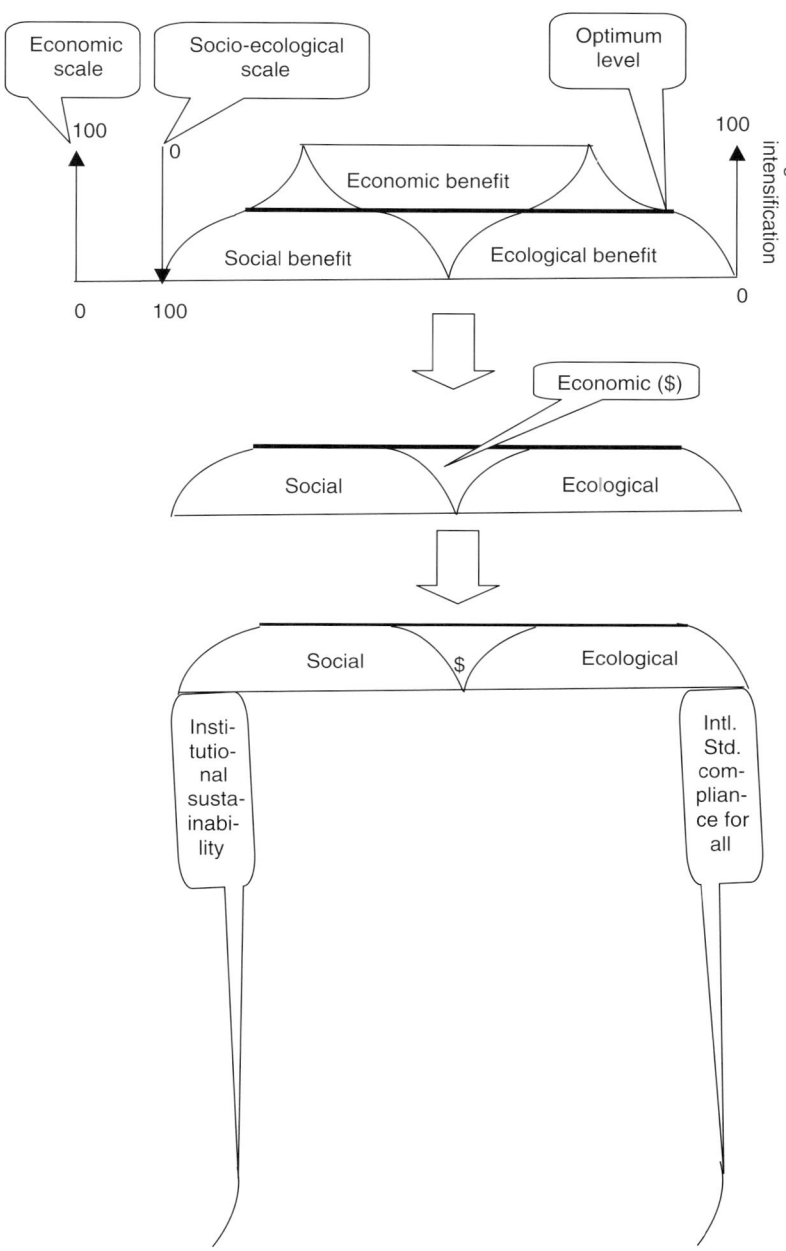

Figure 9.3 Sustainability vision of industrial aquaculture (for long-term vision).

9.4.1 Ecological sustainability

A successful aquaculture system does not have wastes, only by-products, to be used as positive contributors to the surrounding ecosystems and economy (Folke and Kautsky, 1992). Biological nutrient recycling or bioremediation of shrimp effluent would be an ideal management system to protect the farm itself from disease outbreaks and the surrounding ecosystem from pollution. An integrated method of culture with eastern oyster, *Crassostrea virginica*, and the macroalga, *Gracilaria tikvahiae*, is used for biofiltering the effluent from a pond used for intensive culture of Pacific white shrimp, *Penaeus vannamei* (Patricia et al., 2001). So, ecological sustainability will incorporate the long-term sustainable concerns of coastal shrimp farming by reducing or zero pollution impact on ecosystem.

9.4.2 Social sustainability

Social sustainability focuses on the 'macro' level, i.e. on maintaining or enhancing overall long-term socioeconomic welfare (Charles, 2001). This socioeconomic welfare is based on a blend of relevant economic and social indicators focusing on the generation of sustainable net benefits, a reasonable distribution of those benefits among all the participants in the shrimp-farming system, and maintenance of the system's overall viability within local and global economies. On the other hand, community sustainability emphasizes the 'micro' level, i.e. focusing on the desirability of sustaining communities as valuable human systems in their own right, and more than simple collections of individuals (Charles, 2001). Bioremediation or reducing the use of chemicals and drugs might also be helpful for the workers' health in the farm. Along with this compliance with the national and international legislation such as wages, International Labor Organization (ILO) rules would increase the satisfaction level of the employees in the shrimp sector, which would ensure the social sustainability.

9.4.3 Economic sustainability

Economic sustainability is one of the main pillars of eco-friendly sustainability in industrial aquaculture since its objective is to maintain a profit margin at both micro- and macrolevels. Simply, if the industry cannot make a profit, the entire array of benefits from it, including employment opportunities, foreign currency earning and satisfying protein demand or food security will be lost. But this sustainability is the complex web of social, ecological and economic factors. The farm itself is responsible to make sure it is producing a product that is compliant with socio-ecological rules and regulations. The last but not least factor is the international market or consumers who actually make industrial aquaculture profitable. Farms should not intensify their production systems to the extent that technology might allow. Therefore, it is the role of the international consumer to ensure a ready market and willingness to the pay the full (i.e. unsubsidized) cost of production. All players must act accordingly otherwise true sustainability will not be attainable. Where this starts is at the level of ministries and departments within Bangladesh to make the shrimp industry sustainable.

9.4.4 Institutional sustainability

It involves maintaining suitable financial, administrative and organizational capability over the long term, as a prerequisite for these components of sustainability (Charles, 2001). In terms of shrimp farming in Bangladesh, financial institutions such as banks, administrative institutions along with legal and enforcement support, technical support from Department of Fisheries and Department of Agriculture and organizational mobilization support from the Department of Social Welfare are required, all working in harmony, to make this industry sustainable.

Technological development is a dynamic process, for which institutional support might help the farmers by updating them with new and innovative information regarding technological developments. Further, farmers should have their own institution or organization nationally and internationally to harmonize compliance equally. The organizational structure embeds the interdisciplinary decision-making arrangement through the existence of the local area management teams whose focus is on the issues and ongoing problems of the various fisheries in specified management areas (Lane and Stephenson, 2000).

9.4.5 *International standard compliance for all*

Lessons can be learned from the global salmon farming experience. Here, each company is in competition with all others in a global market place. Therefore, success relies on a company being able to undercut its competitors in order to secure market share. In short, to produce the most salmon for the least cost. Thus ensuing maximum consumption of available social and ecological subsidies available (Volpe, 2006). However, production costs vary on the basis of the domestic economic context and level of compliance exacted in each country. Production costs increase as compliance with environmental and social rules and regulations increase. The salmon-farming industry on the coast of British Columbia in western Canada is facing a significant challenge to remain viable in the face of competition from other salmon-producing nations with inherent lower costs of production (e.g. Norway) and/or lower compliance to basic socio-ecological standards (i.e. Chile). Since economic sustainability is a vital component of eco-friendly sustainability, an international agreement among aquaculture-producing countries is another option to 'level the playing field' and promote environmentally sound practices (Naylor et al., 2003). This type of agreement should comprise the standard international codes of practices having the scope of flexibilities and adaptabilities within each country context, which will ensure the commitment of the country to comply with these and national socio-ecological standards as well. At present, there is 'Codes of Practices for Shrimp Farming' by FAO. These are nicely written in document form, but they have not been implemented on farms (Boyd et al., 2002). The main disadvantage of codes of conduct is that they are voluntary, there has been little implementation, and benefits are assumed rather than proven (Boyd et al., 2002).

9.5 Cost and benefit of eco-friendly sustainable shrimp farming

Environmental management is considered an economic challenge by many farm operators who see, for example, wastewater treatment as an additional and unnecessary production cost. However, adopting a longer-term view of the expense tells another story. The initial investment and annual operating costs of water treatment not only allow higher production on the farm, but because the surrounding environment is not impacted by pollution, the farms longevity is greatly increased. Therefore, the earning power of the farm over its lifetime is far greater. Further, because alternative income streams are not impacted (e.g. commercial fisheries), the social–ecological–economic fabric of the community within which the farm is imbedded also remains intact. A hypothetical benefit and cost of compliance with eco-friendly sustainability measures is shown in Table 9.2.

9.6 Discussion

Aquaculture contributes significantly to the world food supply, providing approximately 30% of total fisheries production (Aquaculture Asia, 2005). Shrimp farming is a very diverse subsector of aquaculture, economically, in terms of farming systems, geographic locations, environments and social contexts. The development of commercial shrimp aquaculture in Bangladesh started in early 1970s and is an important source of employment and income for hundreds of thousands of people. Employment and income are generated on the farm but also in the supply industries as well as in shrimp processing and distribution. Lack of regulation governing establishment of shrimp farms has led in some places to improper sitting, overcrowding, changes in land use pattern and appropriation of habitat and associated ecosystem services. Not surprisingly in 1995, the shrimp aquaculture industry was severely impacted by a widespread outbreak of viral disease. Sustainability of shrimp farming has been questioned because of shrimp disease outbreaks and significant economic losses. Of these and other possible negative impacts, water pollution by pond effluent is probably the most common complaint, and this concern has attracted the greatest official attention in most nations (Boyd and Gautier, 2000; cited by Boyd, 2003). Concentrated in the southern coastal belts, Bangladeshi shrimp culture has witnessed a threefold increase in the last decade. This export-driven industry brings substantial foreign capital in to Bangladesh, but this has failed to generate much-needed economic returns or tangible assets such as schools, sanitation or health care in the communities. On the contrary, it has increased the gulf between rich and poor and for many the onset of shrimp farming has led to a diminished quality of life. Industry growth is not motivated by the need to provide food for the hungry, but as a cash crop for an export market. As such, profit is the chief objective and in its pursuit, the industry has undermined local food security and limited livelihood options for those human populations that are most at risk (EJF, 2004). Such problems emphasize the importance of identification and promotion of effective on-farm and sectoral management practices (Phillips et al., 1993). A more significant challenge is to develop new indicators that can differentiate shrimp farm discharges from other loads, and to identify ecosystem attributes that relate to the community-derived environmental values (Burford et al., 2001). Many shrimp farmers often seek to maximize their short-term gain at the expense of the environment. Shrimp farming is sustainable if it is in harmony with other economic activities in using common natural resources. It should produce a stable net financial benefit for both the farm and the community in which it is embedded without degrading the environment. Its development has to be balanced among production, marketing and other supporting services such as hatchery, feed mill and legal measures. Therefore, a sustainable shrimp-farming system has to be biotechnically feasible, environmentally sound and socioeconomically viable in Bangladesh. It is also believed that the country will benefit not only environmentally but also economically, though such achievements are not

Table 9.2 Hypothetical benefit and cost of compliance with eco-friendly sustainability measures in the shrimp aquaculture in Bangladesh.

Sustainability indicators	Farm level	National level
	Benefit	
Environmental/ ecological	• Can improve environmental performance and management • Can facilitate compliance with national and international environmental requirements • Can earn good reputation as green-labeled industry	• Can help the country to fulfill commitments under different international agreements • Can hold a good position in international market as green product • Can strengthen the institutional capacity • Can create new employment opportunity
Economical	• Can create opportunity to add value to the unit price • Can increase export competitiveness • Can reduce the cost of risk such as disease out breaks • Can decrease the production cost for not using any chemicals/drugs. • Can reduce the opportunity cost by increasing the life span of the farm • Can induce innovation of more environment friendly production system	• Can create more opportunities for capital investment • Can increase foreign exchange earnings
	Cost	
Social	• Can increase the satisfaction of employee • Can increase the gender ratio in the recruitment • Can increase the local employee • Can increase the spill over effect to the community • Can increase the corporate image to the community • Can facilitate compliance with national and international labor/wages requirements	• Can help country to eliminate the gender differentiation in employment • Can help increasing regional development • Can improve image of the country globally
Environmental/ Ecological	• Can increase initial cost for effluent treatment system though it can reduce the cost through recycling of nutrients • Can increase training cost for the staff	• Can increase cost to strengthen institutional capacity through research, training and recruitment • Can increase annual expenditure for recruiting new staff for monitoring or enforcement • May require some extra money to produce compliance mechanism such as standard guideline, awareness development
Social	• Can increase the labor cost, though it might be minimized through satisfactory output from the staff and the excess unit price for green product	• Can increase cost to monitor or enforce the wage system in this industry

Source: Reprinted from Shang et al. (1998), with permission from Elsevier.

without some costs (Khatun, 2004). Ultimately, reduction in nutrient discharges is most likely to ensure the environmental sustainability of the industry, and an understanding of the ecological processes affected by aquaculture both in ponds and in downstream provides an important step along with this path (Burford et al., 2001). Bioremediation is recommended as a way to begin to minimize negative environmental impacts.

On the other hand, insufficient institutional capacity leading to poor governance poses serious problems as regulations and legislations are applied partially and unevenly across the entire sector. The lack of any clear national policy or planning and effective uniform implementation of laws compromises the industry at all levels, further encouraging corrupt, unsustainable and inappropriate practices (EJF, 2004). Reform is required where environmental security and social equity must be put at the heart. Development support within the sector needs to be much more carefully directed toward assisting the poorest, especially women, children and the landless. Greater and more effective consultation with local communities should be undertaken. Alternative income-generating activities need to be protected in order that communities in which farms are located are dynamic and therefore resilient. All these things may happen in identifying and execution of best management practices with an appropriate governance model.

9.7 Acknowledgments

The authors express their deep gratitude and regard to Dr Erlend Moksness for his magnificent arrangements of the symposium in Arendal, Norway, and offering financial support to make MALSs participation possible. Two reviewers and Dr Erlend as well provided valuable comments that helped improve the manuscript.

References

ADB/NACA (1996) Aquaculture Sustainability and the Environment. Report on a Regional Study and Workshop on Aquaculture Sustainability and the Environment. ADB Series. Asian Development Bank and Network of Aquaculture Centers in Asia–Pacific, Bangkok, Thailand.

Alongi, D.M., Johnston, D.J. and Xuan, T.T. (2000) Carbon and nitrogen budgets in shrimp ponds of extensive mixed shrimp-mangrove forestry farms in the Mekong delta, Vietnam. *Aquaculture Research* **31**, 387–399.

Aquaculture Asia (2005) Sustainable aquaculture. *Aquaculture Asia Magazine*, **3**, July–September 2005, p. 48.

Banks, R. (2002) Brackish and marine water aquaculture. Dhaka, Bangladesh, Government Printing Office. Government of the People's Republic of Bangladesh.

Barkat, A. and Roy, P.R. (2001) Marine and Coastal Tenure. Community-Based Property Rights in Bangladesh: An Overview of Resources, and Legal and Policy Developments. Prepared for presentation at the Marine and Coastal Resources and Community-Based Property Rights: A Philippine Workshop Organized by Tambuyong Development Centre Tanggol Kalikasan, Centre for International Law and the CBCRM Resource Centre.

Boyd, C.E. (2003) Guidelines for aquaculture effluent management at the farm-level. *Aquaculture* **226**, 101–1112.

Boyd, C.E. and Gautier, D. (2000) Effluent composition and water quality standards. *Global Aquaculture Advocate* **3** (5), 61–66.

Boyd, C.E., Hargreaves, J.A. and Clay, J.W. (2002) 'Codes of Practice and Conduct for Marine Shrimp Aquaculture'. Report prepared under the World Bank, NACA, WWF and FAO Consortium Program on Shrimp Farming and the Environment. Work in Progress for Public Discussion. Published by the Consortium, 31 pp.

Burford, M.A., Jackson, C.J. and Preston, N.P. (2001) Reducing nitrogen waste from shrimp farming: an integrated approach. In: Browdy, C.L. and Jory, D.E. (eds). *The New Wave. Proceedings of the Special Session on Sustainable Shrimp Culture, Aquaculture 2001*. The World Aquaculture Society, Baton Rouge, pp. 35–43.

Charles, A.T. (2001) Sustainable fishery systems. In: Pitcher, T.J. (ed.). *Fish and Aquatic Resources Series 5*. Blackwell Science, UK, p. 370.

DoF. (2002) *Shrimp Aquaculture in Bangladesh: A Vision for the Future*. Government of the People's Republic of Bangladesh, Office of the Director General, Department of Fisheries, Dhaka. An output of the Shrimp Action Plan, sponsored by DFID, Bangladesh through the Fourth Fisheries Project and the Global Environment Facility, October 2002.

EC. (2002) *Policy Research for Sustainable Shrimp Farming in Asia: A Comparative Analysis of Bangladesh, India, Thailand, and Vietnam with Particular Reference to Institutional and Socio-Economic Aspects*. Literature Review on Bangladesh Shrimp. European Commission: PORESSFA, Project No. IC4-2001-10042. Retrieved from http://www.port.ac.uk/research/cemare/publications/pdffiles/sustainableshrimpfarminginasia/filetodownload,28777,en.pdf on 15 April 2006.

EJF. (2004) *Desert in the Delta: A Report on the Environmental, Human Rights and Social Impacts of Shrimp Production in Bangladesh*. Environmental Justice Foundation, London.

Fleming, C. (2004) Challenges facing the Shrimp Industry in Bangladesh. American International School, Dhaka, Senior Project 2004, p. 24. Retrieved from http://www.ais-dhaka.net/School_Library/Senior%20Projects/04_Fleming_shrimp.pdf on 15 April 2006.

Folke, C. and Kautsky, N. (1992) Aquaculture with its environment – prospects for sustainability. *Ocean and Coastal Management* **17**, 5–24.

Funge-Smith, S.J., Matthew, R. and Briggs, P. (1998) Nutrient budgets in intensive shrimp ponds: implications for sustainability. *Aquaculture* **164**, 117–133.

Hopkins, J.S., Devoe, M.R. and Holland, A.F. (1995) Environmental impacts of shrimp farming with special reference to the situation in the continental United States. *Estuaries* **18**, 25–42.

Jackson, C., Peterson, N, Thompson, P. and Burford, M. (2003) Nitrogen budget and effluent nitrogen components at an intensive shrimp farm. *Aquaculture* **218**, 397–411.

Karim, M. (2003) *A Gross Oversimplification: Stop Blaming Shrimp*. Shrimp Seal of Quality Organization, Dhaka.

Khatun, F. (2004) Fish trade liberalization in Bangladesh: implications of SPS measures and eco-labelling for the export-oriented shrimp sector. In: Bostoc, T., Greenhalgh, P. and Kleih, U. (eds). *Policy Research-Implications of Liberalization of Fish Trade for Development Countries*. Natural Resources Institute, Chatham.

Krkošek, M., Ford, J.S., Morton, A., Lele, S., Myers, R.A. and Lewis, M.A. (2007) Declining wild salmon populations in relation to parasites from farm salmon. *Science* **318**, 1772–1775.

Krkošek, M., Lewis, M.A. and Volpe, J.P. (2005) Transmission dynamics of parasitic sea lice from farm to wild salmon. *Proceeding of the Royal Society of London Series B* **272**, 689–696.

Lane, D.E. and Stephenson, R.L. (2000) Institutional arrangements for fisheries: alternative structures and impediments to change. *Marine Policy* **24**, 69–77.

MoWR (2005) *Coastal Zone Policy, 2005*. Ministry of Water Resources, Government of the People's Republic of Bangladesh, p. 12.

MoWR (2006) *Coastal Zone Policy, 2006*. Ministry of Water Resources, Government of the People's Republic of Bangladesh, p. 12.

Naylor, R.L., Eagle, J. and Smith, W.L. (2003) Salmon aquaculture in the Pacific Northwest: a global industry with local impacts. *Environment* **45**, 18–38.

Naylor, R.L., Goldburg, R.J., Mooney, H., Beveridge, M., Clay, J., Folke, C., Kautsky, N., Lubchenco, J., Primavera, J. and Williams, M. (1998) Nature's subsidies to shrimp and salmon farming. *Science* **282**, 883–884.

Patricia, N.K., Samocha, M.T.M., Jones, E.R. and Browdy, C.L. (2001) Characterization of intensive shrimp pond effluent and preliminary studies on biofiltration. *North American Journal of Aquaculture* **63**, 25–33.

PDO-ICZMP (2003) Living in the Coast – Problems, Opportunities and Challenges, Living in the Coast Series 2. Program Development Office for Integrated Coastal Zone Management Plan, Dhaka, Bangladesh, p. 54.

Phillips, M.J., Lin, C.K. and Beveridge, M.C.M. (1993) Shrimp culture and the environment – lessons from the world's most rapidly expanding warm water aquaculture sector. In: Pullin, R.S.V., Rosenthal, H. and Maclean, J.L. (eds). *ICLARM Conference Proceedings*, Manila, Philippines, pp. 171–197.

Primavera, J.H. (1998) Tropical shrimp farming and its sustainability. In: De Silva, S.S. (ed.). *Tropical Mariculture*. Academic Press, San Diego, pp. 257–289.

Rosenbery, B. (1995) *World Shrimp Farming*. Shrimp News International, San Diego.

Shang, Y.C., Leung, P. and Ling, B. (1998) Comparative economics of shrimp farming in Asia. *Aquaculture* **164**, 183–200.

Volpe, J.P. (2006) 'Salmon Sovereignty' and the dilemma of intensive Atlantic salmon aquaculture development in British Columbia. In: Parish, C.C., Turner, N.J. and Solberg, S.M. (eds). *Resetting the Kitchen Table: Food Security, Culture*. Nova Science Publishers Inc., Hauppauge.

WARPO (Water Resources Planning Organization) (2003) *Integrated Coastal Zone Management Plan Project: A Systems Analysis of Shrimp Production*. Dhaka, p. 39.

WWF – The Global Conservation Organization (2004) Position of WWF regarding the 14th Special Meeting of ICCAT, 24 February 2005. Available from http://assets.panda.org/downloads/iccat_tuna.doc.

Chapter 10
Bioshields and Ecological Restoration in Tsunami-Affected Areas in India

Nibedita Mukherjee, Muthuraman Balakrishnan and Kartik Shanker

Abstract

There has been considerable interest in activities concerning 'bioshields' in India following the December 2004 tsunami. There is an ongoing debate about the effectiveness of these bioshields with respect to the tsunami within both scientific circles and local communities. Despite the lack of concrete evidence of their role in protecting the coast and the ambivalence in data and opinion, there have been numerous post-tsunami initiatives that have established and promoted plantations as bioshields in India. In this chapter, we present a brief overview of such initiatives by the government and by nongovernmental organizations (NGOs). We have collated information on the work done on bioshields and attempted to evaluate them from an ecological perspective. The results indicate that such plantation efforts are not new (post-tsunami) but have been practiced for decades by the government. Scant attention has been paid to the ecology of the species being planted in such 'restoration efforts' and very little science has been used either in the formulation of policies governing bioshields or in the implementation of the same. We propose long-term monitoring of these 'restored sites' or shelterbelts to evaluate their effectiveness and sustainability in the long run.

10.1   Introduction

The Indian Ocean tsunami that was caused by an earthquake of intensity of 8.9 on the Richter scale hit the coastline of peninsular India on 26 December 2004 (http://www.eeri.org/lfe/clearinghouse/sumatra_tsunami/reports/EERI-Report-Combined-Yeh-India.pdf, accessed on 15 January 2008). This mammoth natural disaster affected 2260 km of the coastline in mainland India and led to death of 12 405 people in three states (namely Tamil Nadu, Andhra Pradesh and Kerala) in India (UN Report, 2005). Subsequently, the protection of the coast by both natural and artificial structures gained added impetus. It is not surprising therefore that there has been considerable interest in activities concerning 'bioshields' after the tsunami. Bioshields, as the name indicates, are coastal vegetation structures (both natural and planted) that are supposed to contribute to the protection of the coast from storms, cyclones and even tsunamis to varying extents. In the majority of cases along the Tamil Nadu coast, they consist of mangroves and *Casuarina* plantations.

There has been an ongoing debate about the effectiveness of these plantations and other coastal forests in providing protection from the tsunami (Kar and Kar, 2005; Kathiresan and Rajendran, 2005; Kerr et al., 2006; Vermaat and Thampanya, 2006). Kar and Kar (2005) were the first to point out the need for research on the beneficial role of mangroves in mitigating the effects of the tsunami. Kathiresan and Rajendran's (2005) work in 18 tsunami-affected hamlets located along 25 km of the coastline in Tamil Nadu indicated that hamlets protected by (i.e. located on the landward side of) mangroves and other coastal vegetation suffered less human death and damages. Danielsen et al. (2005) also pointed out that in Cuddalore district, Tamil Nadu, the impact of the tsunami was significantly lesser in mangrove- and *Casuarina*-protected villages than in others. However, Kerr et al. (2006) reanalyzed the data of Kathiresan and Rajendran (2005), and came to the conclusion that mangroves and other vegetation did not have a significant effect on the impact of the tsunami. Rather, topography and distance from the shoreline were the major factors determining impact of the tsunami. Following this, Kathiresan and Rajendran (2006) pointed out that the study was limited to 18 hamlets of a particular area and the tsunami run-up there was only 2.8 m. Vermaat and Thampanya (2006) also reanalyzed the data of Kathiresan and Rajendran (2005), supporting the original conclusion that mortality and property loss were actually less behind mangroves. Thus, at least in the scientific literature, there appears to be no clear consensus about the effectiveness of bioshields, particularly *Casuarina*, in coastal protection and tsunami mitigation.

Local communities too have diverging opinions about coastal plantations. While some communities depend on mangroves and other vegetation for their subsistence, there is also evidence that in many hamlets, local fishing communities have opposed and even uprooted *Casuarina* saplings (Rodriguez, 2007). The main causes for this conflict are rights to the coastal land and accessibility to the sea, both of which are affected by *Casuarina* plantations. However, works supporting plantations as bioshields have often been cited to further plantation efforts in various places and substantial work has been done on bioshields in India after the tsunami. Both government and nongovernmental agencies have contributed significantly to the planting of bioshields. In India, most of the plantations along the coast are under the jurisdiction of the Forest Department, which is a government organization. There are also a large number of nongovernmental organizations (NGOs), which work either independently or in close collaboration with the Forest Department in raising these plantations. International bodies are either directly involved or fund bioshields in India.

Past studies elsewhere have demonstrated that monoculture stands may not be sustainable in the long term and provide fewer ecosystem services (Field, 1999; Ellison, 2000). For instance, Walters (2000) found no post-planting recruitment of different mangrove species (other than those used for plantation) into 50–60-year-old 'restored' sites in the Philippines. Often such rehabilitation/restoration efforts move directly into the planting stage without any assessment of stress factors affecting natural regeneration, the suitability of the site for restoration activities or assessment of the socioeconomic dependence of local communities on mangrove resources at the restoration sites (Ellison, 2000). A sizeable proportion of such efforts therefore end in failure. Very often, restoration programs end with the final activity of planting, and very few plantations are assessed with respect to their functionality.

Bosire et al. (2008) propose 'ten commandments' for mangrove restoration. In this study, we focused on these questions in an attempt to evaluate bioshield plantation initiatives that are currently underway in three states in southern India:

(a) What is the scale and extent of bioshields in tsunami-affected states in India?
(b) Did ecological science play a role in the decision-making process?
(c) What are the flaws in such practices from an ecological perspective?
(d) Were local people involved in the process of raising plantations?

10.2 Methods

In India, the Forest Department is the governmental organization, which is primarily responsible for raising plantations along the coast. We conducted structured interviews with Forest Department officials (from the head of the department, the Principal Chief Conservator of Forests, in each state to local Range Forest Officers) in Tamil Nadu, Andhra Pradesh and Kerala states in southern India. We also procured maps and details of the post-tsunami micro-plans of the Forest Department (see Table 10.1 for an illustration of this administrative setup in Kerala state). We collected information regarding various local NGOs working on plantations through repeated field visits and interactions with local communities between November 2006 and June 2007. We also contacted the regional representatives of major funding agencies to get contact details of NGOs raising plantations. Structured interviews were conducted with representatives of NGOs in Andhra Pradesh and Kerala. We also visited some of their plantation sites to gain a better understanding of ground realities.

The following four objectives were focused on to answer the research questions:

- Scale and extent: In the three different states, we tried to estimate the following: How much area was planted? What was the major source and amount of funds?

Table 10.1 Administrative setup of Kerala Forest Department.

Category	Number of circles	Number of divisions	Number of ranges	Number of Sections	Number of beats
Territorial	5	23	74	177	303
Wildlife	3	12	19	48	49
Working Plan and Research	1	9	14	23	–
Vigilance	2	8	19	25	90
Social Forestry	3	14	37	223	–
Nature Study Centre	1	–	1	–	–
Training	1	2	.	–	–

Source: http://www.kerala.gov.in/dept_forest/forest.htm, accessed on 10 October 2007.
Category: Refers to the various administrative divisions of the Forest Department.
Circle: Each state is divided into a number of circles for better administration.
Division: Each circle is subdivided into divisions headed by a Divisional Forest Officer.
Range: With each division are nested several ranges each headed by a Range Officer.
Section and beat: Each division in turn composed of sections (headed by Section Officer) and each section is subdivided into beats (headed by Beat Officer). The origin of the term 'beat' dates back to colonial times when a beat officer's job was to beat the bushes (along with forest guards further down in the hierarchy) to scare away wild animals when a higher official was accompanying him in the forest.

- Ecology:
  (a) Was the plantation a monoculture/polyculture?
  (b) Was any site selection carried out before the process of planting?
  (c) Was any monitoring of the plantation done post-plantation?
- Gaps or flaws: What are the gaps or flaws in the plantation process from an ecological perspective on the basis of the above-mentioned parameters in ecology?
- Involvement of local people: Were local people directly involved in the process of plantation? What are their perceptions toward plantations and were these perceptions heeded to in the policy-making process for bioshields?

10.3 Results

10.3.1 Scale and extent

10.3.1.1 Forest department initiatives

Tamil Nadu: The Tamil Nadu Forest Department has been raising shelterbelt plantations along its coastline since 1960 (Table 10.2). As a result, about 2239 ha had been covered under shelterbelt plantation prior to the tsunami under various plantation schemes such as, Tamil Nadu Afforestation Programme and National Afforestation Programme. Immediately after the tsunami, the Forest Department conducted a rapid assessment to identify sites for further plantations. An area of about 17 754 ha was found to be available along the coast for raising shelterbelts 'after leaving room for encroachment'. Out of this total area, about 11 500 ha of area was found suitable for shelterbelt (*Casuarina* spp.) and 6254 ha for mangroves. Currently, 2000 ha of *Casuarina* spp. plantation and 400 ha of mangroves are being planted in this state. With the assistance of the World Bank, the Forest Department implemented two schemes: (a) Emergency Tsunami Reconstruction Project (ETRP) and (b) National Cyclone Risk Mitigation Project (NCRMP). Under ETRP alone, US$ 1.75 million were sanctioned for *Casuarina* shelterbelt plantation and US$ 250 000 for mangrove plantations.

Andhra Pradesh: The Andhra Pradesh Forest Department also started planting bioshields after the tsunami. A project titled Andhra Pradesh Community Forest Management (APCFM) is currently being implemented. The duration of this project is 5 years and the estimated budget for this project is approximately US$ 162 million (http://forest.ap.nic.in/JFM%20CFM/CFM/A%20P%20CFM%20Index.htm, accessed on 10 October 2007). International Development Agency and Government of Andhra Pradesh are financing this project. Although the main agenda of this project is to reduce the pressure of natural resource extraction on existing forests and poverty alleviation, coastal bioshields are being planted as a part of this project after the tsunami (see Figure 10.1 for flow of funds in this project).

Kerala: In Kerala, the state government had already constructed sea walls along 550 km of their 600 km coastline. Thereby the emphasis on bioshields after the tsunami has been less than the other two states. The Forest Department has, however, played an active role in raising mangrove plantations along the extensive backwaters in this state to prevent rapid erosion of the riverbanks.

Table 10.2 Details of shelterbelt plantations raised by Forest Department of Tamil Nadu.

S. no.	Coastal districts	Year of planting	Area (ha)
1	Villupuram/Cuddalore	1978	54.3
		1985	110
		1988	31.4
		1991	11.7
		1992	19.7
		2000	10
2	Thanjavur and Pudukottai	1988	30
		1999	16.5
		2001	450
		2004	150
3	Nagapattinam and Thiruvarur	1989	59
		1994	15
		1998	5.5
		2000	250
4	Ramanathapuram	1969	200
		1974	23
		1975	200
		1986	10
		1990	33
		1991	23
		1999	17.5
		2001	300
5	Tuticorin	1998	15
		1999	15
6	Tirunelveli	1974	30
		1975	30
		1990	2.5
		1992	5
		1997	0.5
7	Kanyakumari	1960–1973	28

10.3.1.2 Nongovernmental organization initiatives

In addition to the government, several NGOs have also been involved in establishing plantations after the tsunami (see Table 10.3 for a summary of the work done by various NGOs; Figure 10.2 provides a representative map of plantation activities in Tamil Nadu).

In sharp contrast to the government, the majority of the NGOs started plantation activities only after the tsunami. In a sizeable number of instances, many NGOs themselves have been established post-tsunami, or have included plantations within their institutional themes after the tsunami. With the exception of those NGOs that work in close association of the Forest Department and get funded by the department, most of the NGOs interviewed by us are dependent on international funding. The amount received by each organization ranged from a minimum of US$ 12 000 to about US$ 200 000.

140 Integrated Coastal Zone Management

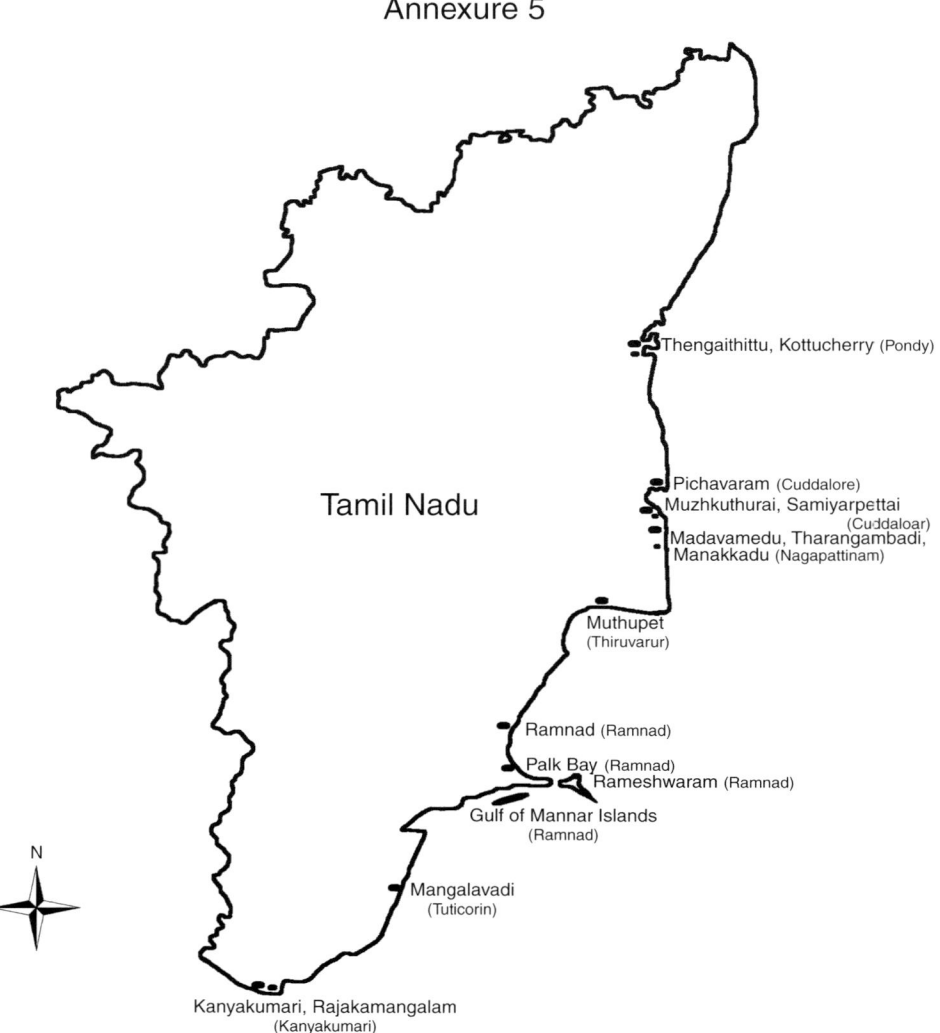

Figure 10.2 Map showing plantation activities in Tamil Nadu.

Soil suitability certificate

It is to certify that I have inspected the _____

_____

Site selected for raising _____
Plantations of _____ Km _____
(Species) in _____ village of _____ Mandal in Ongole (SF)
Range, and after ascertaining soil and rainfall requirements, I am satisfied that the site
selected is fit for raising of _____species.

                                            Forest Range Officer,
                                            Social Forestry, Ongole

Counter Singned

Divisional Forest Officer,
 Social Forestry Division
      Ongole.

Figure 10.3 A copy of the soil suitability certificate emphasizing site selection before plantation in Andhra Pradesh.

Site selection: In almost all instances, we found that no prior site selection was done before the plantation activity. Interestingly, however, there exists a government policy according to which it is mandatory to conduct site selection before plantation (Figure 10.3).

Monitoring: In the majority of cases, plantations are not monitored for growth rate, regeneration status, colonization of nonplanted species, etc., owing to lack of funds and skilled labor.

10.3.2.2 *Nongovernmental organizations*

Species planted: Almost all NGOs are raising monocultures. *A. marina* and *A. officinalis* are widely used in mangrove plantations. *Rhizophora* spp. is also used occasionally. In the case of non-mangrove plantations, only *C. equisetifolia* is planted. In some places, *Palmyrah* is also being planted.

Site selection: Similar to the Forest Department, most NGOs have not done any prior site selection before the plantation activity.

Monitoring: The status of monitoring is even more dismal in the case of NGOs. Several of theses plantations are fairly recent and there was no component of monitoring in their working plans.

10.3.2.3   Gaps/Flaws

Based on species: According to most government working plans in the east coast, a mixture of 90% *Casuarina* sp. with 10% mixed species are supposed to be planted. This consists of *Acacia planifrons, Syzygium cumini, Holoptelea integrifolia, Bambusa arundinacea, Dalbergia sissoo, Thespesia populnea, Jatropha* spp., *Borassus flabellifer, Acacia curticulliformis* and *Cocos nucifera*. However, in most of the circles, only *A. planifrons* is being used sparingly in Tamil Nadu. In Andhra Pradesh, *Casuarina* is planted exclusively. Almost all NGOs are planting monocultures. The coastline is progressively becoming dotted with nonnative uniform *Casuarina* plantations.

Site selection: Despite a government policy, which makes it mandatory to conduct site selection before plantation (Figure 10.3), it is never practiced on the ground. The NGOs too seldom followed any site selection protocol. The importance of remnant natural coastal ecosystems such as sand dunes and sandy beaches, which might be altered by such exotics, is also being ignored.

Monitoring: The government policy states that *Casuarina* should to be monitored for the succeeding 5 years after plantation and mangroves for the next 3 years. This is seldom practiced as noted by our field surveys. As mentioned earlier, the situation is worse in the case of NGOs who seldom have any funds allocated to monitoring.

10.3.3   Involvement of local people

Based on the questionnaire surveys and interviews, we found that in Tamil Nadu, the involvement of the local communities was significantly less than in Andhra Pradesh. The Tamil Nadu Forest Department had to bring in migrant labor form neighboring districts during the plantation season. A considerable portion of the resident local communities was opposed to such plantation activities in the state. In Andhra Pradesh, the Forest Department is raising plantations with the village conservation committees in four coastal circles as documented in Forest Department working plans.

There is greater involvement of local people in the plantation process in the case of NGOs. Most NGOs are engaged with the local communities in other initiatives such as rural development, education programs and livelihood enhancement. This enables them to interact with the communities at a deeper level than with the Forest Department.

10.4   Discussion

A fruitful collaboration between management and ecological research is required for effective forestry activities. As a beginning, we offered suggestions from ecological and socioeconomic perspectives. In the case of mangroves, the importance of the hydrologic regime and soil characteristics of a site cannot be overemphasized during site selection (see Ellison, 2000). There is ample documented evidence to suggest that plantation/afforestation need not be the only approach to 'restoration' of mangroves (see Lewis, 2004). On the contrary, according to some researchers, it may be the last option that should be put into practice (Stevenson et al., 1999). Often, altering the hydrology or soil characteristics that facilitate natural regeneration of mangroves may be the most cost-effective and self-sustaining method to bring back mangroves in a degraded mangrove site. If afforestation is the only option, then mixed plantations should be practiced rather than monoculture of *Avicennia*. After establishing the plantations, the sites should be monitored on a long-term basis to understand the sustainability of such efforts.

In the case of *Casuarina* plantations, the area where such plantations are being carried out needs to be carefully evaluated. Critical turtle nesting sandy beaches and sand dunes should be left untouched. Moreover, the ratio of other mixed species can be raised. A host of tropical dry deciduous or dry evergreen species can be used (e.g. *Manilkara hexandra*, *Mimusops elengi*, *Diospyros ebenum*, *Strychnos nux-vomica*, *Eugenia* spp., *Drypetes sepiaria* and *Flacourtia indica*). In this context, the flora in the recent past in the plantation sites could be of importance. At a more generic level, emphasis should shift from raising artificial plantations to preserving the natural dune ecosystems found in these regions.

The natural resource use of the local people in the vicinity of the plantation site should be borne in mind before the choice of species is made. Their perception toward plantations and their need for the particular plantation (e.g. fuel, firewood or bioshields) should be considered. Their involvement is crucial to the long-term sustainability of such plantations as about two thirds of the population lives close to the coast. Thus, it is critical to weave local support into restoration and management plans.

At a much larger scale, it is vital to analyze policies at local and regional levels, which regulate bioshield management (cf. Dahdouh-Guebas and Koedam, 2006; Rist and Dahdouh-Guebas, 2006). This is particularly significant for mangroves because they occur in diverse land use patterns where multiple policies come into play. Quite often, when there are multiple stakeholders, the policies do not overlap. In such cases, it is essential to be able to identify and comprehend those policies that are of relevance to bioshields.

10.5 Acknowledgments

We are grateful to the Principal Chief Conservator of Forests (Tamil Nadu, Andhra Pradesh and Kerala) for granting us permission to conduct the interviews. We thank all the Forest Department officials and representatives of NGOs whom we met, for giving us their precious time and sharing information with us. We are especially thankful to the DFO, Social forestry Division, Tuticorin and DFO, Territorial Division, Kannur, for logistic support. We are thankful to United Nations Development Programme for funding this project and Ashoka Trust for Research in Ecology and Environment (ATREE) for institutional support.

We are thankful to Aarthi Sridhar and Sudarshan Rodriguez for their help and valuable comments. The fisher folk who became part of our journey in this endeavor are specially thanked. We also thank Anusha, Shivani, Murugan, Gokul, Terenia, Anjana, Naveen, Mamatha and all others at Coastal and Marine Programme at ATREE for their help. We are grateful to Dr Farid Dahdouh-Guebas for his inputs on mangrove restoration. We also thank the three reviewers for their comments on an earlier version of this manuscript. This chapter is dedicated to the countless local people whom we encountered in our journey across the coast. Their warmth and hospitality has been a constant lifesaver in difficult times.

References

Bosire, J., Dahdouh-Guebas, F., Walton, M., Crona, B.I., Lewis III, R.R., Field, C., Kairo, J.G. and Koedam, N. (2008) Functionality of restored mangroves: a review. *Aquatic Botany* **89**(2), 251–259.

Dahdouh-Guebas, F. and Koedam, N. (2006) Coastal vegetation and the Asian tsunami. *Science* **311**, 37.

Danielsen, F., Sørensen, M.K., Olwig, M.F., Selvam, V., Parish, F., Burgess, N.D., Hiraishi, T., Karunagaran, V.M., Rasmussen, M.S., Hansen, L.B., Quarto, A. and Suryadiputra, N. (2005) The Asian tsunami: a protective role for coastal vegetation. *Science* **310**, 643.

Ellison, A. (2000) Mangrove restoration: do we know enough? *Restoration Ecology* **8**, 219–229.

Field, C.D. (1999) Mangrove rehabilitation: choice and necessity. *Hydrobiologia* **413**, 47–52.

Kar, R. and Kar, R.K. (2005) Mangroves can check the wrath of tsunami. *Current Science* **88**, 675.

Kathiresan, K. and Rajendran, N. (2005) Coastal mangrove forests mitigated tsunami. *Estuarine, Coastal and Shelf Science* **65**, 601–606.

Kathiresan, K. and Rajendran, N. (2006) Reply to Comments of Kerr et al. on "Coastal mangrove forests mitigated tsunami (*Estuarine Coastal Shelf Science*, 65 (2005) 601–606)". *Estuarine, Coastal and Shelf Science* **67**, 542.

Kerr, A.M., Baird, A.H. and Campbell, S.J. (2006) Comments on 'Coastal mangrove forests mitigated tsunami' by Kathiresan, K. and Rajendran, N. Estuarine (*Estuarine, Coastal and Shelf Science*, **65**, 601–606). *Estuarine, Coastal and Shelf Science*, **67**, 539–541.

Lewis III, R.R. (2004) Ecological engineering for successful management and restoration of mangrove forests. *Ecological Engineering* **24**, 403–418.

Rist, S. and Dahdouh-Guebas, F. (2006) Ethnosciences – a step towards the integration of scientific and traditional forms of knowledge in the management of natural resources for the future. *Environment, Development & Sustainability* **8**, 467–493.

Rodriguez, S. 2007. A preliminary socio-ecological review of post-tsunami ecosystem-derived livelihoods and rehabilitation efforts. UNDP- PTEI Project Report.

Stevenson, N.J., Lewis, R.R. and Burbridge, P.R. (1999) Disused shrimp ponds and mangrove rehabilitation. In: Streever, W. (ed.). *An International Perspective on Wetland Restoration*. Kluwer Academic Publishers, Dordrecht, pp. 277–297.

UN Report (2005) Tsunami: One Year After. A Joint UN Report, India.

Vermaat, J.E. and Thampanya, U. (2006) Mangroves mitigate tsunami damage: a further response. *Estuarine, Coastal and Shelf Science* **69**, 1–3.

Walters, B. (2000) Local management of mangrove forests in the Philippines: successful conservation or efficient resource exploitation? *Human Ecology* **32**, 177–195.

Chapter 11
The Impact of Population Density and Urbanization in the Coastal Northeast Regions of the Baltic Sea: A Time-Spatial Description

Merle Looring

Abstract

The Baltic Sea is a semi-enclosed ecosystem, which is sensitive to human activities and key anthropogenic impacts such as urbanization and population density affect the ecosystem of the Baltic Sea. The present research deals with the time-spatial analysis of population density and urbanization at the coast in the northeastern region of the Baltic Sea and possible impacts to the coastal ecosystems. The investigation area covers the coastal regions of Finland, Leningrad region of Russia, Republics of Estonia and Latvia, as well as the coastal regions of the Gulf of Finland and the Gulf of Riga. This chapter describes the population density and urbanization dynamics during the past 50 years. The main objects of the research are to find out relative dynamics of population density in the coastal regions compared to the total density figures in the countries mentioned. The general migration trends in the area are analyzed. The demographic magnetism of the cities in northeast Europe, such as St. Petersburg, Riga, Tallinn and Helsinki, has been intense and the process of urbanization relatively rapid. Owing to the fact that the capitals are located on the coast, nutrient loading to coastal ecosystems doubled since around 1950, but the loading has been reduced since 1990s, possibly due to new waste treatment technologies and stabilization of population growth in coastal cities.

11.1 Introduction

The coastal zone is an important area for all kinds of human development. An estimated 60% of the world population lives on the coast. There are many reasons for settling in coastal areas such as transport services provided by the sea, protein from fish and recreation. The coastal zones contribute more than a proportionate share to the respective gross domestic product and to the well-being of local and tourist populations. At the same time, the population density is ever increasing in most coastal areas. The coastal population growth has generally reflected the same rate of growth as the entire nation, but in the limited space of coastal counties (Crossett et al., 2004). The coastal regions have a wide diversity of coastal landscapes configuring a complex territory with regards to different aspects of environmental, social, cultural and economic conditions. The urbanization and population density have a significant influence to affect the ecosystem of the Baltic Sea. This chapter presents the following aspects and viewpoints:

(1) Describing population density and urbanization dynamic and in the coastal regions in the northeast of the Baltic Sea.
(2) A comparison of the causal factors of population dynamics and urbanization.
(3) A description of the anthropogenic impacts affecting the coastal ecosystems.

Coastal population growth is one important component of the total coastal environment as it relates to integrated coastal zone management (ICZM). A particular problem of coastal regions is that they are transit areas for natural and anthropogenic flows of substances from the hinterland. Impacts of anthropogenic factors, which overlap and therefore often reinforce each other along coastlines, turn coastal regions into an ecologically and economically vulnerable system (Eggers and Irmisch, 2002).

11.2 Materials and methods

11.2.1 Description of the study area

The investigation area covers the coastal regions of Finland, Leningrad district of Russia, Republics of Estonia and Latvia, as well as the coastal regions of the Gulf of Finland and the Gulf of Riga.

The Gulf of Finland is the easternmost gulf of the Baltic Sea, surrounded by Estonia, Finland and Russia. The Gulf of Finland (area 30 000 km^2, average depth ca. 37 m) is geomorphologically a direct continuation of the main basin of the Baltic Sea (Baltic Marine Environment Bibliography, 2007a). The total length of the coastline of the mainland and archipelago of the Gulf of Finland is approximately 2000 km (Mälkki, 2003). Leningrad district is situated in the northwest of the European part of Russia over an area of 85 900 km^2 (0.5% of the total area of Russia). The biggest city of the region, St. Petersburg, is located at the east coast of the Gulf of Finland, in the mouth of the Neva River.

The Gulf of Riga is a semi-enclosed gulf of the Baltic Sea, surrounded by the Kurzeme peninsula on the Latvian side in the south and the Isle of Saaremaa and Estonian coastland in the east and north (Berzinsh, 1995). The coastland of the Gulf of Riga provides a rich variety of wildlife and biodiversity, featured with coastal meadows and sand dunes. The shallow sea, reed beds and coastal meadows harbor an excellent biotope for numerous species of birds. The Gulf of Riga is separated from the Baltic proper by the two islands of Hiiumaa and Saaremaa. These islands are important nature reserves and are protected as such (Baltic Marine Environment Bibliography, 2007b). The length of the Latvian coastline at the Gulf of Riga is about 244 km (Eberhards et al., 2006). The capital of Latvia, Riga, is situated in the mouth of the Daugava River. The length of the Estonian coastline is about 470 km. The country is bounded to the north by the Gulf of Finland, to the west by the Baltic proper and to the southwest by the Gulf of Riga. The capital Tallinn is situated on the North-Estonian coastal plain. The land area of the coastal regions of Finland, Leningrad district of Russia, Republics of Estonia and Latvia is 76 895 km^2, which comprises 15.5% of their total land area.

Table 11.1 The population censuses in the countries of the northeast of the Baltic Sea.

Country	Years of the census
Finland	1950, 1960, 1970, 1980, 1985, 1995, 2000
Estonia	1959, 1979, 1989, 2000
Latvia	1959, 1979, 1989, 2000
Russia (Leningrad region)	1959, 1979, 1989, 2002

Source: Central Statistical Bureau of Latvia (2007), Federal State Statistics Service of Russia (2007), Statistics Estonia (2007) and Statistics Finland (2007).

11.2.2 Origin of data

This chapter has used data given by Estonian, Latvian, Finnish and Russian statistics institutions, as well as atlases, maps and other sources. The population census is the major source, until 1991 Estonia and Latvia were under the Soviet rule. After World War II, there were four population censuses: 1959, 1970, 1979 and 1989. After the collapse of the Soviet Union, the first census in Russian Federation was carried out in 2002. In the Republic of Estonia and the Republic of Latvia, the last census was in 2000. In Finland, there have been numerous censuses: 1950, 1960, 1970, 1975, 1980, 1985, 1995 and 2000 (see Table 11.1).

Owing to the reliability of the censuses, the population dynamics has been analyzed between the years 1950 and 2005. The data about nutrient loading are presented during the period 1950–2000.

11.3 Results

Table 11.2 presents the population dynamics of the northeast Baltic Sea coastal regions (excluded cities) between 1950 and 2005. The population density in the periods 1950–1960 and 2000–2005 is shown in Figures 11.1 and 11.2. Apart from the cities, the population in the coastal counties (in the rural areas) has increased by 1.35 million (42.1%); the whole population in the area (including cities) has risen by 3.47 million (49.1%). In the former Soviet areas (Leningrad region, Estonia, Latvia) urbanization decreased, whereas in Finland the process constantly kept its pace. The population in Finnish coastal cities had increased. The population growth in major cities (with more than 20 000 inhabitants) is given in Table 11.3. The population has increased about 2.1 million, which is more than 30% in the cities of the area. The percentage coastal population of Estonia, Latvia, Finland and Leningrad district and Russia compared to the total populations has increased from 57.4 to 70.3% during the period, although coastal areas take up only 15.5% of the territories.

The nutrient load during the period 1950–2000 is presented in Tables 11.4 and 11.5. The combination of a high population density, a well-developed agricultural sector, and other human activities, such as emission from energy production and transport, has resulted in large inputs of nutrients to the northeast of Baltic Sea (Helsinki Commission, 2006).

Table 11.2 Population dynamics between 1950 and 2005 at the coastal regions in the northeast of the Baltic Sea (thousands, at the end of the period, excluded population of cities).

Region/county	Period					
	1950–1960	1961–1970	1971–1979	1980–1989	1990–1999	2000–2005
Ida-Virumaa	12	28	26	23	24	22
Lääne-Virumaa	32	78	78	80	80	68
Harjumaa	72	83	89	106	106	122
Läänemaa	32	31	32	33	33	28
Pärnumaa	36	45	43	43	43	39
Hiiumaa	11	10	10	11	11	10
Saaremaa	42	38	39	39	39	36
Uusimaa	660	675	1033	1135	1291	1359
Eastern Uusimaa			80	84	89	93
Kymenlaakso	364	410	199	194	187	185
Southwest of Finland			406	423	445	456
Åland Islands	21	21	23	24	26	27
Kingisepp district	55	64	72	74	74	81
Lomonosovsky district	51	59	66	68	68	67
Vsevolozhk district	160	184	204	215	214	214
Vyborg district	85	98	108	113	114	110
Limbazi district	29	38	40	41	40	39
Riga district	106	120	127	152	144	150
Talsi district	42	46	49	50	50	48
Tukums district	47	53	55	59	55	55
Total	1857	2081	2779	2967	3133	3209

Source: Central Statistical Bureau of Latvia (2007), Federal State Statistics Service of Russia (2007), Statistics Estonia (2007), Statistics Finland (2007) and according to the calculations by the author.

11.4 Discussion

Population densities are generally higher on the coast than inland. For Europe, population sizes of the coastal regions are on the average 10% higher than inland areas. However, in some countries, this figure can be more than 50%. There are many regions where the coastal population is at least five times the European average density (European Environment Agency, 2006). Most cities of the coastal areas are located in the estuaries of the rivers, connecting the sea with the inner water bodies (Kozlov and Peregudov, 1966). The Baltic Sea is no exception, as St. Petersburg, Riga and Pärnu have a long navigation history. Many of Europe's capital cities are on or close to the coast, including Amsterdam, Athens, Copenhagen, Dublin, Helsinki, Lisbon, London, Oslo, Riga, Rome, Stockholm, Tallinn and Valetta. In total, there are about 280 coastal cities with a population above 50 000. More than 80 cities with more than 50 000 inhabitants are in the area of the Baltic Sea (Pichler-Milanovich, 1997).

Urbanization and increase of townsfolk increased rapidly during the period 1950–1970, whereas from 1980 to 2005 the rate of increase slowed down. During the postwar period, the population of the cities in the area began to increase rapidly. Migration to towns and cities was caused and promoted by various sociopolitical reasons (Marksoo, 2005). In Estonia and Latvia, the Soviet occupation also included changes in socioeconomical

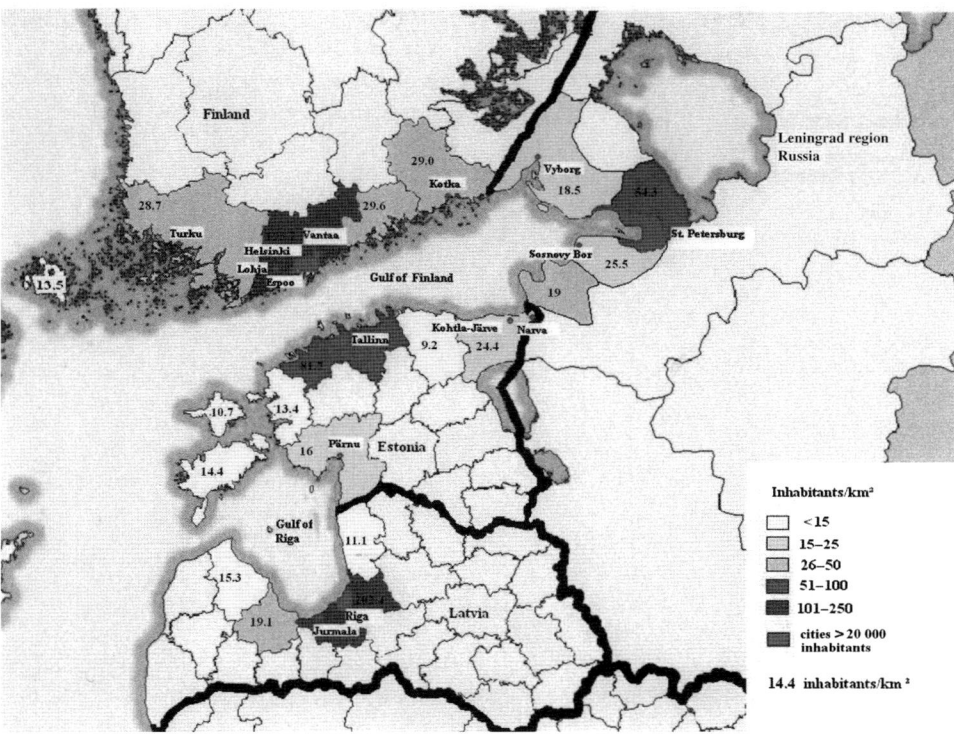

Figure 11.1 Population densities (inhabitants/km^2) in the northeast of Baltic Sea in the period 1950–1960. (To see this figure in color, please see color plate 9.)

politics, the so-called industrialization, creating and promoting heavy industry in areas which traditionally used to be farming or light industry areas. In the postwar period, Estonia and Latvia were restructured to modernize the Soviet industry. The region had sufficiently skilled labor, completed infrastructures, including transportation network (ports, railways and highways). The hidden agenda, however, was to change ethnic composition of the Baltic region. During the industrialization, a large number of Russian workers were employed in the new industries to outnumber local populations. The Kremlin succeeded in Estonia and Latvia, particularly, and the ethnic and demographic situation of these countries underwent drastic changes. As a result, new Russian-speaking townships and industrial cities emerged in the northwest of Estonia. The newcomers did not have any cultural or social connection to their new homeland and ambient environment. The extensive and wasteful development of the economy soon created a damaged and polluted situation. The 1970s, a period of economical stagnation (extensive and wasteful measures in the industry and agriculture), meant a further aggravation in the Soviet economy and in mid-1980s the Soviet Union faced a deep political and economic crisis (Vahtre, 2005). The situation in the Leningrad district was not much different from the Baltic region – the only remarkable feature being the intensity of industrialization. As buildings and infrastructure in Leningrad (St. Petersburg) had been damaged in the war, its rebuilding and reconstruction demanded more labor and

150　Integrated Coastal Zone Management

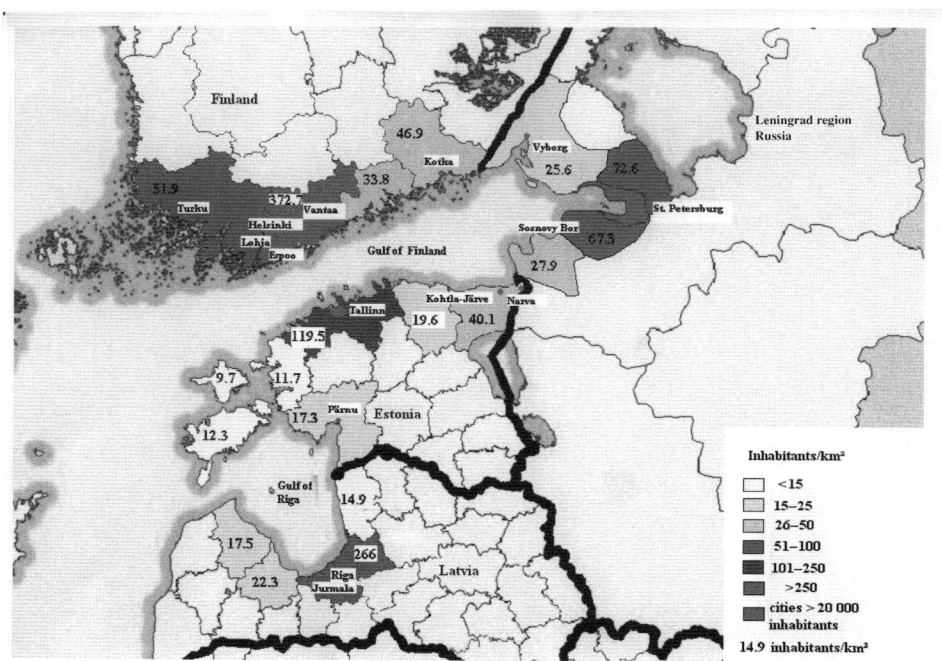

Figure 11.2 Population densities (inhabitants/km^2) in the northeast of Baltic Sea in the period 2000–2005. (To see this figure in color, please see color plate 10.)

resources. People in the neighboring rural areas left the collective farms to escape from poverty and become proletarians – the favored social class. Leningrad and its satellite towns grew rapidly and formed a conurbation. By the end of the Soviet rule, Estonia, Latvia and the St. Petersburg urbanization slowed down and more and more people settled down in the country. Food shortages and constant crop failures resulted in small private business and farming (Marksoo, 2005). After the World War II, Finland was still a typical agrarian society. In 1950, agriculture provided livelihood for 46% of the economically active population, while one-third of the population lived in urban towns. From then on, the industrial structure changed rapidly. Manufacturing, trade and diverse services generated new jobs and migration from rural areas to population centers livened up. At the same time, the fertility rate also began to decline, which partly contributed to the depopulation of the countryside. In 2000, urban population in developed countries in general was estimated as 76% of the total (Hall and Pfeiffer, 2002).

　　The Finnish economy grew in the postwar period all the way to the late 1980s (Statistics Finland, 2008). Urban spread has been very intense around the countries surrounding Tallinn, Helsinki and Riga between 2000 and 2005. Migration process is affected by the Baltic countries – firstly, people move to the economically attractive urban areas; secondly, well-to-do townsfolk move to the green outskirts. For example, the population of the Uusimaa county around Helsinki had grown by 68 000 and Harjumaa around Tallinn attracted 16 000 new inhabitants (between the years 2000 and 2005). Thus, the total population growth in the Helsinki and Tallin was 85 000 in 5 years, which led to greater

Table 11.3 Population dynamics in the cities northeast of the Baltic Sea (thousands, at the end of the period).

City/Country	Period					
	1950–1960	1961–1970	1971–1979	1980–1989	1990–1999	2000–2005
Finland (whole country)	4446	4598	4788	4998	5171	5255
Helsinki	453	524	485	492	551	559
Espoo	57	97	134	170	213	232
Kotka	57	60	61	57	55	55
Turku	130	154	164	160	173	175
Lohja	20	25	30	32	35	37
Vantaa	44	80	120	144	178	187
Leningrad Region	4567	5386	6107	6685	6289	6254
St. Pertersburg	3390	4033	4588	4350	4596	4601
Vyborg	51	65	76	81	79	78
Sosnovy Bor	–	–	42	56	63	67
Estonia (whole country)	1196	1356	1464	1565	1372	1347
Tallinn	282	370	442	499	400	396
Kohtla – Järve	40	83	87	92	48	46
Narva	30	61	76	85	69	67
Pärnu	41	50	54	57	45	44
Latvia (whole country)	2080	2352	2503	2666	2399	2306
Riga	575	729	828	910	766	735
Jurmala	35	48	53	61	56	56
Total in cities	5205	6379	7250	7254	7327	7325
Total in coastal regions	7062	8460	10029	10221	10460	10534

Source: Central Statistical Bureau of Latvia (2007), Federal State Statistics Service of Russia (2007), Statistics Estonia (2007), Statistics Finland (2007) and according to the calculations by the author.

anthropogenic impact in the western part of Gulf of Finland. Last but not least, the navigation between Tallinn and Helsinki (both cargo and tourism) has grown rapidly. Thus, the various migration processes with the economic factors have increased the anthropogenic impact and effects on the coastal areas and sea life.

Urbanization is directly influenced by the economic growth of a country. Over the last 10–15 years across Europe, under the growing pressure of globalization and the apparent decline of the nation-state, the city region (or metropolitan region) has increasingly been defined as the natural focus for economic development policies (Atkinson et al., 2007).

Table 11.4 Nitrogen loading in the northeast of Baltic Sea (thousand tons per year).

Subregion	1950[a]	1980	1992–1994	2000
Gulf of Finland	70 000	200 700	152 300	129 361
Gulf of Riga	40 000	52 000	95 000	81 476
Total	110 000	252 700	247 300	210 837

Source: Andrushaitis et al. (1995), Pitkanen et al. (1997) and Helsinki Commission (2006).
[a]Estimated amounts.

Table 11.5 Phosphorous loading in the northeast of Baltic Sea (thousand tons per year).

Subregion	1950[a]	1980	1992–1994	2000
Gulf of Finland	3 000	10 350	8 490	6 029
Gulf of Riga	1 100	2 000	4 500	2 209
Total	4 100	12 350	12 990	8 238

Source: Andrushaitis et al. (1995), Pitkanen et al. (1997) and Helsinki Commission (2006).
[a] Estimated amounts.

The Baltic Sea and its coastal areas have been strongly affected by anthropogenic activities such as nutrient loading, pollution and mechanical impact during the history of human occupation.

The input of nutrients to the Baltic Sea has increased greatly since about the 1940s, with nitrogen rising by about three times the 1940s level and phosphorous rising by about five times. The nutrient load during the 1940s was probably only slightly higher than at the start of the twentieth century. As a result of this increased nutrient load, the concentration of nitrogen in the sea has doubled since about 1950. To reduce the nutrient load to the levels of the 1940s, a reduction by 65% for phosphorus and 80% for nitrogen is needed (Jansson and Dahlberg, 1999). A range of anthropogenic activities contribute to significant nutrient inputs to the sea, which enter the sea either via runoff and riverine input or through direct discharges. In 2006, nutrient inputs from point sources such as industries and municipalities were cut significantly, the total input of nitrogen to the Baltic Sea is still over 1 million tonnes per year, of which approximately 25% enters as atmospheric deposition to the Baltic Sea and 75% as waterborne inputs (Helsinki Commission, 2007). In the early 1990s, the nutrient input decreased, possibly due to new wastewater treatment technologies and stabilization of population growth in cities. Also, agriculture in Estonia and Latvia underwent rapid changes in the period.

Coastal erosion has a major impact on European coasts and affects the environment and human activity. It is largely caused by human activity in the form of river dams, intensive development and the use of sand for construction and engineering purposes. Due to the irreversible nature of land cover change from natural to urban and infrastructure development, these changes are seen as one of the main threats to the sustainability of coastal zones. The leading cause of land uptake by housing is residential urban sprawl, which on average is responsible for more than 45% of the coast's land conversion to artificial surfaces in Europe (European Environment Agency, 2006). Transport infrastructure development, however, does not consume large areas, but still acts as a main driver of artificial sprawl along the coast, which is associated with the expansion of a new transport infrastructure along the coast.

11.5 Conclusions

The coastal area of the northeast of the Baltic Sea is densely populated and widely developed. The coastal population of Estonia, Latvia, Finland and Leningrad district, relative to the total population in these countries, has increased from 57.7 to 69.5% during the period, although coastal areas take up only 15.5% of the territories. Urban spread has been very intense around the countries surrounding Tallinn, Helsinki and Riga between 2000 and 2005. Thus, the total population growth in the Helsinki and Tallinn was 85 000 inhabitants in 5 years, which led to greater anthropogenic impact in the western Gulf of Finland. The various migration processes combined with economic factors have increased the anthropogenic impact and effects on the coastal areas and sea life. The combination of a high population density, a well-developed agricultural sector and other human activities, such as emission from energy production and transport, resulted in large inputs of nutrients to the northeast of Baltic Sea. In the early 1990s, the nutrient input decreased, possibly due to new wastewater treatment technologies and stabilization of population growth in cities.

The impacts of population growth and urbanization on coastal areas and marine environment discussed in this chapter are summarized as follows:

- Pollution: surface water, ground water, soil, air and marine environment.
- Physical changes: erosion and groundwater depletion.

Other impacts, not discussed in this chapter, may include:

- Habitat degradation: habitat destruction, isolation of habitats, noise disturbance and destruction of fauna and flora.

As the coastal population continues to grow, there is a need to move toward an ecosystem-based management approach to population growth, urban sprawl and their interactions with the vulnerable coastal environment.

Acknowledgment

The current study was partly supported by the fundamental science project of the University of Tartu (contract PP2PC07906).

References

Andrushaitis, A., Seisuma, Z., Legdzina, M. and Lenshs, E. (1995) River load of eutrophying substances and heavy metals into the Gulf of Riga. In: Ojaveer, E. (ed.). *Ecosystem of the Gulf of Riga Between 1920 and 1990*. Estonian Academy Publishers, Tallinn, pp. 32–41.

Atkinson, R., Smith, I. and Sweeting, D. (2007) *The Changing Nature of Urban Governance in England: The Paradox of Decentralization and (Re)centralization*. Cities Research Centre, Bristol, pp. 3–4.

Baltic Marine Environment Bibliography (2007a) Baltic Sea Web. 3rd Periodic Assessment of the State of the Marine Environment of the Baltic Sea. Marine environment of the Baltic Sea. Gulf of Finland. Available at http://www.vtt.fi/inf/baltic/balticinfo/index.html.

Baltic Marine Environment Bibliography (2007b) Baltic Sea Web. Gulf of Riga. Available at http://www.baltic.vtt.fi/demo/baltmap.htm.

Berzinsh, V. (1995) Hydrological Regime. In: Ojaveer, E. (ed.). *Ecosystem of the Gulf of Riga Between 1920 and 1990*. Estonian Academy Publishers, Tallinn, pp. 7–31.

Central Statistical Bureau of Latvia (2007) Databases. Population. Available at http://www.csb.gov.lv.

Crossett, M.K., Culliton, T.J., Wiley, P.C. and Goodspeed, T.R. (2004) *Population Trends Along the Coastal United States: 1980–2008*. Coastal Trends Report Series. U.S. Department of Commerce,

National Oceanic and Atmospheric Administration, National Ocean Service, Management and Budget Office, Special Projects, 1.

Eberhards, G., Lapinskis, J. and Saltupe, B. (2006) Hurricane Erwin 2005 coastal erosion in Latvia. *Baltica* **19** (1), 10–19.

Eggers, H. and Irmisch, A. (2002) Scientific research: the Germen point of view. In: Schernewski, G., Schiewer, U. (eds). *Balitic Coastal Ecosystems. Structure, Function and Coastal Zone Management*. Springer, Berlin, pp. 187–191.

European Environment Agency (2006) The changing faces of Europe's coastal areas environment, EEA Report, 4/2006, 8. Available at http://reports.eea.europa.eu/eea_report_2006_6/en/eea_report_6_2006.pdf.

Federal State Statistics Service of Russia (2007) History Russia's state statistics. Available at http://www.gks.ru.

Hall, P. and Pfeiffer, U. (2002) *Urban Future 21. A Global Agenda for Twenty-First Century Cities*. Spon Press, New York, pp. 3–5.

Helsinki Commission (2006) Development of tools for assessment of eutrophication in the Baltic Sea. *Baltic Sea Environmental Proceedings* **104**, 6–7.

Helsinki Commission (2007) Indicator Fact Sheets for 2006. The Marine Environment. Available at http://www.helcom.fi.

Jansson, B.O. and Dahlberg, K. (1999) The environmental status of the Baltic Sea in the 1940s, today, and in the future. *Ambio* **28**, 312–319.

Kozlov, V.I. and Peregudov S.P. (1966) Rahvastiku paiknevus. In: Kozlov V.I., Peregudov, S.P. (eds). *Maailma Rahvastik*, Eesti Raamat, Tallinn, pp. 78–110.

Mälkki, E. (2003) Groundwater flow conditions in the coastal bedrock area of the Gulf of Finland. *Geological Quarterly* **47** (3), 299–306.

Marksoo, A. (2005) Linnastumine ja ränne nõukogude perioodil. In: Kulu, H. and Tammaru, T. (eds). *Asustus ja ränne Eestis*. Tartu Ülikooli Kirjastus, Tartu, pp. 59–78.

Pichler-Milanovich, N. (1997) The role of Baltic Cities in the European Urban System: forgotten cities or important regional actors. In: Åberg, M. and Peterson, M. (eds). *Baltic Cities: Perspectives on Urban and Regional Change in the Baltic Sea Area*. Nordic Academic Press, Lund, Sweden, pp. 16–42.

Pitkanen, H., Kondtratyev, S., Lääne, A., Gran, V., Kauppila, P., Loigu, E., Markovets, I., Pachel, K. and Rumyantsev, V. (1997) *Pollution Load on the Gulf of Finland from Estonia, Finland and Russia in 1985–1995*. Proceedings of the Final Seminar of the Gulf of Finland Year 1996. Suomen Ympäristökeskus, Helsinki, pp. 9–19

Statistics Estonia (2007) Population census. Available at http://www.stat.ee/582.

Statistics Finland (2007) Population census. Available at http://www.stat.fi.

Statistics Finland (2008) From slash-and-burn fields to post-industrial society – 90 years of change in industrial structure. Available at http://www.stat.fi.

Vahtre, L. (2005) Sulaaeg. In: Vahtre, L. (ed.). *Eesti ajalugu VI. Annekteeritud Eesti 1944–1991*. Kirjastus Ilmamaa, Tartu, pp. 289–302.

Section 3
Integrated Coastal Zone Management (ICZM)

Chapter 12
Future Challenges in Environmental Policy Relative to Integrated Coastal Zone Management

Svein Jentoft

Abstract

From the perspective of local communities, environmental policy-making is typically perceived as something that occurs 'elsewhere' and as resulting from some crisis experienced and defined at the national and even global level. Such crises tend to trigger legislation that is subsequently imposed from the top administrators onto coastal zone managers and local stakeholders, thus turning local policy-making into a predominantly reactive process. I argue that although it is important that global environmental challenges are being met with adequate local response, there is a risk that rules and regulations imposed from the outside will erode the capacity of local communities to adapt to environmental change. It is therefore important that political processes are working not only from the top-down but also from the bottom-up. How to mediate between the global and the local level is a key governance issue.

12.1 Introduction[1]

Let me start with a comment on the key concept appearing in the title. A challenge can be a problem that needs our attention. If we do not act upon it, the problem might go away after some time by healing itself. There is, however, a risk that the problem will remain, that we will continue to suffer from it, or that it will get worse, and perhaps develop into a crisis. A challenge can also manifest itself in the form of an opportunity, something to take advantage of, but perhaps only for a limited time. Again, we may choose whether we want to act upon it or not. If we do, we expect to somehow be rewarded, that something good may come about, although we cannot always be sure, since opportunities often come with risks. Moreover, a problem might be transformed into an opportunity. In the process of solving the problem we may have discovered that we possess capacities that we were not

---

[1] This chapter was originally presented as a keynote speech at the *International Symposium on Integrated Coastal Zone Management*, Arendal, Norway, 10–14 June 2007. The title was given to me by the symposium organizer. I am grateful for the challenge.

aware of. A problem resolved between two parties may have improved their relationship by building mutual trust, or in other words by creating social capital which may prove useful.

The title, notably, is about future challenges. A challenge is here and now, and once we know what it is, it can be addressed intellectually and politically. A future challenge, on the other hand, is not here yet, and its implication are not always clear we can at best only catch glimpses of it. It is therefore impossible to be sure how to act, the rational response is to be cautious. We should not do anything now that might reduce our capacity to deal with it in the future. Take global warming, for example; predictions are that we will be affected, but we do not yet know exactly how. This is partly not only due to the magnitude of the anticipated global changes, but also because we cannot know what policy measures will be initiated and how effective they will be in dealing with these changes. Experts, such as those who form the Intergovernmental Panel on Climate Change (IPCC) have stated that global warming is real, is largely man-made and is going to bring about significant environmental changes (IPCC, 2007). The IPCC's predictions of what is going to happen in different parts of the world with, the polar ice, the sea level, aquatic and terrestrial ecosystems and the weather can no longer be ignored. But how will this transform our society, our political institutions, and how we live? That is more difficult to predict because there is no direct relationship between global warming and social changes. What will happen to our society, to our institutions and to our lifestyle depends on what we, and those who come after us, decide in the future, which is something we cannot know in advance.

This is because we cannot know today what we, or the next generation, will know in the future. We can be sure that we and they will learn, but we cannot know what precisely it is that we and they will learn. After all, innovations are in principle unpredictable (cf. MacIntyre, 1984). With global warming, as well as with many other issues confronting coastal zone management, we are not only dealing with the known and the unknown, we are also challenged by what is unknowable (cf. Gomory, 1995). However, one thing is certain it is now impossible to talk about future challenges in environmental policy relative to integrated coastal zone management (ICZM) without also talking about global warming and climate change. As the European Environmental Agency stated in a 2006 brief, climate change will only exacerbate pressures from demographic changes, economic restructuring, increased living standards and leisure time and global trade patterns (EEA, 2006). Hence, the IPCC endorses ICZM as a 'policy' response to climate change.

This leaves us with three questions: What do we do with what we know about the environmental challenges facing coastal zones? How do we deal with what is unknown? How do we prepare for those challenges that we cannot know? In the case of global warming and ICZM, the answer may well be the same for all three of these questions.

I will touch on the issue of global warming and climate change throughout my talk, as some of my presentation inspired by the reports of the IPCC (2007). My main concern, however, is of a more general nature, which relates to policies that arise globally but must be dealt with locally, such as those pertaining to global warming and climate change, biodiversity conservation and human rights legislation as they affect people's access to coastal resources. What influence do these global initiatives have on ICZM? I argue that there is a risk that they will turn ICZM into a predominantly *reactive process*, thereby eroding its inherent 'adaptive capacity', a term frequently used by the IPCC (2007). I contend that for ICZM to function it must be proactive, which requires the ICZM to work as a social process. In other words, ICZM cannot be reduced to a top-down engineering exercise, but requires local communities to become involved by and through their own creative initiatives.

12.2 Globalization: from challenge to issue

Environmental issues reach far and wide; when they hit us, we do not always know precisely where they came from and how they got there. This seems equally true for policy instruments intended to handle these problems. Not only are environmental problems spreading across the globe, but also the policy and management tools designed to address them. All of a sudden, we have policy measures, but we do not know who exactly we are supposed to do with them.

This is an experience that comes with globalization; O'Riordan and Curch (2001, p. 3) argue that 'there is little, and maybe nothing, that is global that does not have some sort of local manifestation.' In order to understand how globalization is changing the world, we need only look at what is happening in our own community. Think of terrorism for instance, it has become a global concern, and we feel its impacts in our daily lives wherever we are. In my town, people are asking: Who got this crazy idea to fence-off the harbor to the public? The point here is that when global policy measures reach us, they become local and we are forced to implement them at that level, not necessarily by choice, but as conditions given, demands raised and as prescriptions imposed. Thus, such policies are experienced locally as an exogenous force. Policy-making is something that occurs elsewhere, high up and far away. The old slogan 'think globally and act locally' is turned on its head. Someone has been acting globally and it is up to us at the community level to figure out what it really means and what to do.

Global policies, such as those pertaining to terrorism, world trade, culture and the environment, all merge together in the coastal zone and place demands on management institutions. They bring additional challenges to an already complex agenda. As Glantz and Feingold (1992) maintain, global warming is no exception to this rule. ICZM must handle problems that are operating not only at the level of a particular locality, but also elsewhere and at higher levels of governance. Some of these problems have a short time-span and they occur, as catastrophic events, such as when a hurricane hits. Others occur over a much longer time frame, such as with pollution. ICZM systems must be prepared for and function within both kinds of time frames.

ICZM systems must also be embedded in systems of governance that operate at different geographic scales, such as nationally and internationally. With issues such as global warming, biodiversity loss and human rights violations, this becomes increasingly pertinent, not only because these challenges are present but also because authorities at higher societal scales raise demands vis-à-vis local ICZM efforts.

There is also another important dynamic beyond, and is largely triggered by, globalization, which contributes to this influence. Rosenau's concept of 'distant proximities' alludes to this phenomenon (Rosenau, 2003). As globalization makes the world more compact, there is a parallel, partly unrelated, partly compensating and countervailing dynamic that occurs at the local pole. According to Rosenau, as we speak of 'globalization', we must also speak of 'localization'. Thus, globalization and localization are two parts of the same phenomenon that go hand in hand, and we should look at them as interconnected forces that are playing each other out. This implies that in order to understand how globalization is changing the world we do not have to travel far or to be in some global metropolis, we need only to look around at what is happening in our own community, and even at ourselves.

Certainly, many environmental problems require more effective global governance. ICZM cannot prevent climate change and worldwide overfishing on its own. But these and many other global environmental issues end up in the coastal zone and alter the very conditions on which management institutions work. They bring new concerns and challenge the way we make decisions. Thus, Nicholls and Klein (2005, p. 215) argue that: 'given the additional challenge of climate change in coastal zones, the purpose and design of coastal management will have to be revisited'. ICZM institutions are used to deal with dilemmas and trade-offs such as those of conservation versus use, but they may not yet be prepared for the challenges that lie ahead.

Many of these new global demands do not leave coastal managers and stakeholders with much of a choice. They are largely nonnegotiable. You do not argue with nature, as nature does not listen to what you say (unless you believe it holds spiritual forces), it only listens to what you do. Similarly, with human rights, we either comply or we do not because these rights are codified in international law.[2] Thus, regardless of what lies ahead, management institutions have to deliver. It is not enough to support policies in principle by ratifying international conventions and codes; it must be demonstrated in practice, since international institutions audit the implementation of conventions from time to time. Still, at the local level, people often drag their feet or refuse to adhere. Thus, global policy implementation is rarely unidirectional and streamlined; implementation often meets fierce local resistance if such policies are felt to go against particular interests.

Before an environmental challenge is met it must be acknowledged. Some environmental issues facing coastal communities are not all that noticeable. Global warming is a gradual process; marine species do not disappear overnight, pollution occurs gradually. Behavioral patterns, such as those involving natural resource uses, are often so much part of everyday life that it is difficult to perceive them as problematic. Environmental degradation is often acknowledged only when there is a tangible crisis, or when communities are hit with catastrophic events. Sometimes awareness results from comparing the current conditions with the past or with situations elsewhere. But even then, degradation often goes unnoticed. Even when noticed, we ask ourselves: What does this degradation mean, what difference does it make? Is this caused by human or natural fluctuations? Is this permanent or temporary?

Environmental issues are also neglected for other reasons. For instance, fisheries biologist Daniel Pauly (1996) talks about the 'shifting baseline syndrome', overfishing trends do not look apparent when this year's landings are compared with those from last year. If that is also the comparison next year, long-term trends can be over looked and we may quickly find ourselves on a slippery slope toward disaster without even being aware of it. Then, to paraphrase Al Gore, there are those 'inconvenient truths' that are too worrisome to even

---

[2] For instance, the ILO Convention 169 concerning the rights of indigenous peoples (cf. Jentoft et al., 2003).

contemplate. Certain parties may have an interest in suppressing particular issues, like energy industries have allegedly been doing with global warming. Furthermore, there are the 'the politics of blame': Who is causing the problem, who should be held accountable and who should solve it? Here, we get to the core of both the political issue of global warming and overfishing. Since we cannot agree on who should be held accountable, we have difficulties acting upon them.

Let me add another factor. In a now classic article, political scientist David Easton (1957) pointed out that a political demand does not automatically convert into a political issue, i.e. something politicians are prepared to discuss and act upon. The demand may be registered perhaps, but still left untouched, and thus the politics of nondecision can be as interesting as the politics of decision. Easton mentions as one hindrance the location of the initiators of the demand relative to those who have to act upon it. The farther away initiators are situated, the less inclined actors are to relate to them. The outcome may be as follows: the more pressing the demand, the more likely it is to be converted into a political issue and to be turned into *a command*: 'If you do not respond, we will make you comply'.

Therefore, there is a real danger that global environmental policy demands – be they related to global warming, biodiversity loss, or human rights, will transform ICZM institutions at the local and regional level into mere instruments of higher regulatory authority, thus turning ICZM into a reactive process. At the level of the local community there will be nothing more to do than to implement what has been decided elsewhere. As Easton reminds us, the upshot might be even worse because: once a political demand instigated far away is converted into a local decree, it runs the risk of becoming a nonissue because people will turn their back on it.

Environmental management is not a technical fix, but it is fundamentally a political process because mixed interests, power, and social values all have a critical bearing on decisions and outcomes. At the end of the day, it is this process that determines how effective ICZM can possibly be. ICZM is therefore confronted with the challenge of providing an institutional framework that will enable constructive participation of all stakeholder groups in a way that allows interests to come forward, powers to be balanced, values to be acknowledged and principles to be mutually agreed upon. This cannot be facilitated through top-down command and control that puts stakeholders in an inferior position where obedience is the only action required. As Jim Whitman insists:

> Obedience has been the axial principle of task execution in the traditional environment of imperative control. The logic of that environment is reproduced when technology is used only to automate. When tasks require intellectual effort, however, obedience can be dysfunctional and can impede the exploitation of information. (Whitman, 2005, p. 52)

Interestingly, the IPCC also emphasizes the need for decentralization and stakeholder participation in management:

> Changes in development paths emerge from the interactions of public and private decision processes involving government, business, and civil society, many of which are not traditionally considered as climate policy. The process is most effective when actors participate equitably and decentralized decision making processes are coordinated. (IPCC fourth assessment report, Working Group III, 4 May 2007, p. 33)

This statement will find support in any ICZM text or instructions manual, and it is increasingly becoming recognized as an essential governance principle in international agreements, conventions, and development goals. The question is whether it has real implications or not. The crucial issue of stakeholder involvement is a tricky one, particularly if it is imposed from above. Participation usually works better if it is initiated and governed from below. Still, stakeholders are often a conservative force – they hold back if policies go against their individual interests. This tends also to be the case with nation-states, as demonstrated with some international agreements, such as the Kyoto Protocol and the ILO 169. However, the urgency of the problem is often such that action is needed.

There is no easy answer to this dilemma, but it is essential to enhance our ability to effectively address environmental challenges. The key is to be found in the way relationships are built and interactions occur across scales, ranging from the global to the local community.

12.3 Adaptive capacity

The IPCC is concerned with ensuring that social conditions required to live with climate change are identified and strengthened. One such essential condition is 'adaptive capacity', which the panel defines as 'the ability of a system to adjust to climate change (including climate variability and extremes) to moderate potential damages, to take advantage of opportunities, or to cope with the consequences' (IPCC, 2007, p. 22).[3]

Importantly, we should not think of adaptive capacity as implying only marginal and spontaneous adjustment of current processes and practices. Adaptive capacity also entails the aptitude to cope with, and indeed to initiate radical change, innovation and new beginnings. In the face of global warming, the task for ICZM would not be very different from what it is already trying to do, i.e. building resilient coastal ecological and human social systems, protecting coastal habitats, alleviating poverty, and furthering sustainable development through holistic and proactive management approaches. Climate change makes it even more imperative for ICZM to succeed.

Marine systems, including coastal communities, have always possessed a degree of adaptive capacity because of their exposure to climatic variability. McGoodwin (2007) argues that this may help them prepare for more permanent and longer-term climatic change. Still, building further adaptive capacity is critical. This is, in part, an issue concerning technology because adaptive capacity requires better engineering and more research and development. In the context of ICZM, adaptive capacity is not a natural quality but a social one, involving collective choice, decision-making and interactive learning. Therefore, adaptive capacity calls for effective institutions and incentives: it depends on collaboration among stakeholders, mechanisms for resolving use and user conflicts. Again, these are properties that cannot be established through a command-and-control apparatus. They can

---

[3] The same concept is applied in other global exercises such as the Millennium Ecosystem Assessment (www.maweb.org). See also the special issue of *Ecology and Society* **12**(1), 29 (2007), and the synthesis article by Fabricius et al. (2007). For an overview of the literature on adaptability and adaptive capacity of human systems within the context of global climatic change, see Smit and Wandel (2006) and Smith et al. (2003).

surely benefit from exogenous stimuli, sometimes even in the form of a command. But first and foremost they rely on a process that is endogenous, a process situated at the level where conflicts actually occur, as in a local community or a particular industry. To build adaptive capacity, management systems must therefore be granted a certain degree of autonomy and self-governance. The challenge is simply too complex for one agent, such as a government organization, to deal with single-handedly through usual command-and-control mechanisms. As Pahl-Wostl (2007, p. 562) contends: 'Complex adaptive systems are characterized by self-organization, adaptation and heterogeneity across scales and distributed control'. In other words, complex systems, whether they are natural or social, cannot easily be managed exogenously; rather, control must be decentralized, shared and interactive.

Adaptive capacity requires – the ability to learn to manage by managing to learn (Bormann et al., 1994, in Pahl-Wostl, 2007). Notably, learning is not a top-down process, as it always requires active participation of the learner; learning is about empowerment as a mutually reinforcing process. But neither learning nor empowerment can be forced on the learner. Benjamin Barber says: 'Give people some significant power and they will quickly appreciate the need for knowledge, but foist knowledge upon them without giving them responsibility and they will display only indifference' (quoted by Fisher, 2000, p. 29). For this reason, ICZM has to be more than an instrument at the disposal of some centrally situated authority. It must involve both empowerment and learning among all stakeholders.

In many instances, management systems erode rather than build adaptive capacity. They make coastal systems less flexible and less resilient than they should be for short- and long-term environmental changes. For instance, McGoodwin's assessment of the preparedness for climatic change of the commercial fishing economies of Iceland and Alaska concludes that overall these economies 'are operating within very tightly wound and complexly structured management regimes, which confer low degrees of flexibility, autonomy, and adaptability to changing conditions', and 'their potential resilience to the sort of radical changes that are forecast for high-latitude regions seems dangerously low' (McGoodwin, 2007, p. 53). Therefore, these systems are more vulnerable than they need to be. With new environmental challenges, ICZM systems could benefit from a similar assessment: How resilient are they? How much adaptive capacity do they actually possess? How can these qualities of ICZM systems be enhanced? Before such an assessment can be done, we need some yardsticks. What do ICZM systems with adaptive capacity look like? What are their inherent features and strengths? What exactly do they do? What follows is an attempt to answer these questions. Drawing on so-called interactive governance theory of Kooiman et al. (2005) and Jentoft (2007), I will present some ideas regarding adaptive capacity and ICZM institutional design.

12.4 ICZM – conceptual model

Let us think of ICZM as a relationship between two systems, what could be called a 'governing system' and a 'system to be governed'. The governing system is social: it is made up of institutions and steering instruments and mechanisms of human design and implementation. The system to be governed, on the other hand, is partly natural and partly social: it consists of an ecosystem and a diversity of resources, as well as a system of users and stakeholders who among themselves form social relations, such as institutions and political coalitions. Fundamentally, the adaptive capacity of ICZM depends on the inherent qualities of both the governing system and system to be governed as well as the relationships and the interactions between them. Thus, in order to build such capacity, you can target (a) the system to be governed, (b) the governing system and/or (c) the interactions that occur between them.

As previously mentioned, there might be limits to how effective the ICZM system can be. There may be knowledge gaps, both in terms of 'unknowns' and 'unknowables'. Moreover, some realms of the system to be governed may not lend themselves to be as effectively managed when they are inherently chaotic or rule-free. On the other hand, ICZM may get better at doing what it does, for instance, through acquiring new knowledge, improving its institutional design and/or adopting new management tools. So the question is as follows: If the adaptive capacity of ICZM systems is insufficient, how can it be improved? How can it get from where it is now to where it should be, particularly given the new environmental challenges that are literally landing on their shores? And where should it begin?

Governance theory would suggest that we should begin by recognizing what ICZM must be able to do, i.e., the basic challenges that it has to address and its institutional requirements. Section 12.4.1 describes the essential properties of the system to be governed.

12.4.1 *System to be governed: properties*

As shown in Figure 12.1, the systems to be governed present some fundamental qualities that must be taken into account since they establish conditions for the way that ICZM works.

Diversity relates to spatial variability in natural, social and cultural conditions. The number and quality of components (habitats, species, stakeholders, knowledge systems) may vary from system to system, as with small versus large scale or with tropical and temperate marine ecosystems. Human social systems are as diverse as natural systems. This is to be expected given that they draw from and adapt to the natural system. Thus, when people live by and from the sea, their communities are structured accordingly. But there is no direct relationship between the two, as human communities also derive their characteristics from their cultural and social environments. *Complexity* refers to the fact that actions, concerns, needs and interests are overlapping and interdependent, and actually or potentially synergistic or in conflict. Fish and other marine species and organisms feed

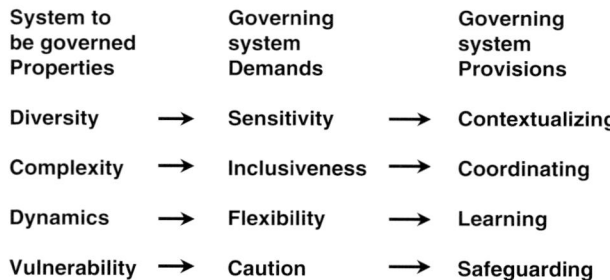

Figure 12.1 Governance model for ICZM and its attributes.

on one another and together they constitute of a food chain. When we add humans to the top of this chain, and we have an intricate, multi-scale 'human-in-nature system' that is inherently difficult to know and control. *Dynamics* is about fluctuation and change that occur as a consequence of the interaction within a system and/or between systems. The situation may change rapidly and unpredictably. Incidents in one part of the system may trigger processes that spread and magnify. The result is uncertainty and shock for those affected. *Vulnerability* refers to the fact that systems to be governed are fragile and therefore easily and sometimes irreversibly harmed. Consequently, things may go wrong, damaging both nature and society.

12.4.2 Governing system: demands

Next are the challenges that each of these properties poses relative to the ICZM system. Adaptive capacity is a function of the degree to which ICZM is able to respond to each of these demands.

As Figure 12.1 illustrates, diversity demands that ICZM be sensitive, whereas complexity calls for it to be inclusive. Furthermore, dynamics lead logically to a need for flexibility, and vulnerability to cautiousness. *Sensitivity* is an attitude as well as an approach on the part of management and a methodology for gathering information and for making decisions. As an attitude, it involves appreciation of variations, a perceptiveness of distinctions, and a compassion for uniqueness. Diversity, then, is viewed as a positive thing – an opportunity more than a problem, recognizing that even small things may matter because they can be part of a bigger whole. *Inclusiveness* is about taking many things into consideration at the same time. Thus, ICZM must address several concerns, among them are ecosystem health, economic efficiency, and social justice. Inclusiveness is partly analytical and partly organizational. Participation of stakeholders in the governance process is a way to ensure that these concerns are considered and that they are balanced. *Flexibility* is about the ability of ICZM to adapt promptly and swiftly to dynamics and change. A flexible ICZM system is able to recover from shocks and to learn from mistakes. Flexibility requires a degree of opportunism, spontaneity and preparedness for the unexpected, and is therefore both a structural capacity and a mindset. *Cautiousness* is about risks. It relates to what we do when we do not know for sure where to tread and what ecological or social footprints we will leave. Cautiousness means to go slow and to be vigilant in order to be safe. (This is now established as a basic standard in the 'precautionary principle' in the UN Fish Stock Agreement, the Biodiversity Convention and in FAO's Code of Conduct for Responsible Fisheries.)

12.4.3 Governing system: provisions

Figure 12.1 also depicts which institutional mechanisms are equipped to meet the four demands. *Contextualization*: If diversity calls for sensitivity, then the one-size-fits-all management approach must be abandoned. Instead, a differentiated, contextualized model should be adopted, i.e., one that takes specific ecological and social factors into consideration. Therefore, we need a decentralized management model that allows us to deal with details and subtleties that can only be seen from up close. *Coordination*: When matters are complex, as they are in the coastal zone, it is not possible for people 'to do their own thing' no matter how desirable that may be from the perspective of the individual. Some form of action coordination is needed to avoid conflict and create symbiosis; for example, the relationship between fisheries, coastal tourism, marine aquaculture, coastal transport and offshore oil exploration does not always have to be antagonistic. *Learning*: Human social systems adapt because people make them do so based on what they experience, learn and decide. When lacking this ability, systems tend to make the same mistakes over and over again. Governance theory argues that learning must be interactive to be efficient at both the individual and system levels. Although crucial for adaptive capacity, learning is never complete (Jentoft, 2007). The complexities are such that neither successes nor failures can be understood in details (Rycroft, 2003). ICZM systems therefore have to learn to cope with the unknown and unknowable. *Safeguarding*: The governing system should not put the system to be governed in jeopardy, its survival must be guaranteed. Thus, ICZM is faced with some absolute, nonnegotiable demands, and has to deliver no matter what. For instance, the Kyoto Protocol, the Biodiversity Convention and human rights legislation put heavy restrictions on what ICZM can do. They also address concerns that cannot be ignored or played down.

Herein is the problem: ICZM would have it rather easy if it could restrict itself to delivering on only one of these items; in reality, it has to handle them all – if not simultaneously, then at least consistently and sequentially. The problem, however, is that these items do not easily add up, but tend to be in conflict. For instance, the institutional response to diversity, sensitivity and contextualization may not only be different from the institutional response to complexity, inclusiveness, and coordination, it may also reduce the ability to deliver on the latter. There are also some things that are better dealt with at the local level than at the central level, and vice versa. Global issues require legislation and other initiatives that do not easily lend themselves to decentralization, yet as argued above, they cannot and should not exclude local involvement and initiative.

Another complicating factor is that institutional design is not only a technical issue, where performance is measured in functional terms, it is also a normative and, hence, a political issue, where moral codes and principles are relevant. The precautionary principle is one example, the subsidiarity principles another – both they are frequently referred to in relation to marine and coastal management.[4] These are principles that gain general support not only because they help ICZM do its job more effectively, but also because they are perceived as normatively and ethically 'right'.

12.5 Discussion and conclusion

Environmental change requires policy initiatives at the global level. But since no community will be unaffected by global environmental change and the policy initiatives to handle it, the response at the level of the community is not less important. At the end of the day, the local response and initiative will largely determine how effective we are at handling environmental change. Global and local environmental policy-making and management

---

[4] Compare, for example, the so-called Malawi Principles for the Eco-system Approach, which talks about both (http://www.fao.org/docrep/006/Y4773E/y4773e0e.htm).

must therefore find mechanisms to connect constructively. This is generally known as the problem of 'scaling up' and 'scaling down' (see, for instance, Keohane and Ostrom, 1995).

I have argued that ICZM must be able to deliver on a broad range of issues and that it cannot work by only using command-and-control mechanisms. Adaptive capacity is not only a capacity to react but also a capacity to initiate, to learn and to be innovative. It consists of multiple elements and mechanisms that require institutional backing, constructive interactions and legitimacy among concerned stakeholders (representing government, coastal industries and civil society, such as coastal communities). As a consequence, and in line with interactive governance theory, ICZM must be a mix of different governing modes that must work on different scales while linking them both as a governing system and as a system to be governed (Kooiman et al., 2005). ICZM systems are therefore deemed to become diverse, complex, dynamic and multi-scale, and to end up as a vulnerable hybrid of tools, competencies, rules, regulations, organizations, partnerships, events and arenas (cf. Figure 12.1). However, there are always going to be limits to their governability, partly due to knowledge constraints and reluctance and resistance among coastal stakeholders. Therefore, ICZM is deemed to be a continual process of learning by trial and error and one of 'muddling through'. More science would be helpful, but it is not sufficient to build adaptive capacity.

As a governance issue, adaptive capacity to environmental change is a relationship between politics and power. ICZM systems represent and organize power. Their mandate, authority and design are backed up by government authorities and by the law. But ICZM systems are also exposed to formal and informal pressure from more or less powerful stakeholder groups with an interest-based agenda. ICZM may provoke resistance when it goes against the interests of powerful stakeholders. Thus, just as ICZM systems are instruments of power, they are themselves the outcome of power; they are established and shaped by powerful interests. Consequently, the urgency of the demand and the seriousness of the environmental challenges do not always provide sufficient tailwind to convert ICZM into an issue to be acted upon. That is why ICZM raises issues of democracy, due process and legal rights (Lafferty and Meadowcroft, 1996).

When regulatory frameworks within which coastal actors operate are becoming increasingly strict and demanding, social relationships at the local level are affected, often leading to conflict and struggle. Environmental politics are therefore as antagonistic as human social relations are competitive and conflictive. These relationships are potentially symbiotic: conflicts may be converted into cooperation if there is trust or an understanding that trust can be built.[5] For this reason, participatory democracy plays a constructive role in ICZM because it brings people together, not because they are forced to, but because they see it to be in their own best interest and because they learn from it. Neither trust nor learning can be imposed upon people; it has to be built from within and through experience. A well-known proverb states that: 'if you do it alone, you can do it faster, but if you do it with others, you can do more'. With ICZM, however, one also has to become more efficient in order to do more. Therefore, government has a pivotal role to play in pushing issues and facilitating interaction; external pressure in this case can be productive. As an example, in Norway, coastal zone planning at the municipal level picked up speed when the Ministry of Fisheries

---

[5] Davos offers interesting insights into this aspect – see Davos (1998) and Davos and Lajano (2001).

made such planning a requirement for the allocation of new aquaculture licenses (Buanes and Jentoft, 2005).

Global environmental changes that are now confronting coastal stakeholders will affect no one in particular while posing threats to everyone. Curbing vested interests and power, turning conflicting, potentially destructive relationships into collaborative, constructive collective action is a key environmental policy issue. Global environmental policies and principles are not developed in a vacuum totally removed from real-life experiences. They synthesize lessons learned from local situations and events. When these experiences are aggregated at the global level and turned into policy frameworks and initiatives, they have to be averaged and decontextualized. But when these policies are implemented at the level of coastal communities, they must again be recontextualized. They must be translated and applied to the actual local situation. The latter is a process as political as the former. If there is little room for maneuver, with everything prescribed and nothing left for local decision-makers, then managers and user-groups rebel rather than obey and comply. Thus, perhaps the greatest political challenge ahead is to find ways to connect the de- and recontextualization processes that work across geographic and temporal scales. How this process of scaling up and down actually works is an issue begging for empirical research. It is here, in the interaction between what is happening globally and locally, that the adaptive capacity of coastal systems to environmental change will be decided. This, I conclude, is the ultimate future environmental challenge confronting ICZM.

References

Bormann, B.T., Cunningham, P.G., Brookes, M.H., Manning, V.W. and Callopy, M.W. (1994) *Adaptive Ecosystem Management in the Northwest*. Gen. Tech. Rep. PNW-GTR-341, USDA Forest Service, Pacific Northwest Research Station, Portland, 22 pp.

Buanes, A. and Jentoft, S. (2005) Challenges and myths in Norwegian coastal zone management. *Coastal Management* **33** (2), 153–167.

Davos, C.A. (1998) Sustaining co-operation for coastal sustainability. *Journal of Environmental Management* **52**, 379–387.

Davos, C.A. and Lajano, R.P. (2001) Analytical perspectives of cooperative coastal management. *Journal of Environmental Management* **62**, 123–130.

Easton, D. (1957) An approach to the analysis of political systems. *World Politics* **9** (3), 383–400.

EEA (2006) The continuous degradation of Europe's coasts threatens European living standards. Briefing 3/2006. Available at http://reports.eea.europa.eu/briefing_2006_3/en.

Fabricius, C., Folke C., Cundill G. and Schultz, L. (2007) Powerless spectators, coping actors, and adaptive co-managers: a synthesis of the role of communities in ecosystem management. *Ecology and Society* **12** (1), 29. Available at http://www.ecologyandsociety.org/vol12/iss1/art29/.

Fisher, F. (2000) *Citizens, Experts, and the Environment: The Politics of Local Knowledge*. Duke University Press, Durham.

Glantz, M.H. and Feingold, L. (1992) Climate variability, climate change, and fisheries: a summary. In: Glantz, M.H. (ed.). *Climate Variability, Climate Change and Fisheries*. Cambridge University Press, Cambridge, MA, pp. 417–438.

Gomory, R.E. (1995) An essay on the known, the unknown and the unknowable. *Scientific American* **272**, 120.

IPCC (2007) Contribution of working group II to the fourth assessment report of the intergovernmental panel on climate change. Available at http://www.ipcc.ch/pdf/assessment-report/ar4/wg2/ar4-wg2-spm.pdf.

Jentoft, S. (2007) Limits of governability: institutional implication for ocean and coastal governance. *Marine Policy* **31**, 360–370.

Jentoft, S., Minde, H. and Nilsen, R. (eds) (2003) *Indigenous Peoples: Resource Management and Global Rights*. Eburon Academic Publishers, Delft.

Keohane, R.O. and Ostrom, E. (1995) *Local Commons and Global Interdependence*. Sage Publications, London.

Kooiman, J., Bavinck, M., Jentoft, S. and Pullin R. (eds) (2005) *Fish for Life: Interactive Governance for Fisheries*. Amsterdam University Press, Amsterdam.

Lafferty, W.M. and Meadowcroft, J. (eds) (1996) *Democracy and the Environment: Problems and Prospects*. Edward Elgar, Cheltenham.

MacIntyre, A. (1984) *After Virtue: A study in Moral Theory*. University of Notre Dame Press, Notre Dame.

McGoodwin, J.R. (2007) Effects of climatic variability on three fishing economies in high-latitude regions: implication for fisheries policies. *Marine Policy* **31**, 40–55.

Nicholls, R.J. and Klein, R.T. (2005) Climate change and coastal management on Europe's coast. In: Vermaat, J.E., Bouwer, L., Turner, K. and Salomons, W. (eds). *Managing European Coasts: Past, Present, and Future*. Springer-Verlag, Berlin, pp. 199–225.

O'Riordan, T. and Curch, C. (2001) Synthesis and content. In: Riordan, T. (ed.). *Globalism, Localism, and Identity: Fresh Perspectives on the Transition to Sustainability*. Earthscan, London.

Pahl-Wostl, C. (2007) The implications of complexity for integrated resources management. *Environmental Modelling and Software* **22**, 561–569.

Pauly, D. (1996) Anecdotes and the shifting baseline syndrome in fisheries. *Trends in Ecology and Evolution* **10** (10), 43.

Rosenau, J.N. (2003) *Distant Proximities: Dynamics Beyond Globalization*. Princeton University Press, Princeton.

Rycroft, R. (2003) Innovation networks and complex technologies: policy implications of the unknown, and the unknowable. *Occasional Paper Series*. Elliot School of International Affairs, The George Washington University, Washington, DC.

Smit, B. and Wandel, J. (2006) Adaptation, adaptive capacity and vulnerability. *Global Environmental Change* **16**, 283–292.

Smith, J.B., Klein, R.J.T. and Huq, S. (eds) (2003) *Climate Change, Adaptive Capacity and Development*. Imperial College Press, London.

Whitman, J. (2005) *The Limits of Global Governance*. Routledge, London.

Chapter 13
Integrated Coastal Zone Management and Sustainable Development of Coastal Area: A Short Overview of International Legal Framework

Md. Saiful Karim and Ridwanul Hoque

Abstract

After the Rio Earth Summit, 'sustainable development' turned out to be a major concern and area of focus of the global community. Principally through state practices, this concept of development has over the years become a principle of international law. Adoption of Agenda 21 brings the issue of sustainable development of coastal zone to the forefront of global agenda. Increased threat of sea-level rise due to climate change is another factor that has made the issue so important. Integrated coastal zone management (ICZM) is now recognized as a major process for ensuring sustainable development of the coastal area. Set against this background, this chapter briefly analyzes the international legal framework for ICZM.

13.1 Introduction

Integrated coastal zone management (ICZM) can be considered the most modern management process for the harmonization of the interests of industrial development and conservation of natural resources in coastal areas. ICZM can be defined as a decision-making and management process with which to achieve the sustainable use, development and protection of coastal and marine areas as well as of their resources. Thus, ICZM stands for an enviro-economic regulatory system for numerous competing interests (Eremina and Stetsko).

To date, however, ICZM has been seen as belonging more to the dominion of policy makers, scientists and economists than to the dominion of lawyers. Yet a comprehensive legal framework is sine qua non for the proper implementation of ICZM process. More importantly, where 'nonstatutory means are employed to achieve the objectives of ICZM, they must function within an established legal framework that already defines the powers and duties of the many public and private stakeholders involved in the administration and use of coastal areas' (Gibson, 2003). In making ICZM policy, policymakers must consider international environmental legal norms so that coastal zone policies will be designed and carried out with a critical understanding of international laws and principles that have a bearing upon sustainable ICZM process. Moreover, at the same time, the states will have to create a national legal framework to facilitate the operation of the ICZM process in a fashion that will conform to their special local conditions including, inter alia, natural characteristics of coastal areas, geopolitical situations, economic conditions and historical and cultural traditions (Beckman and Coleman, 1999). In the discussion below, we focus on, and explain international and national legal imperatives toward the recognition of, ICZM as a tool for achieving sustainable development.

13.2 ICZM and sustainable development of coastal zone

ICZM refers to 'a dynamic, continuous and iterative process designed to promote sustainable management of coastal zones' which 'covers the full cycle of information collection, planning, decision making, management and monitoring of implementation' (European Commission, 1999). Also, ICZM requires 'the informed participation and cooperation of all interested and affected parties to assess the societal objectives in a given coastal area at a given time, and to initiate the actions necessary to move toward meeting these objectives' (European Commission, 1999). The World Bank defines ICZM as follows:

> ICZM is a process of governance and consists of the legal and institutional framework necessary to ensure that development and management plans for coastal zones are integrated with environmental (including social) goals and are made with the participation of those affected (Post and Lundin, 1996).

This definition essentially describes some basic elements for a successful and viable ICZM process in any country. It is clear from this definition that an ICZM seeking to achieve sustainable development of coastal zones needs to be predicated upon at least three essentials: (a) an enabling legal framework, (b) an enabling institutional framework, and (c) stakeholders' participation.

A well-defined legal framework is a must for establishing an enabling institutional framework for ICZM, which will clearly set out the rights and duties of all stakeholders. Moreover, integration of development and management plans with environmental goals can be identified as an implementation of international law principle of sustainable development in coastal zone. As a member of global community, it is now an international legal obligation for every state to implement the 'sustainable development' objective in all parts of its territory, be it in its coastal zone, marine area or an area in far inland.

Defining sustainable development is a difficult task. Over the years, researchers from varied backgrounds have tried to define the term, while the most commonly cited definition comes from the Report of the United Nations World Commission of Environment and Development, popularly known as 'Our Common Future Report'. The report states that sustainable development is the development that 'meets the needs of the present without compromising the ability of future generations to meet their own needs'. In 1987, this report, also known as Brundtland Report, brought the concept of sustainable development to the forefront of international agenda. Sustainable development was the central feature of the 1992 Rio Declaration on Environment and Development. Moreover, it has been a focus of attention in all discourses on development (Tay, 1999; Karim, 2007).

Sustainable development does not solely focus on environmental issues. It equally deals with economic, environmental and social issues. The elements of sustainable development may include sustainable utilization of natural resources, integration of environmental protection and economic development, the right to development, intergenerational equity, intragenerational equity, and some procedural elements such as environmental impact assessment and public participation in decision making (Boyle and Freestone, 1999).

Over the years, through its universal recognition, sustainable development has become a globally accepted norm of international law. The former vice president of the International Court of Justice Judge Christopher G. Weeramantry observed in his celebrated separate opinion in the case concerning *Gabčíkovo–Nagymaros Dam* (Hungary/Slovakia):

[...] I consider it [sustainable development] to be more than a mere concept, [it is]. [...] a principle with normative value (International Court of Justice, 1997; Lowe, 1999)

He also observed:

After the early formulations of the concept of development, it has been recognized that development cannot be pursued to such a point as to result in substantial damage to the environment within which it is to occur. Therefore development can only be prosecuted in harmony with the reasonable demands of environmental protection. Whether development is sustainable by reason of its impact on the environment will, of course, be a question to be answered in the context of the particular situation involved. (International Court of Justice, 1997)

It is now well established that sustainable development is a fundamental principle of international law, widely accepted in multilateral treaties, international declarations, establishment documents of international organizations, practices of international financial and other institutions and, importantly, in state practices (International Court of Justice, 1997).

Undoubtedly, therefore, ensuring sustainable development is an international law obligation which states owe to the entire humankind. ICZM may be identified as a potential key process for ensuring sustainable development of coastal zone, and hence as means of discharging this international obligation by states. Below is a brief assessment of the existing international legal framework vis-à-vis ICZM.

13.3 International legal framework for ICZM

A large set of international laws covertly or overtly influences the process of ICZM. Although not directly steered toward achieving the ICZM, the Law of the Sea Convention (1982) and other conventions relating to marine environmental conservation and pollution control have an influential role to play in the ICZM process. Moreover, international environmental treaties such as the Convention on Wetlands of International Importance Especially as Waterfowl Habitat (1971), the Convention Concerning the Protection of World Cultural and Natural Heritage (1972), the Convention on International Trade in Endangered Species of Wild Fauna and Flora (1973), the Convention on Biological Diversity (1992), the

United Nations Framework Convention on Climate Change (1992) and the United Nations Agreement on Stocks of Transboundary and Highly Migratory Species (1995) may also play a vital role in achieving the ICZM process. Apart from these hard law instruments, soft law instruments such as the Programme of Action for the Sustainable Development of Small Island Developing States (1994), the International Coral Reef Initiative (1995), the World Bank Guidelines for Integrated Coastal Zone Management (1993), UNEP Guidelines for Integrated Management of Coastal and Marine Areas (1995) and Global Programme of Action on Protection of the Marine Environment from Land-Based Activities (1995) are also important for the process of ICZM.

More importantly, Chapter 17 of 'Agenda 21', the blueprint action plan for sustainable development for the twenty-first century, exclusively deals with coastal and marine environment, addressing the issues of protection of oceans, seas and coastal areas, and the protection, rational use and development of their living resources. In Agenda 21, the States Parties committed 'themselves to integrated management and sustainable development of coastal areas and the marine environment under their national jurisdiction' (Quarrie, 1992). Agenda 21 identified following steps as important preconditions for the sustainable development and ICZM:

(i) Adoption of an integrated policy and decision-making process, including all involved sectors, to promote compatibility and a balance of uses.
(ii) The identification of existing and projected uses of coastal areas and their interactions.
(iii) Concentration on well-defined issues concerning coastal management.
(iv) The application of preventive and precautionary approaches to project planning and implementation, including prior assessment and systematic observation of the impacts of major projects.
(v) Providing, as far as possible, an access for concerned individuals, groups and organizations to relevant information, and opportunities for consultation and participation in planning and decisionmaking at appropriate levels (Quarrie, 1992).

The 1992 Convention on Biological Diversity (CBD) can play a vital role in the formulation of national legislation for ICZM. The CBD imposes on the global community certain broad obligations for the conservation of biodiversity, sustainable use of biological resources and equitable sharing of benefits derived from the use of biological resources. The Convention defines sustainable use as 'the use of components of biological diversity in a way and at a rate that does not lead to the long-term decline of biological diversity, thereby maintaining its potential to meet the needs and aspirations of future generations' (UNEP, 1992). This definition can be very instrumental in formulating a national policy for the sustainable use of coastal biodiversity, which is indispensable for a viable planning of coastal zone management. Under Article 8 of the CBD, the parties to the Convention undertake an obligation to respect, preserve and maintain knowledge, innovations and practices of indigenous and local communities embodying traditional lifestyles relevant for the conservation and sustainable use of biological diversity and to develop or maintain necessary legislation and/or other regulatory provisions for the protection of threatened species and populations.

The second meeting of the Conference of Parties (CoP) to the CBD held in Jakarta in 1995 adopted a Declaration, called the Jakarta Mandate, which identified marine and coastal biodiversity as an area for priority action. The Mandate identified ICZM as one of its special attentions. It recommended a number of steps to preserve the coastal and marine ecosystems and that Parties to the CBD introduce in their domestic regimes the process of integrated coastal area management for the conservation and sustainable use of coastal and marine biodiversity (Downes and Fontaubert, 1996).

While making national legislation, many countries will most likely have to consider the issue of climate change and sea-level rise as one of their main priority concerns. The United Nations Framework Convention on Climate Change (UNFCCC) aims to establish a framework for stabilization of greenhouse gas concentrations in the atmosphere at a level that would prevent dangerous anthropogenic interference with the climate system. Parties commit to take precautionary measures to anticipate, prevent or minimize the causes of climate change and mitigate its adverse effects. Parties also pledged to take guidance from the principle of intergenerational equity, intragenerational equity and sustainable development in taking action for combating climate change (United Nations, 1992). The UNFCCC recognizes ICZM as one of the main instruments with which to combat the evils of climate change and sea-level rise. Under its Article 4(e), the convention parties commit themselves to 'develop and elaborate appropriate and integrated plans for coastal zone management' (United Nations, 1992).

The above discussion allows us to conclude that for ensuring sustainable development of the coastal area, the states should essentially consider some emerging international environmental law principles while enacting laws and framing polices for coastal zone management. In addition to the principles as aforementioned, certain other international law principles require legislative, executive and judicial implementation for the purpose of having a development-friendly ICZM. These include, inter alia, the principles of intergenerational equity, intragenerational equity, prior environmental impact assessment, notification of activities having adverse effects, the principles of preventive action, and 'polluter pays', 'precautionary principle' and the no-harm principle. The value of these principles in framing well-organized national policies and domestic legal frameworks with a view to achieving sustainable development of coastal area needs no stressing.

13.4 Implementation of international legal obligations in domestic arena

Although there is a wide participation of states in the above-mentioned international legal instruments which impose a legal obligation on states to implement ICZM process in the domestic arena, many countries are yet to reform their legal framework where changes are due in order to accommodate these international norms in the domestic legal framework.

States' obligation to comply with international obligations while enacting legislations in a particular area is a complex area, and the state practices in this regard are marked by diversity. Many states consider it against the state sovereignty to directly apply, administratively or judicially, international obligations undertaken under international treaties without first having enabling national laws to implement them. Paradoxically, on the other hand, while some states (e.g. the US, the Netherlands) consider international obligations to be of direct applicability, the practices of domestically implementing international law obligations even in such countries are murky. Space constraint prevents a detailed analysis of this issue, but it must be noted that there is now an increasingly strong proposition that states are not only bound to respect international treaty and customary law obligations, they can also do so without even having enabling legislations through, for example, administrative activity and judicial interventions (Vereshchet, 1996; Vagts, 2001; Bahdi, 2002; Alam, 2007).

Lack of national legal and institutional framework incorporating international norms and standards is not the only cause of weak implementation of international legal norms in developing countries. Apart from the lax national enabling legal framework, inherent weakness of international law is also a major cause of noncompliance with these international norms in many countries. Although several international legal instruments clearly endorsed the need for a legal framework for ICZM, no international legal instruments clearly elaborated what legal steps are needed to be taken for the implementation of ICZM process. For example, the UNFCCC calls for development of appropriate and integrated plans for coastal zone management, but fails to prescribe any basis or criteria of 'appropriate and integrated' plans. Not that all international instruments are silent on this issue. For example, Agenda 21, although it is not a legally binding instrument *stricto senso*, elaborated the precise nature of provisions which are needed to be incorporated in the domestic legal framework.

On the other hand, there is a reservation in certain governments in taking a statutory approach to coastal zone management. They consider it more a policy issue than legal. But conflicts among several sector-based laws may frustrate the implementation of the ICZM process in a harmonious way. Further, the term coastal zone is needed to be well defined in the national legislation for implementation of the ICZM process, and there is the need for harmonizing the definitions of coastal zone in various domestic laws of different countries. Moreover, government authorities responsible for the management of coastal areas are also de-coordinated. Apart from functioning and co-ordaining government departments, legislative reform for harmonizing different sectoral laws is also needed for proper implementation of the ICZM process. The umbrella law for ICZM should harmonize laws relating to environmental pollution and conservation; land use, administration and management; fishery; forestry; ocean management; rural and urban planning; and water resources. In some countries, collective land right of indigenous people and artisanal fishing communities' fishing rights are also important factors that need to be considered while harmonizing legislations for ICZM, as these stakeholders are greatly dependent on coastal resources to maintain their livelihood.

Republic of Korea may be a good example for making a short review of a model national legal framework for ICZM. Like many other countries, the ocean management in Korea has been divided among many government agencies. There was no comprehensive coastal zone policy. Against this backdrop, the Korean government in 1996 established the Ministry of Maritime Affairs and Fisheries (MOMAF) to integrate ocean-related activities of ten government agencies (Hong and Chang, 1997). Over the last decade, this ministry has established an integrated process of ocean and coastal area management. The first legal reform of the government was an amendment to the Government Organization Act (GOA). The amended law entrusted to the MOMAF such functions as administering fisheries, shipping and ports, preserving the marine environment, doing oceanographic and marine science technology research, achieving marine resources development and securing the maritime safety. Moreover, the government also integrated the functions of relevant government agencies including the Maritime and Port Administration (MPA), the Fisheries Administration (FA) and National Marine Police Administration (NMPA) under the control of MOMAF (Cho, 2006).

In 1999, the Korean government enacted the Korean Coastal Zone Management Act for the preservation of the coastal environment and sustainable development of coastal zone. According to this act, the coastal zone has to be managed in an integrated way in order to harmonize ecological, cultural and economic values. The act provides a good definition of coastal zone and defines the duties and responsibilities of government agencies in coastal areas. Having provisions for the participation of stakeholders in formulating an integration plan, the Act also elaborates the process of initiating and conducting coastal readjustment projects which include:

- Projects to protect the coastal zone from coastal hazards and erosion and to restore damaged coastline.
- Projects to clean coastal waters and to eliminate abandoned vessels in order to preserve and improve coastal waters.
- Projects to revitalize a waterfront (Korean Coastal Zone Management Act, 1999).

ICZM process in Korea is now operating under a well-defined legal framework where role of the core ministry and other ministries is very clear and coordinated, which makes Korea's overall ocean and coastal governance a success. As mentioned by an expert, over the last decade, the ocean governance in Korea has been successful and strong (Cho, 2006). Korea undoubtedly is a good example to show why a well-defined legal framework is a must for the success of ICZM process and how nations can translate globally recognized legal norms in their national legal systems.

13.5 Concluding remarks

Humans have changed the natural environment from the very beginning of civilization through uncontrolled and unplanned utilization of natural resources. Coastal area is one of the main victims of human interventions in nature, as many major cities in the world are on the banks of the sea. Consequently, for the sake of very existence of human civilization on earth, all concerned must ensure that the development process, particularly in the coastal areas, is carried out in a sustainable manner. Undoubtedly, the legal framework is a very important driving force for ensuring sustainable development of coastal zone. A well-planned legal framework can certainly accelerate the process of sustainable ICZM. By contrast, an ill-planned legal framework may become the major hurdle for the ICZM. Making or drafting a law for ICZM is not a 'cure-all'. Proper implementation of the law and a sensitized compliance with international commitments, therefore, remain the main challenge for the world nations.

References

Alam, M.S. (2007) *Enforcement of International Human Rights Law by Domestic Courts*. New Warsi Book Corporation, Dhaka.
Bahdi, R. (2002) Truth and method in the domestic application of international law. *Canadian Journal of Law and Jurisprudence* **15**, 255–279.

Beckman, R. and Coleman, B. (1999) Integrated coastal management: the role of law and lawyers. *The International Journal of Marine and Coastal Law* **14**, 491–522.

Boyle, A. and Freestone, D. (eds) (1999) *International Law and Sustainable Development: Past Achievements and Future Challenges*. Oxford University Press, Oxford.

Cho, D.O. (2006) Evaluation of the ocean governance system in Korea. *Marine Policy* **30**, 570–579.

Downes, D.R. and Fontaubert, A.C.D. (1996) Biodiversity in the Seas: Conservation and Sustainable Use Through International Cooperation. CIEL Brief No. 4 (1996) (de Fontaubert & Downes) [BW96–1]. Center for International Environmental Law, Washington, DC.

Eremina, T.R. and Stetsko, E.V. (2007) Legal Provision for Integrated Coastal Zone Management. Available at http://www.unesco.org/csi/act/russia/legalpro5.htm, accessed on May 26 2007.

European Commission (1999) *Towards a European Integrated Coastal Zone Management (ICZM) Strategy: General Principles and Policy Options*. Office for Official Publications of the European Communities, Luxembourg.

Gibson, J. (2003) Integrated coastal zone management law in the European Union. *Coastal Management* **31**, 127–136.

Hong, S.-Y. and Chang, Y.-T. (1997) Integrated coastal management and the advent of new ocean governance in Korea: strategies for increasing the probability of success. *The International Journal of Marine and Coastal Law* **12**, 141–161.

International Court of Justice (1997) GabCikovo–Nagymaros Project (Hungary, Slovakia). Judgment, I. C. J. Reports, pp. 88–119.

Karim, M.S. (2007) Sustainable development needs to be environment-friendly. *The Daily Star* **5** (1101), 13.

Korean Coastal Zone Management Act (1999) Available at: http://www.globaloceans.org/icm/resources/laws/koreanczma.html, accessed on 23 February 2008.

Lowe, V. (1999) Sustainable development and unsustainable arguments. In: Boyle, A. and Freestone, D. (eds). *International Law and Sustainable Development: Past Achievements and Future Challenges*. Oxford University Press, Oxford, pp. 19–38.

Post, J.C. and Lundin, C.G. (eds) (1996) *Guidelines for Integrated Coastal Zone Management*. World Bank, Washington, DC.

Quarrie, J. (ed.) (1992) *Earth Summit '92: The United Nations Conference on Environment and Development, Rio de Janeiro, 1992*. The Regency Press, London.

Tay, S.S.C. (1999) Southeast Asian fires: the challenge for international law and sustainable development. *Georgetown International Environmental Law Review* **11**, 241–300.

UNEP (1992) Convention on biological diversity. *International Legal Materials* **31**, 849–873.

United Nations (1992) United Nations Framework Convention on Climate Change. *International Legal Materials* **31**, 818–842.

Vagts, D.F. (2001) United States and its treaties: observance and breach. *American Journal of International Law* **95**, 313–334.

Vereshchet, V.S. (1996) New constitutions and the old problem of the relationship between international law and national law. *European Journal of International Law* **7**, 29–41.

Chapter 14
Lobster Reserves in Coastal Skagerrak – An Integrated Analysis of the Implementation Process

Alf Ring Pettersen, Even Moland, Esben Moland Olsen and Jan Atle Knutsen

Abstract

The Norwegian fishery for European lobster (*Homarus gammarus*) has gradually decreased over the last 50 years. In 2006, four experimental lobster reserves (0.5–1 km^2 in area) were established along the Skagerrak coast in Southern Norway, with only hook- and line-type gear allowed inside the reserves. The objective was to provide knowledge of the effects of small-scale closures on lobster population development. The four lobster reserves were nominated by local fishers following a series of consultations with public officials. The present study analyses the process from early stage planning to legislative establishment. We conducted an integrated analysis of the implementation process in order to see how it has met the selection criteria, how the scientific monitoring is designed to detect changes and how different stakeholder groups have been included in the implementation process. We conclude that the information flow from management to stakeholders has been successful and the local legitimacy for the lobster reserves is high. However, local residents, and especially recreational fishers, who are directly affected by the regulations, were not included in the process. We argue that non-organised stakeholders should be included in future implementation programmes for marine protected areas in coastal areas of Norway in order to ensure a successful outcome.

14.1 Introduction

Norwegian catches of European lobster (*Homarus gammarus*) in Skagerrak have gradually decreased over the last 50 years. Catch per unit effort (lobsters/pot/24 h) is reduced from 0.15–0.2 before the 1950s to 0.05–0.07 over the last decade (Figure 14.1). Official landings from 1945 to 1970 were 500–700 metric tons year^{-1}, decreasing to 59 tons year^{-1} in 2006 (Knutsen et al., 2007). In 2006, European lobster in Norway was listed as near threatened in the national IUCN red list (Oug et al., 2006). At present, lobster is a popular catch for recreational fishers. The Norwegian lobster fishery is regulated through season closures, gear restrictions and size limitations (minimum legal length). Nevertheless, the current management regime has not been sufficient to rebuild the lobster population.

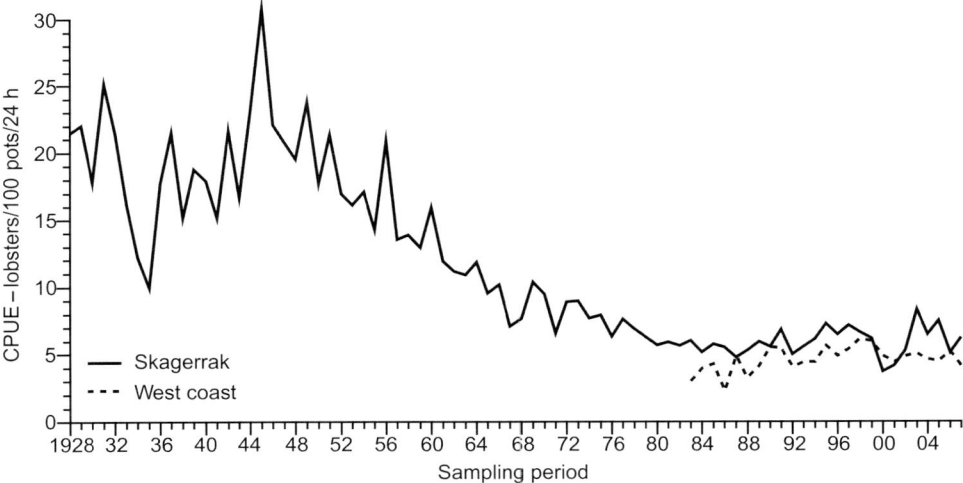

Figure 14.1 Catch-per-unit-effort (CPUE) of lobster in Skagerrak and the North Sea west coast of Norway during 1928–2007, estimated from standardised logbook reports from commercial fishers in collaboration with the Institute of Marine Research.

Marine protected areas (MPAs) have attracted much attention over the latest years as a tool to rebuild and maintain fish stocks and habitats (Russ, 2002). Globally, different species of lobster have generally shown a quick response to protection. Examples of positive effects of protection for lobster are found for spiny lobster (*Jasus edwardsii*) (Edgar and Barrett, 1999; Kelly et al., 2000; Davidson et al., 2002), American lobster (*Homarus americanus*) (Rowe, 2002) and European lobster (Sweeting and Polunin, 2005). However, there is a lack of well-designed data collection efforts on the effects of protection for European lobster in the northeast Atlantic. On 19 September 2006, the Ministry of Fisheries and Coastal Affairs gazetted four lobster reserves in coastal Skagerrak. The objective for the reserves is to understand how lobster populations develop within limited areas when the fishery is excluded and thus test the potential of MPAs in future lobster management.

The four lobster reserves established in Norway in 2006 are Flødevigen (~ 1 km^2), Risør (~ 0.6 km^2), Bolærne (~ 0.7 km^2) and Kvernskjær (~ 0.5 km^2) (Figure 14.2). All four reserves are located along the coastal zone in Skagerrak. These reserves are protected under the saltwater fishery law, regulating the use of fishing gear inside the reserves, prohibiting the use of standing gears such as pots and nets. Hook and line fishing is allowed. The areas are protected for at least 10 years.

Stakeholder participation is found vital to a successful management regime (Mascia, 2003; Fernandes et al., 2005). By participating in the management process, stakeholders are more likely to acknowledge the benefits of a protected area, take credit for the designation and support and enforce the regulations they establish (Leigh Kessler, 2004). Christie et al. (2003) argued that most MPAs have both biological and social goals. These goals can be contradicting between different groups, resulting in controversy and conflict. Christie et al. (2003) further argued that these clashes of goals contribute to a high potential rate of MPA failure.

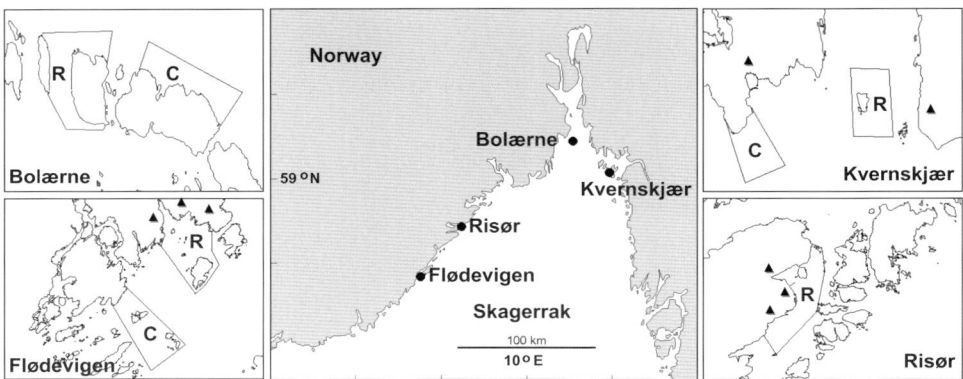

Figure 14.2 Map of the four experimental lobster reserves established in coastal Skagerrak in 2006. Reserve (R), control (C), residential areas (triangles). Risør does not have a control area.

In the event of future planning of a network of reserves along the Norwegian coast, there is a need to analyse the process resulting in implementation of these first lobster reserves, understand the societal issues regarding legitimacy and expectations of their performance, as well as document the biological effects. In this chapter, we analyse both scientific and social processes from early stage planning to legislative implementation as perceived by individuals from different stakeholder groups.

14.2 Materials and methods

In order to evaluate the process towards implementation of the lobster reserves, we selected certain criteria to be tested. First, we analysed the selection criteria developed by management authorities and scientists. The Directorate of Fisheries and the Institute of Marine Research (IMR) developed a set of criteria for the potential locations of the lobster reserves. The areas should (a) have an acceptable lobster population, (b) have a habitat suitable to hold a substantial lobster population, (c) be effective to monitor and (c) be supported by local commercial fishers. Secondly, we analysed the scientific monitoring programme. It is too early to detect biological effects of the lobster reserves, due to the short time since the implementation. However, we can analyse the monitoring programme in order to evaluate if the design is good enough to trace potential biological effects of the lobster reserves. We chose the recommended BACIP (before-after-control-impact-pairs) design described in Russ (2002) and Underwood (1994) as an optimal method for tracing effects of MPAs. Thirdly, we analysed the social factors with regard to the establishment process. Important social factors were inclusion of stakeholders in the implementation process, information flow to the public and level of public support for the lobster reserves (legitimacy). In order to do so, we analysed all the hearing documents, collected publications in media and went through the different steps in the implementation process.

In July to August 2007, nearly a year after implementation of the lobster reserves, a questionnaire survey was sent out to residents and cottage owners in the areas where two of the lobster reserves were sited (Flødevigen and Kvernskjær). We inferred that among these property owners we would find the stakeholders perceiving themselves as most directly affected by the implementation of the lobster reserves, generally having their residencies and/or boats either in areas bordering to or in close vicinity to the two reserves. In Flødevigen, a number of houses are on waterfront properties bordering the lobster reserve. Kvernskjær lobster reserve surrounds a small uninhabited island. No properties are bordering the reserve, but two nearby islands have residential areas (Figure 14.2). Mailing addresses to residents were provided by local authorities. The questionnaire was designed to evaluate (a) information flow from management to stakeholders, (b) stakeholders' fishing activities, (c) potential expectations of the effects of the reserves and finally (d) the stakeholders' involvement in the implementation process. The letter included an anonymous multiple-choice questionnaire and a prepaid return envelope. A total of 186 questionnaires were sent out to property owners around the Flødevigen lobster reserve and 145 to the Kvernskjær reserve area (total of 331).

14.3 Results

The Directorate of Fisheries and the IMR developed a set of criteria for the potential locations of lobster reserves in 2001. The first stakeholder involvement was conducted in 2004, in which local commercial fishers were asked to suggest potential locations for lobster reserves. IMR conducted surveys to evaluate the potential of four of the suggested areas and made a report to the management authorities. IMR concluded that three of the suggested areas were suitable based on the biological selection criteria (Table 14.1). These areas had an acceptable resident lobster population and suitable habitat represented within the area. However, the area that was found to have the poorest potential based on its location in a harbour area and low catch per unit effort data gained a high public interest (Risør), and it was suggested that this area should be included in the public hearing process (J.A. Knutsen, personal communication). For each area, different policing strategies were discussed. The Flødevigen reserve is located outside Flødevigen Marine Research Station and would be easily controlled. The Risør reserve is located around the harbour of Risør city in close vicinity to the Risør aquarium and police. The Bolærne reserve is the area that is furthest away from population centres. However, police and local commercial fishers agreed to look after the reserve. The Kvernskjær reserve is in close vicinity to a fishery harbour and a pilot station, with pilots passing the area daily. For all areas, police and coastguard agreed to put extra effort into surveying the reserves for poaching. The four suggested areas went out to public hearing in November 2005. A total of 21 return forms were received from commercial fishing organisations, research institutions, environmental non-governmental institutions (NGOs) and governmental institutions on different levels from municipalities to ministries (Table 14.2). None of the responses were negative to the suggested lobster reserves. However, there were some suggestions regarding reserve size and boundaries. The bill for implementation of four lobster reserves in coastal Skagerrak was signed on 19 September 2006.

Research fishing, including capture–mark–recapture, has been conducted inside the reserve areas for three consecutive years prior to establishment (since 2004). In 2006, in the last season before implementation of the lobster reserves, adjacent control areas, with habitat structure comparable to the reserves, were included in the survey. The standardised annual fishing effort was 100 pots deployed for 24 h inside reserves and control areas

Table 14.1 Data collected from the proposed lobster reserves during the pre-selection sampling cruises in coastal Skagerrak in 2004 and 2005.

Site	Biological selection criteria		Habitat sampling mode
	CPUE[a]	Habitat representation[b]	
Kvernskjær	0.82	ABCDEF	ROV/SCUBA
Bolærne	0.63	ABCDEF	ROV/SCUBA
Risør	0.27	–	–
Flødevigen	0.32	ABCDEF	SCUBA

[a]Resident lobster population in proposed reserves evaluated by fishing. Numbers represent lobsters per trap per day (CPUE) sampled in 2004 and 2005 (pooled).
[b]Habitat representation: A, rock patches; B, boulder fields; C, kelp; D, sand; E, mud flats and F, rock outcrops and ridges. Habitat representation was not evaluated in the Risør area.

simultaneously during the same week at each site each year (Figure 14.2). Three out of four reserves were included in the monitoring programme (excluding the Risør reserve). Every lobster caught in reserve and control areas was tagged on first capture (T-bar anchor tags inserted ventrally through the thoracoabdominal membrane) and released alive. Data were collected on gender, total length and carapace length. In addition, a tissue sample was collected from each egg-bearing female in order to allow future genetic analyses. Scientific monitoring of the reserves has been planned to be continued for up to 10 years post-implementation.

A total of 186 questionnaires were sent out to property owners around the Flødevigen lobster reserve and 145 to the Kvernskjær reserve area. The response rate was 38% for Flødevigen and 37% for Kvernskjær ($N = 125$). Thirty-eight per cent of the respondents were females. Out of all respondents, 48% had their summer house in the area, while 46% were local residents. The remainder had other connections to the area. There were a higher proportion of respondents living at the border, or closer than 500 m from the reserve, at Flødevigen than at Kvernskjær. Seventy-two per cent of the respondents owned their own boat, while additional 16% had access to a boat. Seventy-three per cent of the respondents knew that there had been established a lobster reserve in their region. Moreover, 84% of those using fishing gear that had been excluded from the reserve (standing gear such as nets, pots and traps) were familiar with the reserve, which were significantly different to the rest (67.5%) ($\chi^2_{(3.698), df=1}$, $p < 0.05$). This group also reported to know rules and regulations

Table 14.2 Institutional responses to the public hearing document regarding the implementation of four lobster reserves in coastal Skagerrak.

Hearing institutions	Number of respondents
Local counties	4
Regional counties	3
Governmental bodies	8
Commercial fisher organisations	2[a]
Environmental organisations	2
Research institutions	1

[a]Speaking on behalf on local commercial fishers.

'well or to some extent' (75%) as well as geographical boundaries (76%) of the reserve. The difference in knowledge of rules and borders was significantly different to the group that did not use standing gear ($\chi^2_{(9.116), df=1}$, $p < 0.05$ and $\chi^2_{(9.929), df=1}$, $p < 0.05$, respectively). Media coverage of the implementation process can be termed high. An internet media news search (www.sesam.no/nyheter) shows that from January 2006 to March 2007, newspapers with web-based services covered the lobster reserves at least 53 times. The Directorate for Fisheries also distributed pamphlets and posters with information about the reserves.

Fifty-nine per cent of the respondents characterised themselves as either commercial or recreational fishers. Within these groups, there was a significantly greater amount of people fishing with standing gear in the Kvernskjær region (54%) than in Flødevigen (22%) ($\chi^2_{(13.071), df=1}$, $p < 0.05$). However, there were no differences in the proportion of fishers in general. A high proportion of the respondents (63%) that fished with standing gear had been fishing inside the lobster reserve previous to its implementation. Consequently, they responded that they had changed fishing area. Nearly half of this group (46%) informed that they had to travel further to fish after the lobster reserve establishment. A smaller amount of the respondents (21%) said that the lobster reserve had made them fishless. Twenty-eight per cent of the respondents within this group informed that they now tended to fish on the borders of the reserve, indicating that fishing effort has increased just outside borders.

The respondents were asked a set of questions regarding their expectations regarding the effect of the lobster reserves. Eighty-three per cent expected a positive effect on the lobster population inside the reserve. More interestingly, 70% expected a positive effect on the lobster population outside the reserve, whereas 58% expected a positive effect on the actual catches of lobsters outside the reserve. In general, the respondents were positive to the idea of using reserves as a tool to manage lobster (88%). They reported a high degree of interest in the management of local resources, where 41% said they were very interested in this issue. Only 6% said they were not interested, while the rest were to some degree interested.

Only three persons had been involved in the implementation process, and they were all commercial fishermen. The rest of the respondents (98%) said they had not been involved at all in the implementation process. However, a high number of respondents wished to participate in future processes. Furthermore, more fishers (46%) were interested in participating in future MPA planning processes, than people that did not fish (28%). The difference was significant ($\chi^2_{(4.298), df=1}$, $p < 0.05$). We asked those who were interested in participating in future planning processes to choose a preferred type of participation. The respondents could choose more than one alternative of participation. The preferred ways of potential future involvement were information letters (72%), followed by information meetings (22%) and public hearing meetings (20%), as well as information through local media (41%).

14.4 Discussion

The a priori evaluation of areas proposed as lobster reserves showed that there were acceptable lobster populations in the areas, and that habitats seemed suitable to hold substantial lobster populations. However, the proposed reserve in Risør did not fulfil these requirements to the same extent as the other three areas. Nevertheless, the location of the area (on the city ocean front, adjacent to the local aquarium) coupled with local enthusiasm regarding the plans to establish reserves ensured inclusion of this area as well. Cooperation between management authorities, coastguard, local police and commercial fishers should be sufficient to ensure an effective policing of the reserves. Reports from the policing authorities indicate few instances of poaching inside the reserves. However, there is a need to evaluate the efficiency of these control measures in the future. Based on the hearing documents and the few responses from commercial fishers received in the questionnaire survey, there are strong indications that local commercial fishers support the lobster reserves. Geographical locations for the areas were suggested by commercial fishers, and their organisation has been involved in the entire implementation process. An interesting question is what type of expectations commercial fishers harbour regarding the effect of the reserves. Their support might be based on expectations of a 'spill over effect' from the reserves, in which fishers expect that the reserves over time might 'leak' lobsters that will be catchable outside reserve borders, as described by Robertson and Caporossi (2003).

Concurrent with the overall aim for the reserve establishment, IMR has launched a long-term scientific monitoring programme in order to document the effects of small-scale fishery closures on local lobster populations. The ongoing scientific monitoring consists of a research fishery conducted annually and is designed as a BACIP study. Before-data inside reserves were sampled in three consecutive years prior to establishment. However, before-data in controls were sampled only in the last year prior to reserve establishment. The limited before-data from controls is insufficient to observe before-establishment variations in controls, but useful when comparing the baseline situation in reserves and controls. Internationally, little research on the effects of MPAs has been found to meet the BACIP design requirements. An optimal design should be balanced and include sampling of data from controls and impacts from the same point in time (Russ, 2002). For the ongoing study designed to monitor the effects of the lobster reserves, the requirement to include sufficient before-data inside reserves is met. However, the lack of sufficient before-data from controls may be partially adjusted for by using data from fishers operating in nearby areas and already part of a long-term IMR reporting scheme (Table 14.3).

Based on the questionnaire, nearly 75% of the respondents knew about the lobster reserves. The group that was directly affected by the lobster reserves was better informed than others, indicating that the target group (fishers using standing gear) was well informed regarding the reserves, their rules and their borders. Media coverage has been high and might have been an important tool for distribution of information. There is reason to argue that the information flow from management to stakeholders has been successful.

Commercial fishing organisations, environmental NGOs and governmental authorities on different levels were active participators in the implementation process. However, commercial fishers were the only non-governmental stakeholder group that were included from the start. Other stakeholder groups were formally included at the time when the public hearing was launched.

Data collection regarding local public attitudes towards the lobster reserves is not necessarily representative to the whole area where the questionnaires were distributed. The main aim of the survey was to understand how local stakeholders and user groups have been involved in the implementation process and how they wish to be involved in the future.

Table 14.3 Evaluation of the implementation process of lobster reserves in coastal Skagerrak, in relation to selection criteria, scientific monitoring and stakeholder involvement.

Selection criteria	Acceptable lobster population	1	3/4 of the reserves
	Suitable habitat	1	3/4 of the reserves
	Effective to monitor	1	
	Support of local commercial fishers	1	
Scientific monitoring	Before-data inside reserves	1	Data from 3 years
	Before-data controls	2	Data from 1 year
	Replication (pair)	1	
	After-data inside reserves	1	
	After-data controls	1	
Stakeholder involvement	Inclusion of commercial fishers	1	
	Inclusion of recreational fishers	3	Not included
	Information to public	1	
	Public support	1	

1, optimal; 2, satisfactory; 3, sub-optimal.

We expect that stakeholders wishing to be involved in future processes did put an effort into answering the questionnaire. We also expect that recreational fishers would answer the survey more frequently than those who do not fish.

Local stakeholders in close vicinity to the lobster reserves had limited involvement in the implementation process. From this group, recreational fishers are most affected, especially those who are fishing with standing gear which is excluded from the lobster reserves. This might represent a potential conflict, since for example the Kvernskjær area is known to be used mostly by recreational fishers. Commercial fishers did not use this area often according to Knutsen (2005). Commercial fishers suggested areas where they did not fish often themselves, but knew that it was used by recreational fishers. However, recreational fishers were not involved at all in the implementation process. Limited level of involvement is confirmed in our stakeholder survey, where no recreational fishers say they have been included in the implementation processes. Arguably, those who mainly fished in the area were not included in the process. International experiences show that this can easily lead to a failure of the management goals with low legitimacy from user groups (Russ and Alcala, 1999; Guenette et al., 2000).

The coastal zone and marine coastal resources have a number of different stakeholder groups. Mikalsen and Jentoft (2001) listed a number of stakeholder groups related to fisheries management in Norway. They divided the stakeholders into definitive, expectant and latent, dependent of the groups' urgency, power and legitimacy for the fishery resources concerned. Mikalsen and Jentoft (2001) identified stakeholders such as commercial fishers, bureaucrats and scientists to be definitive stakeholders, while environmental groups and local communities were defined as expectant stakeholders. Sport fishers (or recreational fishers), on the other hand, were found to be latent stakeholders with low urgency and power, but with an increasing legitimacy. They argued that apart from fishers, very few groups and individuals will be directly affected by, or compelled to obey, the laws and rules for fisheries management. However, fisheries regulations in the coastal zone, such as MPAs, that are excluding all fishing activities will directly affect recreational fishers that used to fish in these areas. Along the Skagerrak coast, recreational fishers outnumber commercial fishers more than in other parts of Norway (Hallenstvedt and Wulff, 2004). On the basis of this, one might argue that the legitimacy of recreational fishers in this area is higher than in other parts of Norway. Conversely, when only 3.5% of the recreational fishers in Norway are organised (Hallenstvedt and Wulff, 2004), the user group might have little power. This is different compared to the US where marine recreational fishers seem to be more organised and powerful and have gained recognition in political processes (Robertson and Caporossi, 2003). Therefore, it seems to be a gap between the number of organised recreational fishers and their potential interest in marine management and planning processes. However, the recreational fishers' desires and level of involvement in MPA planning processes need to be investigated further.

Houde et al. (2001) argues that local residents and other community representatives should be involved in MPA establishment processes for a number of reasons, including that 'they will destroy or undermine the integrity of the MPA if they are not involved in its establishment and management'. Because of the lack of involvement of recreational fishers, a potential for a negative attitude towards the lobster reserves in Norway might have been expected. This is not the case for local stakeholders living near the studied lobster reserves. These stakeholders show a positive attitude towards the use of MPAs for management of lobster, including those who were using fishing gear that was excluded from the reserves. A majority of respondents expect an increased lobster population outside reserves as well as increased catches. There might be an inconsistency between the small sizes of the reserves under study and the stakeholder's expectations.

Examples of MPAs failing with regard to user group support are from societies that are harbouring stakeholders that rely on fisheries for subsistence, mostly in tropical developing countries, such as the Philippines (described by Russ and Alcala, 1999). Nevertheless, it has been shown that recreational fishers tend to be the most vocal against MPAs in western countries, as experienced in the US (Salz and Loomis, 2004). In order to understand why local acceptance of lobster reserves is high, there is a need for further studies including in-depth interviews. Small areas might be less likely to create opposition. In the event of future planning of a large-scale network of MPAs along the Norwegian coast, there is reason to anticipate stronger opposition, or expectancy of involvement, from stakeholders. We conclude that the inclusion of local stakeholders, such as recreational fishers, has not been fulfilled in the implementation process described above.

In the event of using MPAs as a future management tool in coastal areas in Norway, a number of species and habitats might be included in the process, as well as a network design. The protected areas may be bigger and more numerous. A successful MPA establishment is highly dependent on the participation of local stakeholders throughout the process, probably more so than for the experimental lobster reserves described herein. The positive attitudes voiced by the stakeholders in our survey indicate a potentially successful introduction of a network of MPAs along the Norwegian Skagerrak coast. However, public attitudes towards future zoning plans are likely to be affected by the outcome from the current monitoring programme designed to evaluate the performance of small, experimental lobster reserves. Therefore, management should share scientific findings and define realistic expectations when interacting with stakeholders.

Acknowledgements

Financial support for this work was provided by the Research Council of Norway (through the Oceans and the Coastal areas programme) and Institute of Marine Research. Thanks to all stakeholders that contributed to the evaluation process.

References

Christie, P., McCay, B.J., Miller, M.L., Lowe, C., White, A.T., Stoffle, R., Fluharty, L.D., McManus, L.T., Chuenpagdee, R., Pomeroy, C., Suman, D.O., Blount, B.G., Huppert, D., Villahermosa Eisma, R.L., Oracion, E., Lowry, K. and Pollnac, R.B. (2003) Towards developing a complete understanding: a social science research agenda for marine protected areas. *Fisheries* **28**, 22–26.

Davidson, R.J., Villouta, E., Cole, R.G. and Barrier, R.G.F. (2002) Effects of marine reserve protection on spiny lobster (*Jasus edwardsii*) abundance and size at Tonga Island Marine Reserve, New Zealand. *Aquatic Conservation – Marine and Freshwater Ecosystems* **12**, 213–227.

Edgar, G.J. and Barrett, N.S. (1999) Effects of the declaration of marine reserves on Tasmanian reef fishes, invertebrates and plants. *Journal of Experimental Marine Biology and Ecology* **242**, 177–144.

Fernandes, L., Day, J., Lewis, A., Slegers, S., Kerrigan, B., Breen, D., Cameron, D., Jago, B., Hall, J., Lowe, D., Innes, J., Tanzer, J., Chadwick, V., Thompson, L., Gorman, K., Simmons, M., Barnett, B., Sampson, K., De'ath, G., Mapstone, B., Marsh, H., Possingham, H., Ball, I., Ward, T., Dobbs, K., Aumend, J., Slater, D. and Stapleton, K. (2005) Establishing representative no-take areas in the Great Barrier Reef: large-scale implementation of theory on marine protected areas. *Conservation Biology* **19**, 1733–1744.

Guenette, S., Chuenpagdee, R. and Jones, R. (2000) *Marine Protected Areas with An Emphasis on Local Communities and Indigenous Peoples: A Review*. Fisheries Centre Research Reports. Fisheries Centre, University of British Columbia, Vancouver, p. 57.

Hallenstvedt, A. and Wulff, I. (2004) Fritidsfiske i sjøen 2003 (in Norwegian) [Recreational fishing in the sea 2003]. Norwegian College of Fisheries Science/University of Tromsø, Tromsø, p. 66.

Houde, E., Coleman, F.C., Dayton, P.K., Fluharty, D., Kelleher, G., Palumbi, S.R., Parma, A.M., Pimm, S., Roberts, C.M., Smith, S., Somero, G., Stoffle, R. and Wilen, J. (2001) *Marine Protected Areas, Tools for Sustaining Ocean Ecosystems*. National Academy Press, Washington, DC, p. 272.

Kelly, S., Scott, D., MacDiarmid, A.B. and Babcock, R.C. (2000) Spiny lobster, *Jasus edwardsii*, recovery in New Zealand marine reserves. *Biological Conservation* **92**, 359–369.

Knutsen, J.A. (2005) Etablering av hummerreservater på Skagerrakkysten (Eng: Establishment of lobster reserves in coastal Skagerrak). Field report, Institute of Marine Research, Flødevigen, p. 13.

Knutsen, J.A., Agnalt, A.L., Moland Olsen, E., Knutsen, H. and Moland, E. (2007) Hummer (Eng: Lobster). In: Dahl, E., Kupka Hansen, P., Haug, T. and Karlsen, Ø. (eds). *Kyst og Havbruk*. Institute of Marine Research, Norway, pp. 90–91.

Leigh Kessler, B. (2004) *Stakeholder Participation: A Synthesis of Current Literature*. National Marine Protected Areas Center, Silver Spring, Maryland, p. 24.

Mascia, M.B. (2003) The human dimension of coral reef marine protected areas: recent social science research and its policy implications. *Conservation Biology* **17**, 630–632.

Mikalsen, K.H. and Jentoft, S. (2001) From user-groups to stakeholders? The public interest in fisheries management. *Marine Policy* **25**, 281–292.

Oug, E., Djursvoll, P., Aagaard, K., Brattegaard, T., Christiansen, M.E., Halvorsen, G., Vader, W. and Walseng, B. (2006) Krepsdyr – Crustacea. In: Kålås, J.A., Viken, Å. and Bakken, T.S. (eds).

Norsk Rødliste 2006 – 2006 Norwegian Red List (in Norwegian). Artsdatabanken, Trondheim, pp. 197–206.

Robertson, R.A. and Caporossi, G. (2003) New England recreational fishers' attitudes toward marine protected areas: a preliminary investigation. In: Murdy, J.J. and Jameson, S.S. (eds). *Proceedings of the 2003 North Eastern Recreation Research Symposium.* Department of Agriculture, Forest Service, Northeastern Research Station, Newtown Square, pp. 121–127.

Rowe, S. (2002) Population parameters of American lobster inside and outside no-take reserves in Bonavista Bay, Newfoundland. *Fisheries Research* **56**, 167–175.

Russ, G.R. (2002) Yet another review of marine reserves as reef fishery management tools. In: Sale, P.F.S. (ed.). *Coral Reef Fishes: Dynamics and Diversity in a Complex Ecosystem.* Elsevier Science, pp. 421–443.

Russ, G.R. and Alcala, A.C. (1999) Management histories of Sumilon and Apo Marine Reserves, Philippines, and their influence on national marine resource policy. *Coral Reefs* **18**, 307–319.

Salz, R.J. and Loomis, D.K. (2004) Saltwater anglers' attitudes towards marine protected areas. *Fisheries* **29**, 10–17.

Sweeting, C.J. and Polunin, N.V.C. (2005) *Marine Protected Areas for Management of Temperate North Atlantic Fisheries.* School of Marine Science and Technology, University of Newcastle upon Tyne, UK, p. 64.

Underwood, A.J. (1994) On beyond BACI: sampling designs that might reliably detect environmental disturbances. *Ecological Applications* **4**, 3–15.

Chapter 15
Efficiency of Fishing Vessels Affected by a Marine Protected Area – The Case of Small-Scale Trawlers and the Marine Protected Area in Nha Trang Bay, Vietnam

Quach Thi Khanh Ngoc, Ola Flaaten and Nguyen Thi Kim Anh

Abstract

Are marine protected areas (MPAs) positive for adjacent fisheries? This is a study of the technical efficiency of small-scale trawlers in Nha Trang, Vietnam, following the establishment of Nha Trang Bay marine protected area (NTB-MPA) that imposed a trawl ban to protect marine biodiversity and regenerate fish stocks. Data were collected through a survey of small-scale trawler owners. Using a stochastic frontier analysis, this study demonstrates that engine power, household size and operating characteristics of vessels strongly affect technical efficiency. The number of days at sea is the most important factor affecting the output revenue. Understanding these relationships is an essential condition for effective management. Despite the ban on trawling, the vessels operating in the vicinity of the MPA are still more technically efficient than those operating in an unprotected area. Thus, the alternative grounds still sustained the activity of the trawl fleet affected by the ban. However, secondary data from the NTB-MPA project indicate a reduction in fish stocks in this area. Our findings combined with the secondary data may provide some policy implications. An MPA and a trawl ban do not seem to be sufficient to achieve improved management. In addition, it is essential also to deal with the link between poverty and resource management. Alternative income generation and effective education to achieve compliance from local communities are among the measures that are important for the success of an MPA.

15.1 Introduction

Marine protected areas (MPAs) have attracted increasing attention in recent years due to their potential as a tool for both fishery management and biodiversity conservation. The creation of MPAs generates a range of benefits. However, MPAs also create a number of negative effects on user groups, especially fishermen in the short term, due to the spatial reallocation of fishing effort. One consequence may be that the fishermen increase their effort in areas still open to fishing (Alban el al., 2006). Assessing the effectiveness of management of MPAs thus requires analysis of how different groups of stakeholders have been impacted by the zoning established within MPAs, especially concerning how catches and revenue are changed as a result of this regulation.

Theoretical and empirical perspectives have been the concern of previous studies on the effects of MPAs on the capture fisheries. Hannesson (1998) and Anderson (2002) used deterministic bioeconomic models to address the question of the impact of an MPA on fishery under an open-access assumption. They concluded that little would be gained in terms of economics in the open-access fisheries. As Hannesson (1998) indicated, the catch might be larger than under open access to the entire area, but this gain would be nullified by the increased cost of fishing. Anderson (2002) supplemented this assessment with the suggestion that spillover from the marine reserve would simply attract more vessels. Conrad (1999) investigated the benefits of MPAs for fisheries in deterministic and stochastic environments. Like Hannesson (1998) and Anderson (2002), he doubted the role that MPAs played in fisheries management. He stated that in the deterministic model there was no need for a marine reserve under perfect management. In the stochastic model, the reserve with the migration process had the ability to reduce variation in biomass and harvest. However, Conrad (1999) emphasised that whether the value of this ability would exceed the loss in expected present net value from creating an MPA was difficult to determine. Focusing on economic issues related to a marine reserve, Flaaten and Mjølhus (2006) investigated the extent to which the reserve size may be tuned to achieve economic and biological objectives. They concluded that a marine reserve and open access in a harvest zone can ensure a greater total fish population than pre-reserve open access, and could even increase consumer surplus. Certainly, the magnitude of these factors depends on economic, biological and reserve parameters and on management objectives.

Contrary to theoretical research, which is often pessimistic about the role of an MPA, empirical research has suggested strong evidence for the benefits of MPAs to fisheries. Whitmarsh et al. (2003) explored the consequences of a trawl ban on the economic sustainability of artisanal fishery after such a ban was introduced in the Gulf of Castellammare, NW Sicily to deal with overfishing and conflicts between trawlers and small-scale artisanal vessels in the area. They showed that the artisanal fishery appears to be sustainable in terms of economics and that demersal stocks had recovered very well since the trawl ban was introduced. Gell and Roberts (2003), utilising results from case studies in different areas such as the Egyptian Red Sea, New Zealand, South Africa, Canada and the US, concluded that fishery benefits from MPAs developed quickly. The evidence suggested that fishermen tended to move their fishing activities closer to MPA boundaries to obtain the benefit from spillover of resources from MPAs into fishing grounds. Gell and Roberts (2003) also indicated that in some cases, the catches within MPAs had increased after the imposition of protection, despite a reduction in the area of the fishing grounds.

Research on MPAs in Asia as well as in Europe and America is expanding, but relatively little research has been conducted to describe and analyse the effects of MPAs in Vietnam specifically. Do the MPAs in the context of Vietnam achieve the same results as MPAs in other areas? How does an MPA impact fishermen's activities? These questions are investigated in this chapter, using the case of the first MPA established in Vietnam – Nha

Trang Bay marine protected area (NTB-MPA). Two main objectives were identified for this project. The first objective was to conserve a representative sample of internationally significant and threatened marine biodiversity. The second objective was to enable local island communities to improve their livelihoods, working in partnership with other stakeholders to effectively protect and manage the marine biodiversity in Nha Trang Bay as a model for collaborative MPA management in Vietnam.

Recently, there has been a trend towards measuring the technical efficiency of fishing fleets in order to allow fishery managers to assess the effects of management measures on technical efficiency and potential catch and revenue. Our study demonstrates the technical efficiency of trawlers 4 years after implementation of the MPA. We investigated the relationship between technical efficiency and its determinants, and compared it between vessels operating in the vicinity of the NTB-MPA and those operating in unprotected areas. Also, based on secondary data from the reports of the NTB-MPA project covering the period 2002–2005, we have attempted to trace changes in fish abundance of different areas and different kinds of fish species (Tuan et al., 2005a, b). This information plays an important role in evaluating the impacts of the NTB-MPA on the regeneration of fish stocks.

15.2 Materials and methods

15.2.1 Study site and its fisheries

The NTB-MPA, established in 2001, was considered as a pilot project for other MPAs presently being established, as well as for future MPAs in Vietnam. The reason that the NTB-MPA was selected for the first comprehensive preservation project was due to the fact that the area has the highest biodiversity in Vietnam. According to the initial survey of the area, there were 350 species of coral, 250 species of fish, 122 species of crustaceans, 112 species of mollusks, 69 species of seaweed and 27 species of echinoderms (Tuan, 2002).

Figure 15.1 shows a map of Vietnam and the location and boundaries of NTB-MPA, which was created to meet the objectives.

On 11 March 2002, the People's Committee of Khanh Hoa Province issued a Temporary Regulation and Zoning Scheme for the NTB-MPA. The regulation sought to promote a management regime for the protection of marine biodiversity, while providing for the regeneration of fish stocks and balancing the various uses of the areas. The scheme set out a series of management zones to regulate resource use and the extraction of resources within the MPA. This multiple-use zoning system was the key management tool for balancing marine biodiversity conservation and resource use. The scheme applied three zones, with different levels of use and protection: core zone, buffer zone and transition zone.

Figure 15.2 shows the zones of the NTB-MPA in 2005, after some adjustments the same year. The island Hon noc, to the east of Hon tre, was removed from the core zone. The north-facing bays of northeastern Hon tre and the southern corner of this island, known for high coverage of seagrass, were added to the core zone sanctuary. Trawling activities were banned in core zones and buffer zones, and were limited in the transition zone. It should be noted that imposing a ban on trawling activities did not directly diminish the number of trawlers; instead, the fishing effort was reallocated to areas where fishing was allowed. The trawl fishermen acknowledged that the imposition of the MPA had altered the position

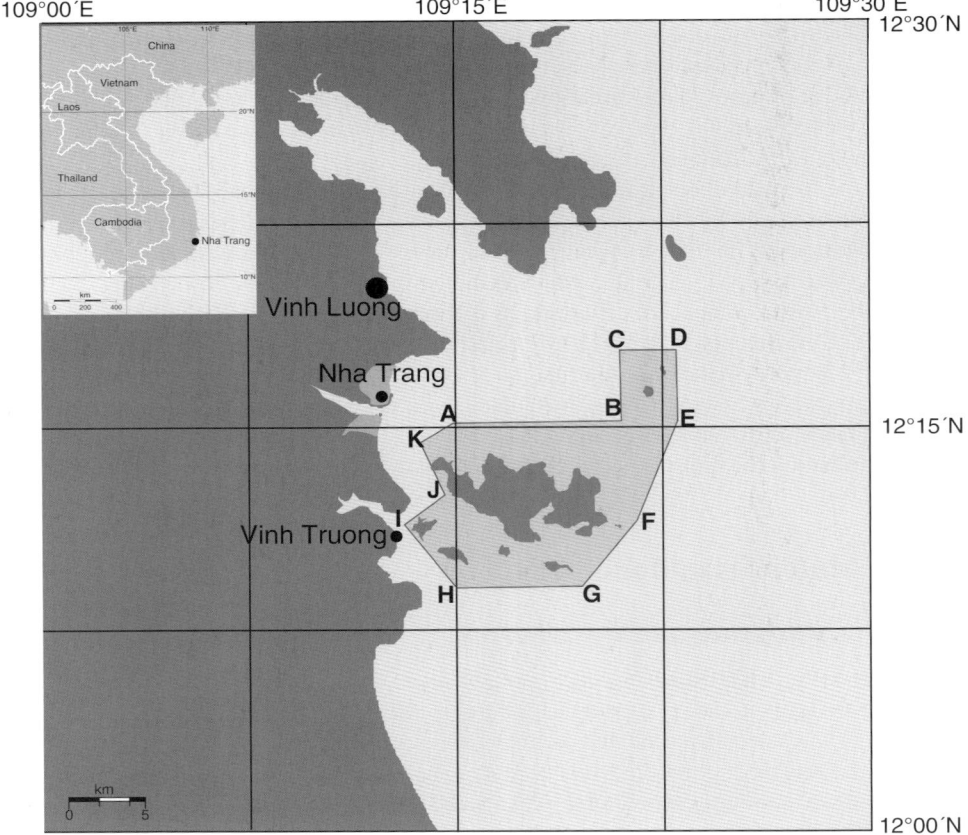

Figure 15.1 The location and boundaries of the Nha Trang Bay marine protected area, Vietnam. (To see this figure in colour, please see colour plate 11.)

of their fishing grounds, thereby negatively influencing their fishing activities. Assessing the effect of the establishment of the NTB-MPA and the ban on trawlers should ideally have been based on data concerning how fishermen chose locations for fishing after the introduction of the ban, and how those choices affected their catches and their earnings. Unfortunately, data necessary for this kind of analysis was unavailable, and as a second best solution, data on costs and revenues of the trawlers in 2005 and 2006 were gathered for this study.

The fisheries of Nha Trang contain a multiplicity of species. They are fished using a variety of fishing gear, including gillnet, longline, trawl and seine nets. The fisheries are regarded as open-access fishery. The use of dynamite and chemicals for fishing is prohibited, as is fishing near the shore at depths less than 10 m, and rules for minimum mesh size are implemented. Of the 2648 registered vessels, 725 (27%) are trawlers, including both single trawl and pair trawlers. Trawl is thus considered to be one of the most important fishing methods in Nha Trang.

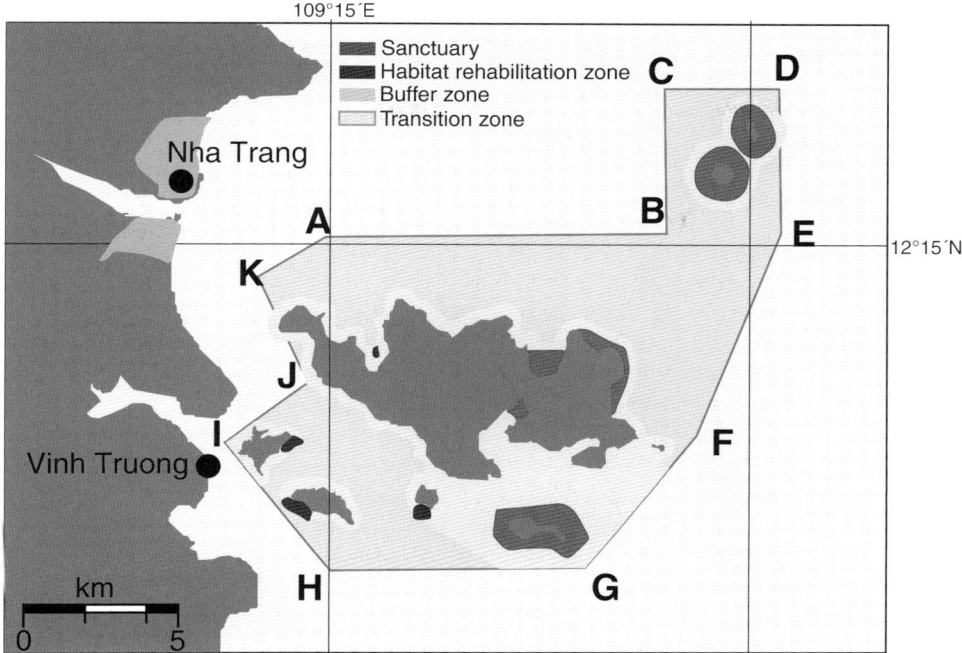

Figure 15.2 The zoning scheme of Nha Trang Bay marine protected area in 2005. (To see this figure in colour, please see colour plate 12.)

Trawl fisheries have existed in Nha Trang since 1975, when fishermen in the area disseminated knowledge of this technology which they had learned from Thai fishermen. An outstanding feature of trawl fishing is its capacity to catch fish at different depth layers. The investments in trawlers are quite low compared to other fleets, mainly due to the small-scale size of the vessels. Consequently, the number of trawlers increased sharply and the expansion of the trawling activities have placed a heavy burden on the marine resources and environment (Yen and Bernard, 2002).

The trawl fleets are primarily located in Vinh Truong and Vinh Luong communes, and they operate in different fishing grounds. Vessels in Vinh Truong often operate in the area outside the buffer zone and in the vicinity of nine islands in the NTB-MPA, while vessels in Vinh Luong operate in Nha Phu Lagoon – a short way north of Nha Trang city. Most vessels are relatively small and owner operated. The fishermen use very simple trawls which are often equipped with no, or very short wings.

Almost all trawlers fish year-round, mostly fishing at depths from 40 to 50 m. An important characteristic of this fishery is that most trips are only overnight, but sometimes trips of 3–4 days duration are made by vessels with larger gear and higher engine power (40–55 HP). Catch preservation techniques are very simple. In the afternoon when the fishermen depart, they take only ice. All of the catch is kept on ice in plastic baskets until they return early the next morning. The catch component of the trawl fleet consists of 12 ecological groups: cephalopods, demersal fish, small pelagic, mixed fish, anchovy, trash fish, coral fish, octopus, crabs, squid, cuttlefish and shrimp, and in which demersal fish, mixed fish, trash fish, crabs and shrimp dominate. More than 80% of the catch (both in terms of volume and revenue) consists of demersal fish, mixed fish, trash fish, crabs and shrimp.

15.2.2 Stochastic production frontier

15.2.2.1 Theoretical framework

The concept of frontier production function for efficiency measurement was first introduced by Farrell (1957). Since then, many authors have applied this to measure technical efficiency, which is considered to be the ability of a firm to produce maximum output possible from a given set of inputs and production technology. Charnes et al. (1978) developed a non-parametric approach (data envelopment analysis – DEA), which is based on a mathematical programming technique and produces a deterministic production frontier as it does not take into account random variations in the data. As a result, the estimates of efficiency may be biased, given that the production process is largely characterised by stochastic elements (Pascoe and Mardle, 2003).

Aigner et al. (1977) based their parametric approach on econometric techniques and introduced the stochastic production frontier (SPF). The parametric approach is considered to be the best method for assessing the technical efficiency of fishing vessels, as it involves stochasticity, which is especially important in natural resource and environmental economics. Differing from DEA, SPF requires the specification of technological characteristics of a production process. This provides information on the level of efficiency and the impacts of key inputs on levels of output. It is also different from the production function approach, which assumes that there are no differences in efficiency in the use of inputs between firms. Production frontier analysis can provide information on the relative efficiency of certain groups from the relationship between observed production and some ideal production (Greence, 1993).

The stochastic frontier model takes the form:

$$\ln q_{it} = x'_{it}\beta + v_{it} - u_{it} \quad (15.1)$$

where q_{it} is the output produced by firm i, x'_{it} is a vector containing the logarithms of input factors and β is a vector of estimated parameters.

The error term includes two components: v_{it} and u_{it}. They are distributed in different ways. The error term v_{it} is commonly assumed to be independently and identically distributed (iid), as $N(0, \sigma_v^2)$, and shows the exogenous stochastic shocks beyond the control of the firms. It can be positive or negative.

The error term u_{it} is a non-negative random variable which is often assumed to be independently and identically distributed as truncations (at zero) of the normal distribution $N(\mu_{it}, \sigma_u^2)$ and captures the technical inefficiency in production. The independent distribution between u_{it} and v_{it} allows the separation of statistical noise and technical inefficiency.

The random variable associated with technical inefficiency was further assumed as a function of firm-specific characteristics and period of observation (Battese and Coelli, 1995):

$$u_{it} = z_{it}\delta + w_{it} \quad (15.2)$$

where z_{it} is a vector of firm-specific characteristics and period of observation that affect technical inefficiency, δ is a vector of estimated parameters and w_{it} is the error term. The frontier model may include intercept parameters and period of observation in both the frontier model and in the inefficiency model if the inefficiency effects are stochastic.

The assumptions concerning the two error terms require using the maximum likelihood (ML) method to estimate parameters in the stochastic frontier model. The concept of ML estimation is based on the idea that the sample has been generated from some distributions rather than others. The ordinary least squares (OLS) can obtain consistent estimators of the slope parameters; however, the OLS estimator of the intercept parameters is biased downwards (Coelli et al., 2005).

The parameters in the stochastic frontier model were estimated by ML with the variance parameterised by: $\sigma^2 = \sigma_v^2 + \sigma_u^2$ and $\gamma = \frac{\sigma_u^2}{\sigma_v^2 + \sigma_u^2}$ (Battese and Corra, 1977). It should be noted that γ lies between zero and one. If $\gamma = 0$, all deviations from the frontier are owing to noise, while $\gamma = 1$ means that all deviations are owing to technical inefficiency.

From the results of the estimated model, the output of each firm would be compared with the best output and the technical efficiency could be specified as (Coelli et al., 2005):

$$TE_i = \frac{q_{it}}{\exp\left(x'_{it}\beta + v_{it}\right)} = \frac{\exp\left(x'_{it}\beta + v_{it} - u_{it}\right)}{\exp\left(x'_{it}\beta + v_{it}\right)} = \exp(-u_{it}) = \exp(-z_{it}\delta - w_{it})$$

(15.3)

when $u_{it} = 0$, the firm i lies on the stochastic frontier and is technically efficient. Otherwise, when $u_{it} > 0$, the firm i lies beneath the frontier, hence the firm is inefficient ($0 < TE < 1$).

15.2.2.2 Testing model specification

The structure of the model needs to be confirmed by a number of tests. These require imposing restrictions on the model and using the generalised likelihood ratio (LR) to determine the significance of the restriction. LR is given by:

$$LR = -2\{\ln[L(H_0)] - \ln[L(H_1)]\}$$

(15.4)

where $L(H_0)$ and $L(H_1)$ are the values of the log-likelihood function under null and alternative hypotheses, respectively. LR has a Chi-squared (χ^2) distribution with the number of degrees of freedom provided by the number of restrictions imposed. This is the difference between the parameters estimated under H_1 and H_0, respectively.

15.2.2.3 Data

The dataset used in this study was obtained from a survey of small-scale trawlers in the Nha Trang area. In this fishery, the imposition of the NTB-MPA and trawl ban was expected to directly affect the trawlers in Vinh Truong. The lack of data on activities of trawlers in Vinh Truong before creation of the MPA obliged us to conduct the survey of trawlers in both Vinh Truong and Vinh Luong. This in order to compare the differences of technical efficiency between the vessels fishing outside the buffer zone, in or near the MPA, and those fishing in unprotected areas.

Table 15.1 Descriptive statistics of the main trawler variables 2005 and 2006 (US$ 1 = 16 015 VND).

Name	N	Mean	Standard deviation	Minimum	Maximum
Revenue (1000 VND)	130	205 830	82 966	48 503	479 700
Length (m)	130	11.6	1.8	8.2	16
Engine power (HP)	130	35.3	12.5	20	56
Crew size (persons)	130	3.1	0.7	2	5
Days at sea (days)	130	210.8	41.1	100	286
Age of vessel (years)	130	8.3	5.4	0.5	28
Fishing experience (years)	130	28.4	8.1	9	50
Household size (persons)	130	6.2	1.7	1	10

The survey was undertaken with independent random samples in order to obtain balanced panel data of 65 trawlers in 2005 and 2006. Of the total 65 vessels, 36 were home ported in Vinh Truong, and 29 were home ported in Vinh Luong. Since the data were collected through a personal household interview, a questionnaire was designed for this purpose. In addition to the information on costs and earnings, information on vessel and skipper characteristics was collected. A summary of descriptive statistics for key variables used in the stochastic frontier and in the inefficiency model is given in Table 15.1.

The revenue was inflated to the 2006 value by the consumer price index (CPI). It ranged from 48 503 thousand VND to 479 700 thousand VND per year, with the mean equal to 205 830 thousand VND. Compared to the average annual revenue of a longliner, which was 568 250 thousand VND (Long et al., 2006), and a gillnetter, which was 851 333 thousand VND (Kim Anh et al., 2006), it can be seen that the revenue of trawlers was relatively small. The main reasons for these differences were that the trawlers were smaller and mostly operated near shore. A number of vessels that had previously operated in longline and gillnet fishing changed gear and switched into trawling, since their technological characteristics could not meet the requirements for offshore fishing.

The majority of trawlers in the two communes were small-scale vessels. The vessel length ranged from 8.2 to 16 m, with an average length of 11.6 m. Similarly, the engine power, measured in horsepower (HP), ranged from 20 to 56 HP, with the mean being 35.3 HP. Because of this, the fishermen can fish only in waters close to shore. Many fishermen did not remember or know the year of construction of their vessels because they often used second hand vessels which had been bought from other fishermen. Thus, the age of vessel used in this chapter was the duration of use by the present owner, which varied from 0.5 to 28 years with a sample mean of 8.3 years. Crew size was small, averaging 3.1 persons with a range of 2–5. The annual average number of days at sea, including both travelling and fishing time, is 210.8 and ranged from 100 to 286 days. The fishing experience was measured by the number of years the fishermen had been involved in fishing activity. Clearly, the skippers were highly experienced, with an average 28.4 years of fishing experience. Household size was relatively high; on average, there were 6.2 persons per household.

15.2.2.4 Econometric specification

In most studies of efficiency, the output is a physical measure of volume, since individual outputs can be identified in the production process of many industries. Fisheries are often characterised by mixed outputs due to different species in the catch. Thus, the landed weight has been used as the output only for single species fisheries, whereas the value of catch is a common proxy for output when multi-species fisheries are examined (Pascoe and Mardle, 2003).

In fisheries economics, uncertainties from variations in resource abundance cause the production frontier to become stochastic. As a result, the main inputs used in the frontier often represent capital, capital utilisation, labour utilisation and stock abundance (Kirkley et al., 1995; Sharma and Leung, 1999; Grafton et al., 2000; Pascoe and Coglan, 2002). The capital level can be measured in terms of monetary investments or in terms of physical inputs (gross register tonnage, length, HP). The choice between these two measures depends on the availability of data and the expectation to capture all inputs used (economic measure), or some of the inputs used (physical measure). Pascoe et al. (2003) found that physical measures were neither better nor worse than economic measures. They emphasised that physical measures were even more robust, and helped in conducting the efficiency analysis on a larger scale.

In most empirical applications, a translog function is applied since it is flexible and allows more scope for substitution among production inputs. However, an LR test undertaken shows (see below) that the Cobb–Douglas function is in this case the most appropriate function form for SPF. Thus, the functional form of the model is specified as:

$$\ln(\text{Revenue}_{it}) = \beta_0 + \beta_1 \ln(\text{Length}_{it}) + \beta_2 \ln(\text{Crewsize}_{it}) + \beta_3 \ln(\text{Days}_{it}) + v_{it} - u_{it} \qquad (15.5)$$

In this study, physical inputs – vessel length and number of days at sea – are used as proxy measures of capital invested and capital utilisation in the fisheries. The crew size representing labour force is also included in the model as a variable input. Output variables are represented by annual revenue due to the multi-species nature of trawl fisheries.

For the inefficiency model, a number of variables are used to investigate the determinants of technical efficiency of trawlers. The choice of variables depends on data availability and on the statistical significance of parameter γ. The key inputs used in this chapter are (1) engine power measured in HP, (2) age of the vessel in years (the ownership of the vessel by the present owner), representing vessel characteristics, (3) fishing experience in years, representing skipper characteristics and (4) the household size in persons, representing socio-demography. In addition, a dummy variable of 1 for the vessels from Vinh Truong and 0 for Vinh Luong is utilised to distinguish characteristics of each location and resource conditions on the fishing grounds. The operating characteristic of the vessel is introduced by a dummy variable of 1 for the vessel taking a trip of one night and 0 for the vessel taking a trip of 3 or 4 days. Change in technical efficiency over time is covered by a dummy variable for the year 2005.

$$\begin{aligned} u_{it} = &\ \delta_0 + \delta_1 \ln(\text{vessel age}_{it}) + \delta_2 \ln(\text{HP}_{it}) + \delta_3 \ln(\text{fish exp}_{it}) + \delta_4 \ln(\text{household}_{it}) \\ &+ \delta_5 D_{\text{trip}} + \delta_6 D_{\text{fg}} + \delta_7 D_{2005} + w_{it} \end{aligned} \qquad (15.6)$$

Table 15.2 Hypothesis testing.

Null hypothesis	$-2[\ln\{L(H_0)\} - \ln\{L(H_1)\}]$	Number of restrictions	Critical χ^2
$\gamma = \delta_0 = \delta_1 = \delta_2 = \delta_3 = \delta_4 = \delta_5 = \delta_6 = \delta_7 = 0$ (No technical inefficiency)	37.14	9	16.274[a]
$\beta_4 = \beta_5 = \beta_6 = \beta_7 = \beta_8 = \beta_9 = 0$ (Cobb–Douglas frontier)	0	6	12.59
$\delta_1 = \delta_2 = \delta_3 = \delta_4 = \delta_5 = \delta_6 = \delta_7 = 0$ (Truncated normal distribution)	44.52	7	14.07

[a]Kodde and Palm (1986) statistic for a one-sided livelihood ratio test.

15.3 Results

15.3.1 Econometric results

The production frontier and inefficiency models were estimated using FRONTIER 4.1 (Coelli, 1996). Likelihood ratio tests are presented in Table 15.2.

The first test focuses on whether there exists a technical inefficiency in the model. It is equivalent to testing the null hypothesis $\gamma = \delta_0 = \delta_1 = \delta_2 = \delta_3 = \delta_4 = \delta_5 = \delta_6 = \delta_7 = 0$. Since the alternative hypothesis is that $0 < \gamma < 1$, the test has an asymptotic distribution, which is a mixture of χ^2 distribution $0.5\chi^2_{p-1} + 0.5\chi^2_p$ where p is the number of restrictions involved in the null hypothesis. The critical values of the test can be obtained from Kodde and Palm (1986). In this chapter, the test rejects the null hypothesis at 5% of significance.

For other tests, the null hypothesis is tested by using standard χ^2 distribution. The null hypothesis that the assumption of a truncated normal distribution was more appropriate than the inefficiency model, $\delta_1 = \delta_2 = \delta_3 = \delta_4 = \delta_5 = \delta_6 = \delta_7 = 0$, was also strongly rejected.

The results from the LR tests suggest that the Cobb–Douglas production frontier was the most appropriate functional form. OLS could not be used in this study, and ML estimation should therefore be suitable for obtaining efficient estimates.

All parameter estimates of the SPF (Equation 15.5) and the inefficiency model (Equation 15.6) are presented in Table 15.3. Almost all of these parameters are statistically significant at 1 or 5% level of significance.

15.3.2 Technical efficiency

The technical efficiency score ranges from 0.54 to 1, with the mean equal to 0.83. The value of γ in the models is 0.134, and this is statistically significant at the 1% level. It suggests that the variance of technical inefficiency effects accounts for only 13% of the total output variance, or we can say that the output variability of the vessels is dominated by the random variation rather than by technical inefficiency. This is not a surprising result, since fishing is sensitive to weather, resource and environment conditions.

The factors affecting technical inefficiency could be explained by the algebraic sign and the significance of the estimated coefficients. The coefficients in the inefficiency model

Table 15.3 Stochastic production frontier model.

	Coefficient	t-ratio
Stochastic frontier model		
Constant	4.293	5.995[b]
ln(Length)	0.438	2.509[a]
ln(Crewsize)	0.211	1.965[a]
ln(Days)	1.264	12.059[b]
Inefficiency model		
Constant	0.723	1.374
ln(Vessel age)	0.011	0.301
ln(HP)	−0.318	−2.902[b]
ln(Fish exp)	−0.569	−0.562
ln(Household)	0.221	3.283[b]
D_{trip}	0.509	3.849[b]
D_{fg}	−0.214	−3.155[b]
D_{2005}	0.013	0.625
Variance parameter		
σ^2	0.039	7.039[b]
γ	0.134	5.836[b]
Mean technical efficiency	0.833	

[a]Significant at 5% level.
[b]Significant at 1% level.

would express the direction of the effects of corresponding factors on technical efficiency. Note that the positive sign will imply a negative effect and a negative sign will imply a positive one. With the exception of the age of vessel variable, skippers' experience variable and the dummy variable for the year 2005, remaining variables in the inefficiency model are statistically significant. The coefficient for engine power is negative suggests a positive effect of engine power on the vessels' technical efficiency. The negative coefficient for the dummy variable for the commune suggests that trawlers in Vinh Truong are more technically efficient than those in Vinh Luong. The coefficients for household size and the dummy variable for the length of a fishing trip have a positive sign, implying that the household size has a negative impact on technical efficiency and the vessels which took a trip of 3 or 4 days were more efficient than those that took a trip of one night.

15.3.3 The elasticity and returns to scale

The relation between inputs and output can be investigated by elasticity and returns to scale analysis. Let us now analyse the contribution of each input variable to the value of the output by its elasticity, and examine the returns to scale that measures the change in output corresponding with the change in the use of all inputs. As the Cobb–Douglas function was applied for the frontier, the coefficients are information to report the output–input elasticity for each of the three inputs. The input that makes the largest contribution to the value of the output is the number of days at sea. It has elasticity of about 1.26, so an increase of 10% in the number of days at sea can increase revenue by 12.6%. The output/revenue elasticity of length of vessel is 0.43 and crew size makes the smallest contribution to revenue; its elasticity is 0.21.

The returns to scale for trawlers in Nha Trang fisheries was greater than 1, so an expansion of vessel length, crew size and number of days at sea would increase revenue with the increasing returns to scale. In general, an increase of 10% in all inputs will result in a revenue increase of 19%. However, we do not know if costs would increase even more.

15.4 Discussion

15.4.1 *Factors affecting efficiency and the fishing process*

The inefficiency model and the frontier model can help us to investigate the factors affecting efficiency and the production process. From the inefficiency model, the facts that engine power has a positive impact and that the household size has a negative impact on efficiency are not surprising. A trawl is a mobile fishing gear, so greater engine power can allow more gear to be worked. Greater engine power also helps the vessels to access the fishing ground more quickly, which then reduces the fishermen's travelling time and fuel cost. With regard to the impact of household size, as household size increases, the number of dependents increases and the demand for family food also increases. This may suggest that fishermen then reserve a larger part of their own catch for family food, thus reducing the number of fish that can be sold. Consequently, revenue from each fishing trip then also decreases.

The result also shows that technical efficiency was improved if a vessel took a trip of 3 or 4 days instead of a trip of one night. An explanation for this is that the vessels taking the trip of 3 or 4 days often accessed fishing grounds that were further away and often larger, offering more resources for exploitation. The dummy variable for the year 2005 was not significant, so there was no difference in technical efficiency between 2005 and 2006. The effect of age of vessels on efficiency was also found not to be statistically significant. This may appear strange, but it is probably because vessels are regularly repaired and maintained. Regular maintenance can help to improve the conditions of a vessel and to extend its lifetime. Thus, older vessels may still obtain catches similar to the new vessels. However, recall that we have used the time of ownership of vessel as a proxy for vessel age, since exact reliable data are not available. The extent to which this is a source of bias is difficult to determine.

Another interesting finding from the inefficiency model that should be discussed is that the technical efficiency varies with the fishing grounds. The main fishing ground of fishermen in Vinh Truong is around Nha Trang Bay, mainly among the nine islands of NTB-MPA. The fishermen in Vinh Luong often fish in Nha Phu Lagoon (a short way north of Nha Trang city). The sign of area dummy variable suggests that trawlers in Vinh Truong are more technically efficient than those in Vinh Luong. The area dummy variable indirectly accounts for area variations in the resource stocks, fishing practices and labour arrangements. For mixed catch fisheries in developing countries, a direct estimate of the abundance of each resource stock is usually not available. In such cases, variation in relative aggregated catch rates may indicate changes in stock abundances, to be studied by use of dummy variables for different months, years or areas (Pascoe and Mardle, 2003). Then, in our case what is the reason for the higher technical efficiency of trawlers operating near the NTB-MPA? To what extent is this due to the creation of the NTB-MPA? We discuss this issue in the following section.

In the frontier model, all coefficients estimated for parameters were significant and showed their impacts on the production process. The elasticity and return to scale analysis has shown that the output elasticity for the number of days at sea is greater than 1. This result may seem strange, since it is not clear how the output/revenue can increase by a higher rate than the increase in input/fishing effort while resources are considered overexploited (Yen and Bernard, 2002). However, this is mainly a cross-sectional analysis for just 2 years; feedback effects from the stocks are not explicitly included, only indirectly through the dummy variables in Table 15.3. Trawlers use mobile fishing gear, and an increase in the number of days at sea allows them to operate under a wider range of conditions. The NTB-MPA project's survey of fishing vessels fishing in or near NTB-MPA (Dinh et al., 2005) has shown that for trawlers with a home port outside the MPA but which are fishing here, catch per unit of effort (CPUE) for fish and acetes, a small shrimp about 1.5 cm long, has increased considerably. From 2003 to 2005, CPUE for fish increased from 7.91 to 9.28 kg h^{-1} and for acetes from 7.73 to 25.43 kg h^{-1}. For other kinds of fishing gear, such as stick-held dip net and lift net of fishermen living outside the MPA, and stick-held dip net, trammel net, lift net and longline of fishermen living inside the MPA, CPUE values decreased from 2003 to 2005.

The elasticities of vessel length and crew size on the output revenue were less than 1, and they resulted from the characteristics of trawl fisheries. The length of vessels may be considered a proxy for carrying capacity. As mentioned above, each trip of a small-scale trawler often lasts only one night, so the number of hauls is small. Furthermore, preservation methods are quite simple and do not require the use of larger vessels. The elasticity for crew size was characterised by the share system between the crew and owner. Crew members receive 50% of income after deduction of all operating expenses. It is this share system that offers incentives for the crew to work hard for income and helps the vessel owners to use labour more efficiently.

15.4.2 *Technical efficiency, stock abundance and management regime effects*

The results above suggest that there is no difference in the technical efficiency of trawlers between 2005 and 2006. However, examination of technical efficiency of trawlers fishing in different areas indicates that fishermen working in areas adjacent to the NTB-MPA are still more efficient than those working in unprotected areas. Is this a result of the spillover effect from the creation of the MPA? This is not an easy question to answer. Unfortunately, we do not have data on catch and revenue of trawlers before the imposition of the MPA, so a comparison of technical efficiency of trawlers before and after creation of the NTB-MPA could not be made. However, the data on fish stock abundance from the report of the NTB-MPA project can help us to investigate part of this problem.

The benefits of MPAs for adjacent fisheries have been found in several MPAs around the world, and these benefits have also attracted great interest from fishery managers in Nha Trang. It has been reported that in some cases the effects on the fishery developed quickly and could be detected within 5 years of an MPA being established (Gell and Roberts, 2003).

With the zoning scheme and the trawl ban in the NTB-MPA, it was expected that the fish stock abundance in the core zone would increase and that this in turn would benefit the stock in exploited areas due to spillover effects. After 4 years of protection, there were expectations of increasing fish stock abundance, but these seem to have been unrealistic.

The assessment of ecological status at eight locations within NTB-MPA waters in the period 2002 and 2005 concluded that 'assessment of recovery of coral reefs as measured by cover of living corals indicated that the cover remained relatively stable in the NTB-MPA overall' (Tuan et al., 2005a, p. 2). However, fish abundance tended to decline in the MPA overall, and in most of locations, although the declines in some locations were not significant. The abundance of larger fish, from 21 to 30 cm and larger than 30 cm, was at a very low level. Fish larger than 30 cm became even rarer in the area in 2005. The quantity of small fish (below 10 cm) contributed almost all of the total fish abundance. These results were also consistent with those from an assessment of biodiversity of the NTB-MPA (Tuan et al., 2005b). From this assessment, it was revealed that from 2002 to 2005 there had been little overall change in richness and abundance of corals, significant declines in richness and abundance of fish, and non-significant declines in invertebrates, together with increases in macro-algae. Tuan et al. (2005b) further indicated that the decline in richness and abundance of fish were unlikely to be a sampling artefact, and more likely reflected the continuing intense fishing effort within MPA waters.

Focusing on comparisons between core zone and non-core zones, Tuan et al. (2005a) found that slightly higher fish abundance was present in the core zones. These differences were not significant, so this did not support the view that there were spillover effects. Only in the core zone in Hon Mun, where local authority made a great effort in controlling fishing and anchor damage, as well as educating and increasing the awareness of environmental issues to tourists and dive boat operators, did the stock abundance became relatively higher.

No systematic data are available on stock abundance and catch of trawlers in Vinh Luong fishing in the Nha Phu Lagoon to assess changes over time. However, Strehlow (2006) asserted that the introduction of new fishing techniques and increasing fishing pressure had led to a decline in fishery resource in recent years in the Nha Phu Lagoon.

The decrease in fish stock abundance in the non-core zone, and even in the strictly protected core zone of the NTB-MPA, indicates that overfishing has been present in the NTB-MPA. It is likely that this has occurred in the Nha Phu Lagoon as well.

15.4.3 Policy implications

Fishing activities are of significant importance in sustaining livelihoods and supplying food for a large number of people living in coastal areas of developing countries. Fisheries in Nha Trang, Vietnam, are characterised by small scale and fishing households rather than fishing companies. In the late 1990s, about 80% of the households in coastal communities of this city relied on fish capture and aquatic resource use (STREAM, 2000). Thus, changes in economic conditions as well as fishery management policies may not only greatly affect fleet activity and fish stock health but also impact local communities.

The findings of this study may provide fishery managers with some policy implications. Firstly, the results from frontier analysis can help to investigate the factors affecting technical efficiency. In this analysis, the trip length and the skipper's household size are the most significant factors. The results of the analysis also reveal that technical efficiency varies across vessels and that there are opportunities for fishermen in the communes of both Vinh Truong and Vinh Luong in order to raise technical efficiency given an appropriate use of all inputs. The frontier analysis also provides information on the relationship between input and production process. This information is a good guide for fisheries managers when formulating management and regulatory conditions. The increasing returns to scale suggest that any increase in all inputs used, say by 1%, results in an increase in output of more than 1%. The number of days at sea is the most important factor positively affecting the output revenue. This result indicates that from a resource conservation objective point of view, policies aiming to induce the fishermen to reduce the number of days at sea would be effective since they would reduce revenue and catch by a rate higher than the rate of decrease in the number of days at sea. Programmes that help to generate alternative income and to develop the socioeconomic situation could be implemented to mitigate resource overuse.

Secondly, in the short run, the establishment of an MPA in Nha Trang Bay with a trawl ban would impose losses on the trawlers due to the reduction of area for fishing and the increase in distance from the main port to the alternative grounds. However, our analysis has pointed out that even influenced by the trawl ban the trawlers fishing in the vicinity of the NTB-MPA are somewhat more efficient with respect to revenue than those fishing in the Nha Phu Lagoon. It implies that the alternative grounds to the banned grounds can still sustain the activity of the trawl fleet affected by the ban. Nevertheless, a likely decrease in fish stock abundance of both areas, the NTB-MPA and the Nha Phu Lagoon, over the last few years may indicate that stocks have suffered from overfishing. Resolving this will be a challenge for resource management.

The findings involving the trawlers in Nha Trang lead to further discussions related to open-access fisheries and the role of an MPA in fisheries resource management. The mean technical efficiency of the trawlers in Nha Trang is relatively high, with a value of about 0.833; but is an increase of technical efficiency in open-access fisheries always an objective to pursue? Obviously, the case of trawlers in Nha Trang indicates that increases in technical efficiency can raise catches and incomes to fishermen in the short run. However, without a feasible management regime, in the long run, the increased technical efficiency will increase the catch in open-access fisheries and put strains on fish stock capacity, diminishing the numbers. This consequence results to some extent from the link between employment, poverty and resource degradation. The close link between poverty and resource degradation has been raised as a main challenge for natural resource management in developing countries, and this is also the case for Vietnam.

In the period 2002–2005, the NTB-MPA project helped villagers who are affected by zoning within the NTB-MPA by providing financial support, technical skills, market information and generating alternative income for them in agriculture, tourism and aquaculture. A monitoring programme was also established to assess the success of the project. With 5300 inhabitants, in about 1000 households in the area with a relatively low standard of living, the NTB-MPA still faces enormous pressure from human activities. Credit regimes aiming to provide loans to villagers with a flexible payment schedule were recognised as a well-run measure in changing the proportion of occupations. The percentage of men participating in fishing reduced from 79.8% in 2002 to 76% in 2005. The percentage of men working in other activities, such as the tourism service, aquaculture and raising of animals, increased from 5.6 to 14% over those years (Thu, 2005). It is clear that loans from the project helped many households to obtain more income from new employment opportunities. Nevertheless, there will remain a number of fishermen living outside the MPA and exploiting the resources there. Trawl fishermen in Vinh Truong are an example of this. Monitors had the opportunity to travel and observe the fishermen in Vinh Truong and Vinh Luong during the time that this study was conducted. For many households in these areas, fishing is the only livelihood and this imposes serious challenges on fisheries management. Many fishermen would like to try alternative livelihood options, which are less risky than fishing and which also help them to earn more income. However, in reality, the fishermen have very limited opportunities to switch to alternative employment and to access long-term credit and bank loans. As a result, labour continues to be focused into the fishing sector. If in the past perhaps 10% of the population went for fishing, nowadays 50% go for fishing (Strehlow, 2006). The cycle of poverty and resource degradation continues to occur.

The Department of Fishery in the Khanh Hoa province is planning to create an MPA in the Nha Phu Lagoon. This will be the second MPA in this province. Trawlers in Vinh Luong and other vessels operating in the Nha Phu Lagoon will be influenced by this new MPA. Policy planners, of course, cannot avoid the fact that these vessels will work the waters in the area open for fishing and that this could impose a serious burden on fish stocks in the newly designated area. The case of NTB-MPA should be studied as a lesson for the MPA in Nha Phu Lagoon, so that fishery managers can implement the factors which result in a successful MPA. The experience of the NTB-MPA shows that the authority must intervene when necessary to protect resources from external threats, such as illegal fishing, poaching and trawling close to the shore in all locations of the MPA. Additionally, the development of other sectors, such as agriculture, aquaculture and tourism, to attract labour and capital away from fisheries is essential if marine resources are to be sustainably maintained. The success of the MPA also requires that more attention should be paid, as it was in the Hon Mun zone, to the education of fishermen and local communities so as to comply with and support the objectives of the MPA.

15.5 Conclusions

This chapter provides a study of efficiency of trawlers in the NTB-MPA. Although data on costs and revenues of the vessels before the creation of the MPA were unavailable to this chapter, a comparison of efficiency between vessels fishing in the vicinity of the MPA and the vessels fishing in an unprotected area provides valuable insight into how these areas function. Without a spillover effect from the MPA to fishing areas, a higher level of efficiency of vessels in the vicinity of the MPA implies that an alternative ground to the banned ground can still sustain the trawlers affected by the ban. Although in many cases MPAs have shown spillover benefit for fisheries by increasing the abundance and the size of animals (Gell and Roberts, 2003), creation of the NTB-MPA seems not to have provided effective protection for fish resources. It is obvious that an MPA with a trawl ban is not sufficient for the purposes of resource conservation. Thus, to help ensure the success of the MPA, it is necessary to identify the most appropriate enforcement that can contribute to effective management and to the development of strategies that work best for local communities. The success of an MPA hinges on enforcement and the support of local communities, and not particularly on habitat type, geographical location, the kind of fishery involved or the technological sophistication of management. This is the lesson that can be drawn from other successful MPAs (Gell and Roberts, 2003) as well as from this study on the NTB-MPA.

Acknowledgements

The project was funded by the Global Environmental Facility (GEF) through the World Bank, the Danish International Development Agency (DANIDA) – Government of Denmark and the Government of Vietnam. The Ministry of Fisheries, Khanh Hoa People's Committee and IUCN – the World Conservation Union – were responsible for implementing this project. We thank the Norwegian Agency for International Development Cooperation (NORAD), Project SRV2701, for funding the trawler survey, Mr Tran Trong Long for research assistance, colleagues at our two institutions for a cooperative attitude and three anonymous reviewers and the editors for valuable comments.

References

Aigner, D., Lovell, C.A.K. and Schmidt, P. (1977) Formulation and estimation of stochastic frontier production function models. *Journal of Econometrics* **6** (1), 21–37.

Alban, F., Appere, G. and Boncoeur, J. (2006) *Economic Analysis of Marine Protected Areas: A Literature Review*, EMPAFISH Project, Booklet no. 3.

Anderson, L.G. (2002) A bioeconomic analysis of marine reserves. *Natural Resource Modeling* **15** (3), 311–334.

Battese, G.E. and Coelli, T. (1995) A model for technical inefficiency effects in a stochastic frontier production function for panel data. *Empirical Economics* **20** (2), 325–332.

Battese, G.E. and Corra, G.S. (1977) Estimation of a production frontier model: with application to the pastoral zone of Eastern Australia. *Australian Journal of Agricultural Economics* **21** (3), 169–179.

Charnes, A., Cooper, W.W. and Rhodes, E. (1978) Measuring the efficiency of decision making units. *European Journal of Operational Research* **2** (6), 429–444.

Coelli, T. (1996) A guide to frontier 4.1: a computer program for stochastic production and cost function estimation. Centre for Efficiency and Productivity Analysis (CEPA) Working Papers No. 7/96, University of New England, Australia.

Coelli, T., Rao, D.S.P., O'Donnell, C.J. and Battese, G.E. (2005) *An Introduction to Efficiency and Productivity Analysis*. Springer, New York.

Conrad, J.M. (1999) The bioeconomics of marine sanctuaries. *Journal of Bioeconomics* **1** (2), 205–217.

Dinh, H.B., Vu, N.P.U. and Quang, V.V. (2005) Results of fishing monitoring in Nha Trang Bay marine protected area North wind season (12/2004–01/2005). Technical Report to IUCN Hon Mun MPA Pilot Project, Nha Trang, Vietnam. Biodiversity Report No. 14.

Farrell, M.J. (1957) The measurement of productive efficiency. *Journal of the Royal Statistical Society* **120** (3), 253–290.

Flaaten, O. and Mjølhus, E. (2006) Nature reserves as a bioeconomic management tool – a simplified modeling approach. Working Paper Series in Economics and Management. Department of Economics and Management, Norwegian College of Fishery Science, University of Tromsø No. 04/06.

Gell, F.R. and Roberts, C.M. (2003) The fishery effects of marine reserve and fishery closures. Report, World Wild Fund, Washington, DC.

Grafton, R.Q., Squires, D. and Fox, K.J. (2000) Private property and economic efficiency: a study of a common-pool resource. *The Journal of Law and Economics* **43** (2), 671–714.

Greence, W.H. (1993) *Frontier Production Functions, EC-93–20.* Stern School of Business, New York University, New York.

Hannesson, R. (1998) Marine reserves: what would they accomplish? *Marine Resource Economics* **13** (3), 159–170.

Kim Anh, N.T., Flaaten, O., Tuan, N., Dung, P.T. and TramAnh, N.T. (2006) *A Study on Costs and Earnings of Gillnet Vessels in Nha Trang, Vietnam.* IIFET Portsmouth Proceedings, Portsmouth.

Kirkley, J.E., Squires, D. and Stand, I.E. (1995) Assessing technical efficiency in commercial fisheries: the Mid-Atlantic sea scallop fishery. *American Journal of Agricultural Economics* **77** (3), 686–697.

Kodde, D.A. and Palm, F.C. (1986) Wald criteria for jointly testing equality and inequality restrictions. *Econometrica* **54** (5), 1243–1248.

Long, L.K., Flaaten, O. and Kim Anh, N.T. (2006) Economic performance of off-shore long-line vessels in Nha Trang, Vietnam. *IIFET Portsmouth Proceedings*, Portsmouth.

Pascoe, S. and Coglan, L. (2002) The contribution of unmeasurable inputs to fisheries production: an analysis of technical efficiency of fishing vessels in the English channel. *American Journal of Agricultural Economics* **84** (3), 585–597.

Pascoe, S., Hassaszahed, P., Anderson, J. and Korsbrekke, K. (2003) Economic versus physical input measures in the analysis of technical efficiency in fisheries. *Applied Economics* **35** (15), 1699–1710.

Pascoe, S. and Mardle, S. (eds) (2003) Efficiency analysis in EU fisheries: stochastic production frontiers and data envelopment analysis. CEMARE Report 60, CEMARE, University of Portsmouth, UK.

Sharma, K.R. and Leung, P. (1999) Technical efficiency of the longline fishery in Hawaii: an application of stochastic production frontier. *Marine Resource Economics* **13** (4), 259–274.

STREAM (2000) *Poverty and Aquatic Resource in Vietnam: An Assessment of the Role and Potential of Aquatic Resources Management.* DFID-SEA ARMP, Bangkok.

Strehlow, H.V. (2006) *Integrated Natural Resources Management of Coastal Fisheries – The Case of Nha Phu Lagoon, Vietnam.* PhD Dissertation, University of Berlin, Germany.

Thu, H.V.T. (2005) Socio-economic impact assessment of the Hon Mun MPA project on local communities within the MPA. Technical Report to IUCN Hon Mun MPA Pilot Project, Nha Trang, Vietnam.

Tuan, V.S. (2002) National report of Vietnam for GCRMN EA SEA workshop. Regional workshop for East Asian Seas, Ishigaki, Japan, pp. 124–130.

Tuan, V.S., Long, N.V., Hoang, P.K., Ben, H.X. and DeVantier, L. (2005a) Ecological monitoring of Nha Trang Bay marine protected area, Khanh Hoa, Vietnam. Reassessment 2002–2005, Biodiversity Report No. 15. Technical Report to IUCN Hon Mun MPA Pilot Project, Nha Trang, Vietnam.

Tuan, V.S., Long, N.V., Hoang, P.K., Ben, H.X. and DeVantier, L. (2005b) Biodiversity of the Nha Trang Bay MPA, Khanh Hoa, Vietnam. Reassessment 2002–2005. Hon Mun marine protected area pilot project.

Whitmarsh, D., Pipitone, C., Badalamenti, F. and D'Anna, G. (2003) The economic sustainability of artisanal fisheries: the case of the trawl ban in the Gulf of Castellammare, NW Sicily. *Marine Policy* **27** (6), 489–497.

Yen, N.T.H. and Bernard, A. (2002) Hon Mun MPA pilot project on community-based natural resources management. *STREAM Journal* **1**, 1–3.

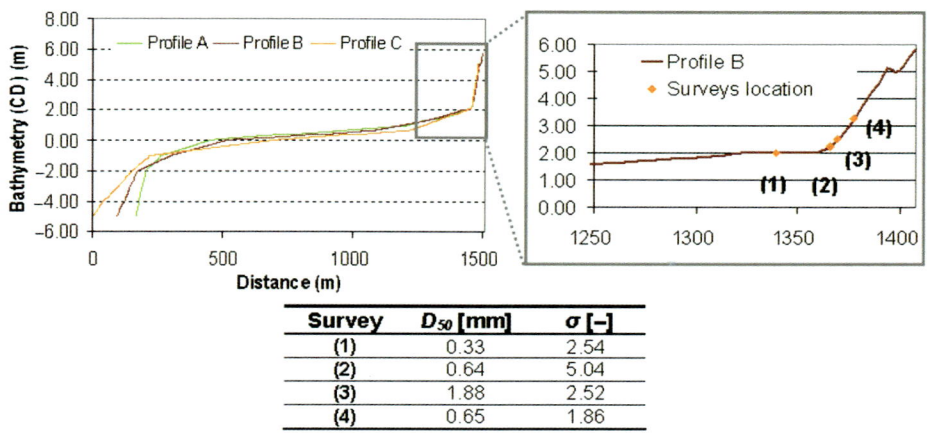

Plate 1 (Figure 3.2) Geometry of the beach profiles A, B and C; detail of profile B with location of the sedimentologic surveys (1) to (4); and sedimentologic parameters, D_{50} and σ, for each survey.

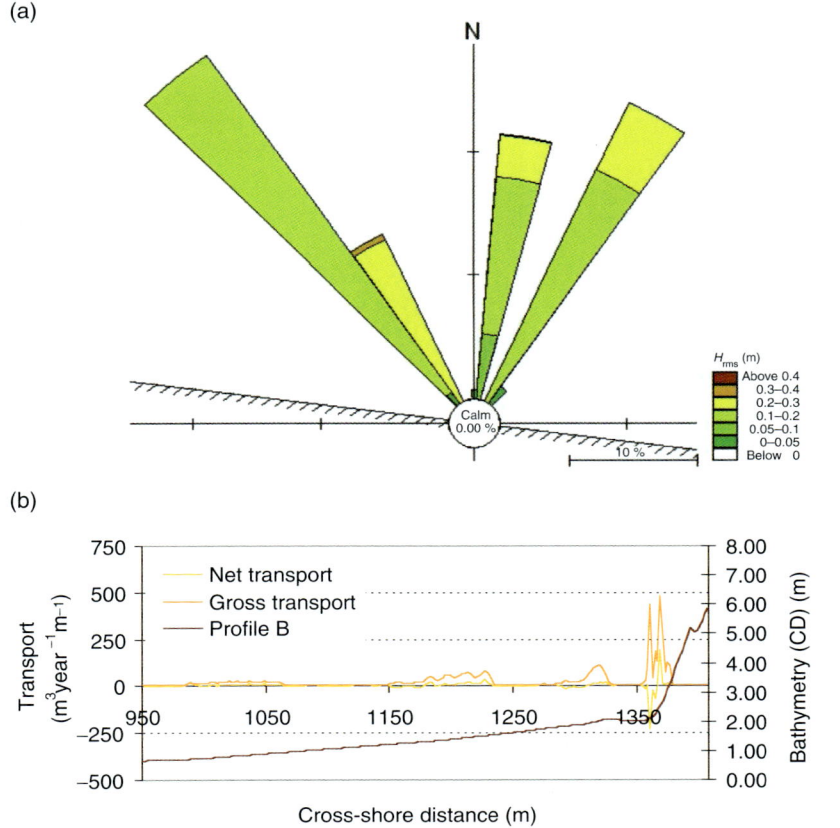

Plate 2 (Figure 3.6) Longshore sediment budget at profile B: (a) discretisation per H_{rms} class and 10° directional sector and (b) cross-shore distribution in the active profile.

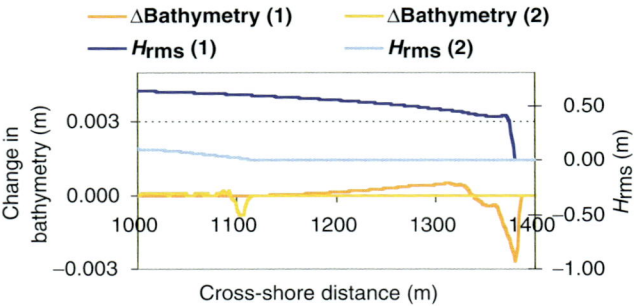

Plate 3 (Figure 3.8) Wave propagation and morphological variation along profile B due to the passage of a catamaran at speed 20 knots, at mean high sea level (1) and mean low sea level (2).

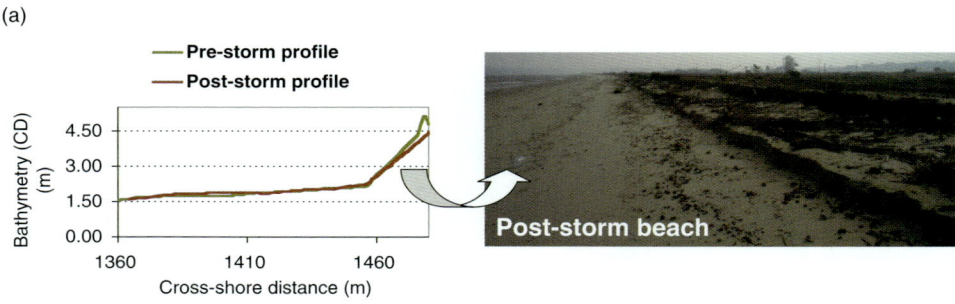

Plate 4 (Figure 3.9) Topo-hydrographic surveys of the pre- and post-storm profile C (a) and the numerical profile C (b).

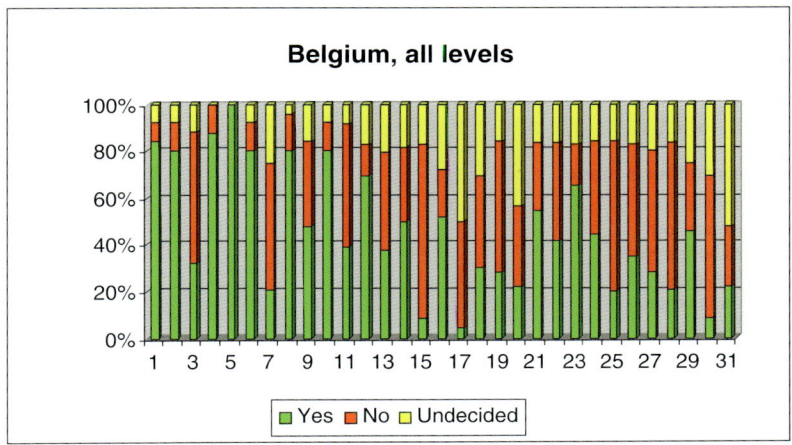

Plate 5 (Figure 6.1) The results from Belgium showing the all the responses for the 31 action levels.

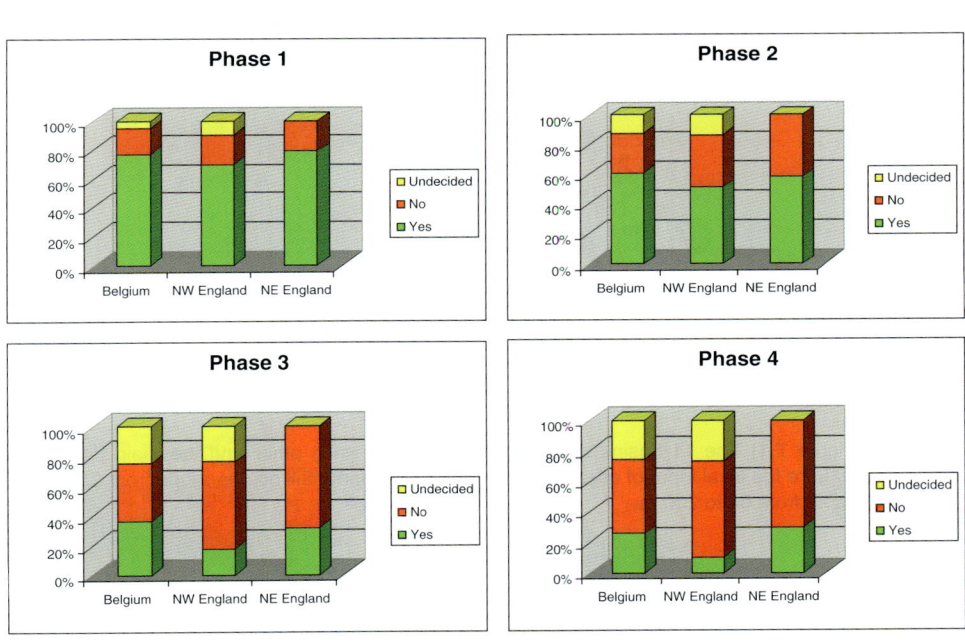

Plate 6 (Figure 6.2) An overview of all results from Belgium, NW and NE England, Phases 1–4.

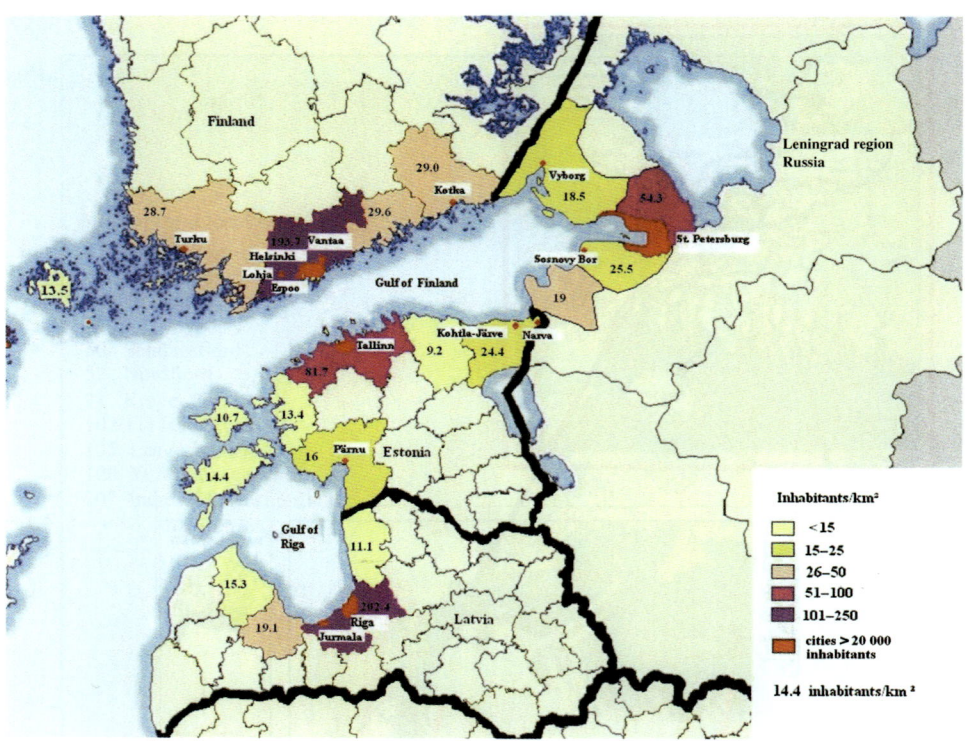

Plate 9 (Figure 11.1) Population densities (inhabitants/km^2) in the northeast of Baltic Sea in the period 1950–1960.

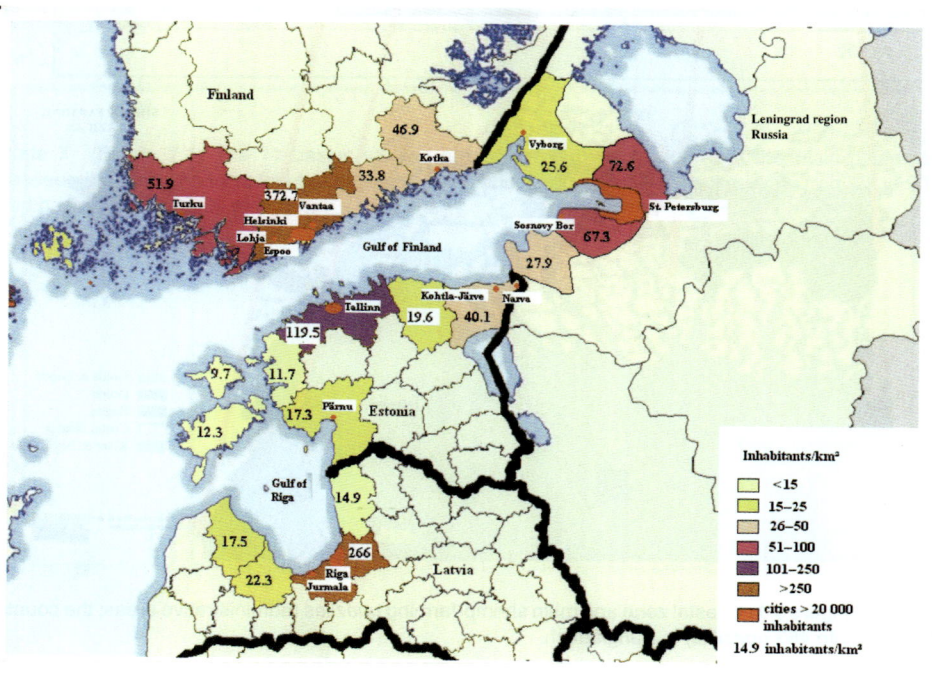

Plate 10 (Figure 11.2) Population densities (inhabitants/km^2) in the northeast of Baltic Sea in the period 2000–2005.

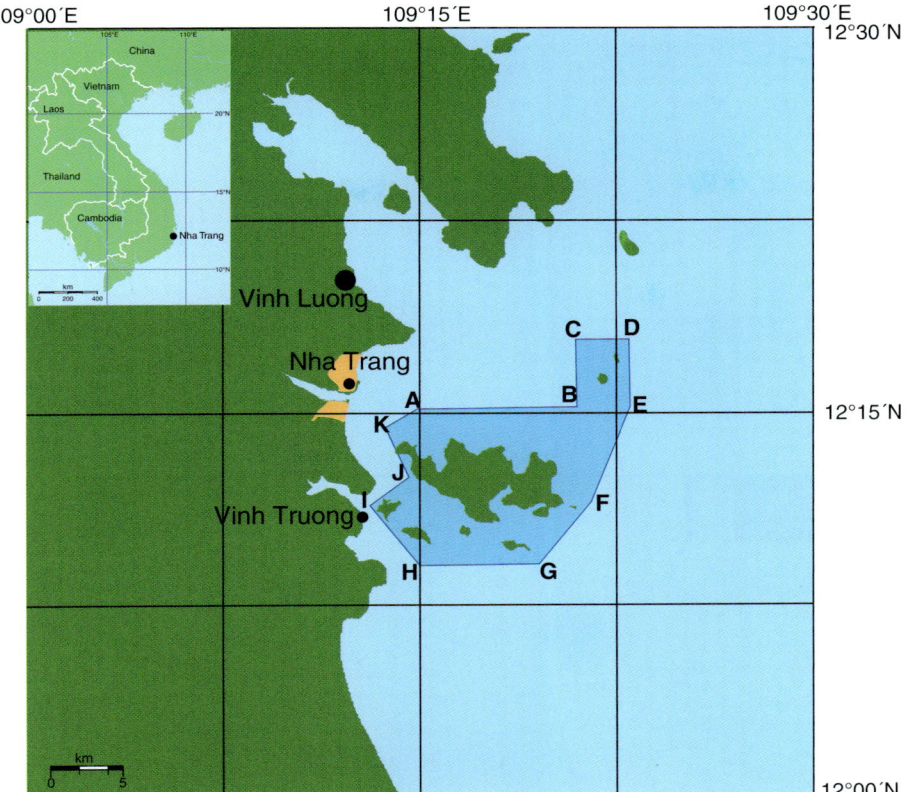

Plate 11 (Figure 15.1) The location and boundaries of the Nha Trang Bay marine protected area, Vietnam.

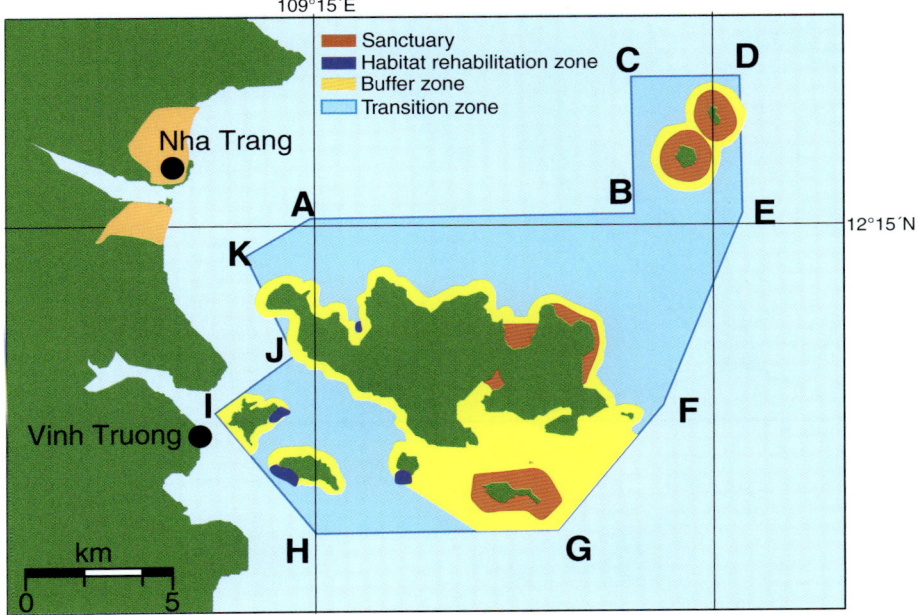

Plate 12 (Figure 15.2) The zoning scheme of Nha Trang Bay marine protected area in 2005.

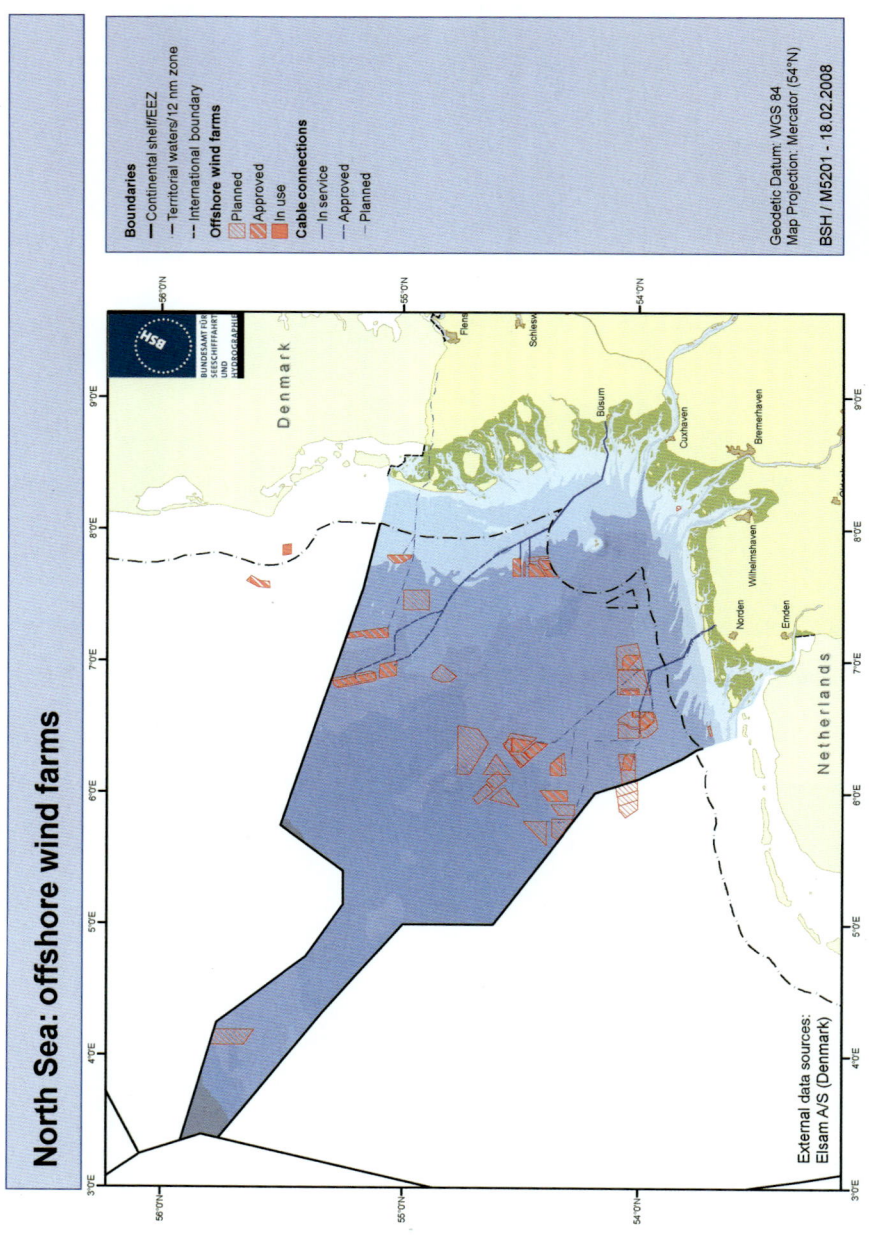

Plate 13 (Figure 16.1) Planned and approved wind farms in the German North Sea. (*Source:* Federal Maritime and Hydrographic Agency CONTIS Information System 2008.)

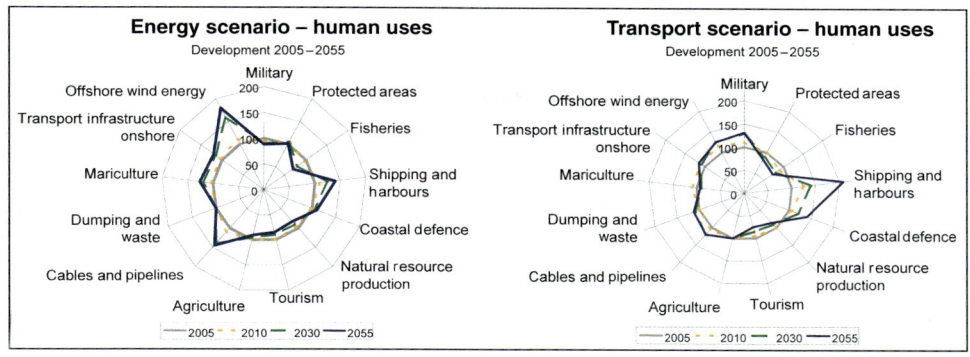

Plate 14 (Figure 16.2) Use patterns in the scenarios 'coast and sea as energy production area' and 'coast and sea as a transport area'.

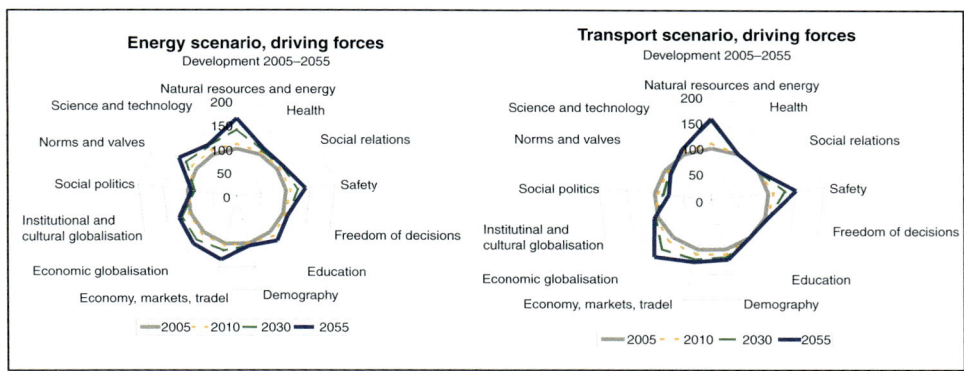

Plate 15 (Figure 16.3) Driving forces in the scenarios 'coast and sea as energy production area' and 'coast and sea as a transport area'.

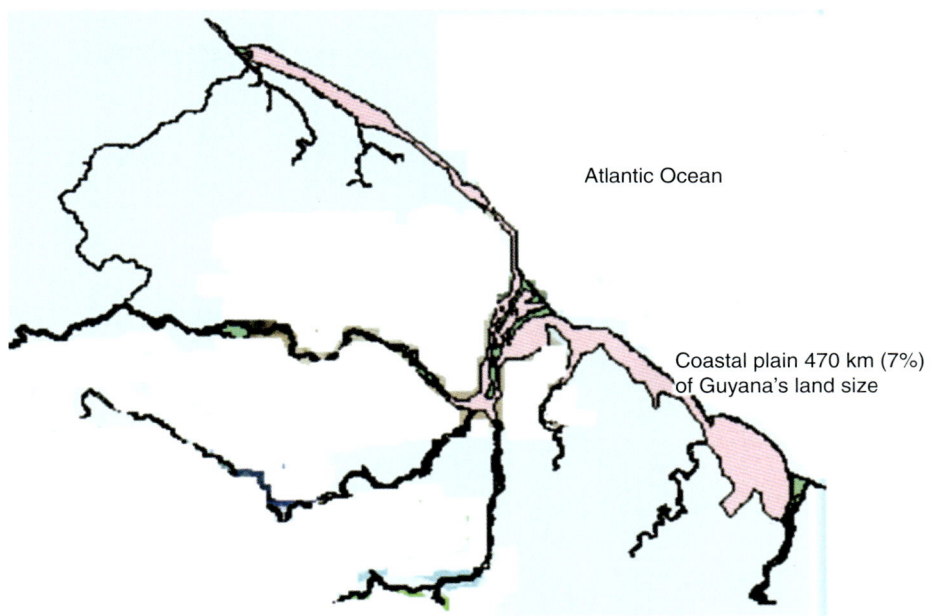

Plate 16 (Figure 17.1) Coastal plain of Guyana. (Adapted from Da Silva, 2007.)

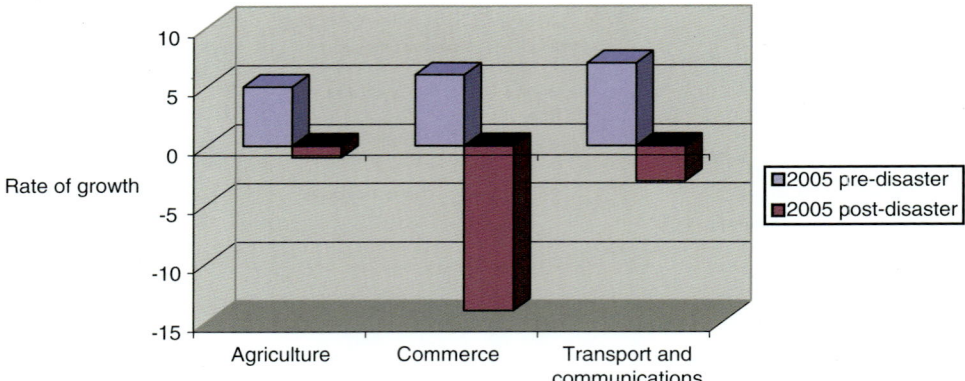

Plate 17 (Figure 17.2) Growth of sectors in Guyana showing pre- and post-disaster scenarios. (Adapted from UNDP, 2006.)

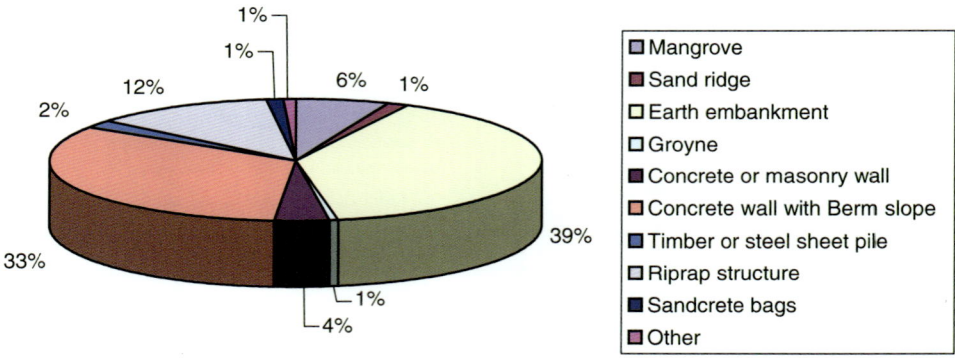

Plate 18 (Figure 17.3) Lengths of Guyana's coastal sea defence in terms of percentages. (Adapted from EU, 2006.)

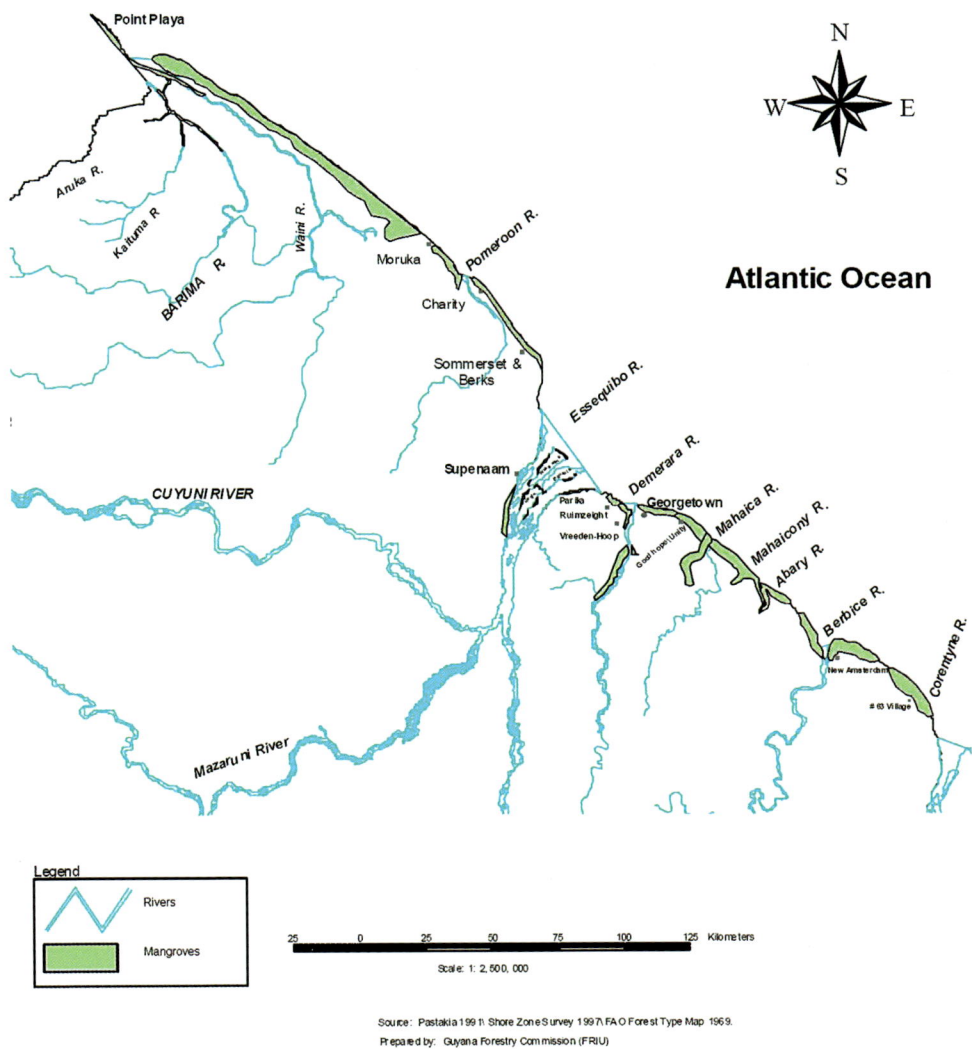

Plate 19 (Figure 17.4) Distribution of Mangroves within the coastal zone of Guyana. (Adapted from Pastakia, 1991.)

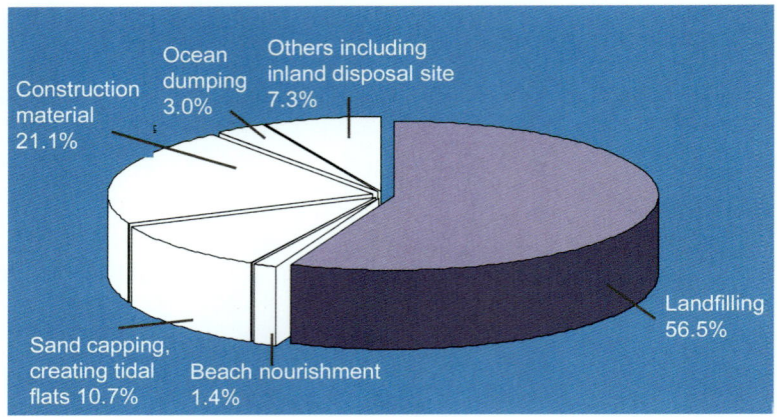

Plate 20 (Figure 18.1) Proportional representation of the use of dredged material in FY 2002.

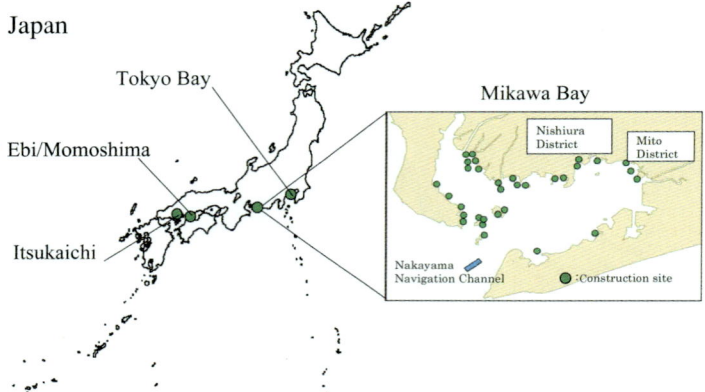

Plate 21 (Figure 18.3) Areas in which tidal flats and shallows construction, as well as sand capping, were carried out in Mikawa Bay. The material used was fine sand dredged from widening projects in Nakayama Navigation Channel. The project started in 1998, and 32 habitats (construction site area approximately 620 ha) had been constructed in the bay by FY 2004.

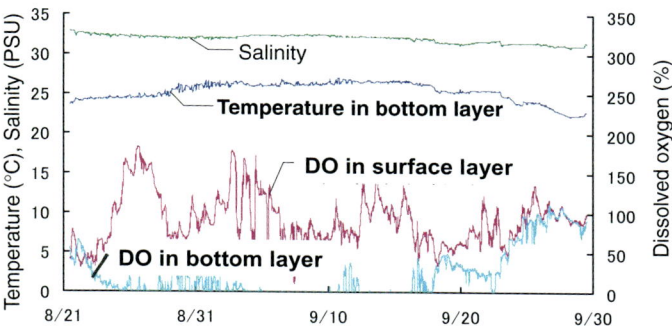

Plate 22 (Figure 18.7) Typical results of continuous monitoring of salinity, temperature and DO in Stn. 4 of a borrow pit before filling.

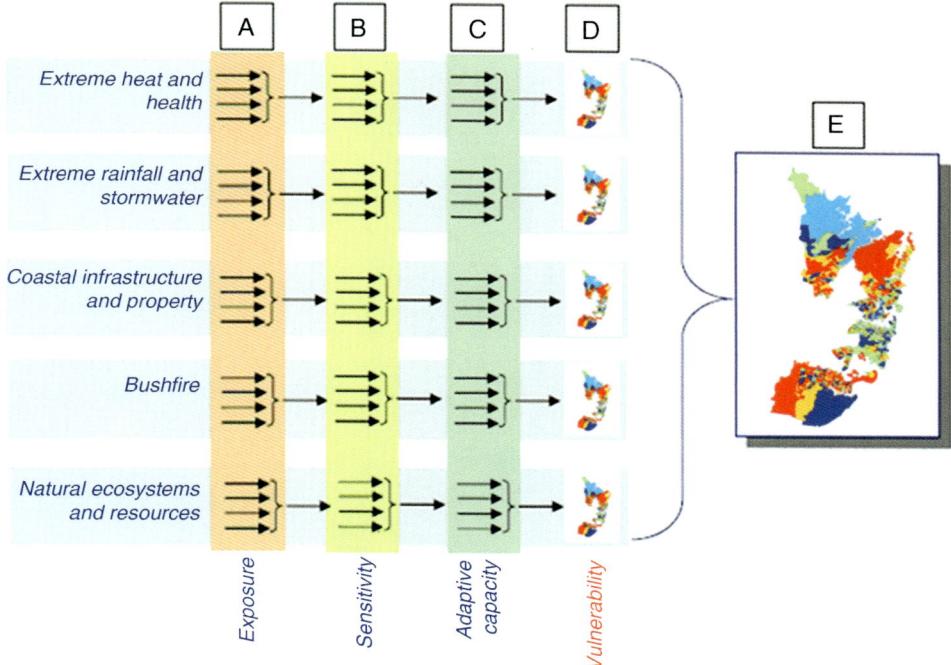

Plate 23 (Figure 25.3) Illustration of the process of determining relative vulnerability for the 15 member local governments of the Sydney Coastal Councils Group (Preston et al., 2007).

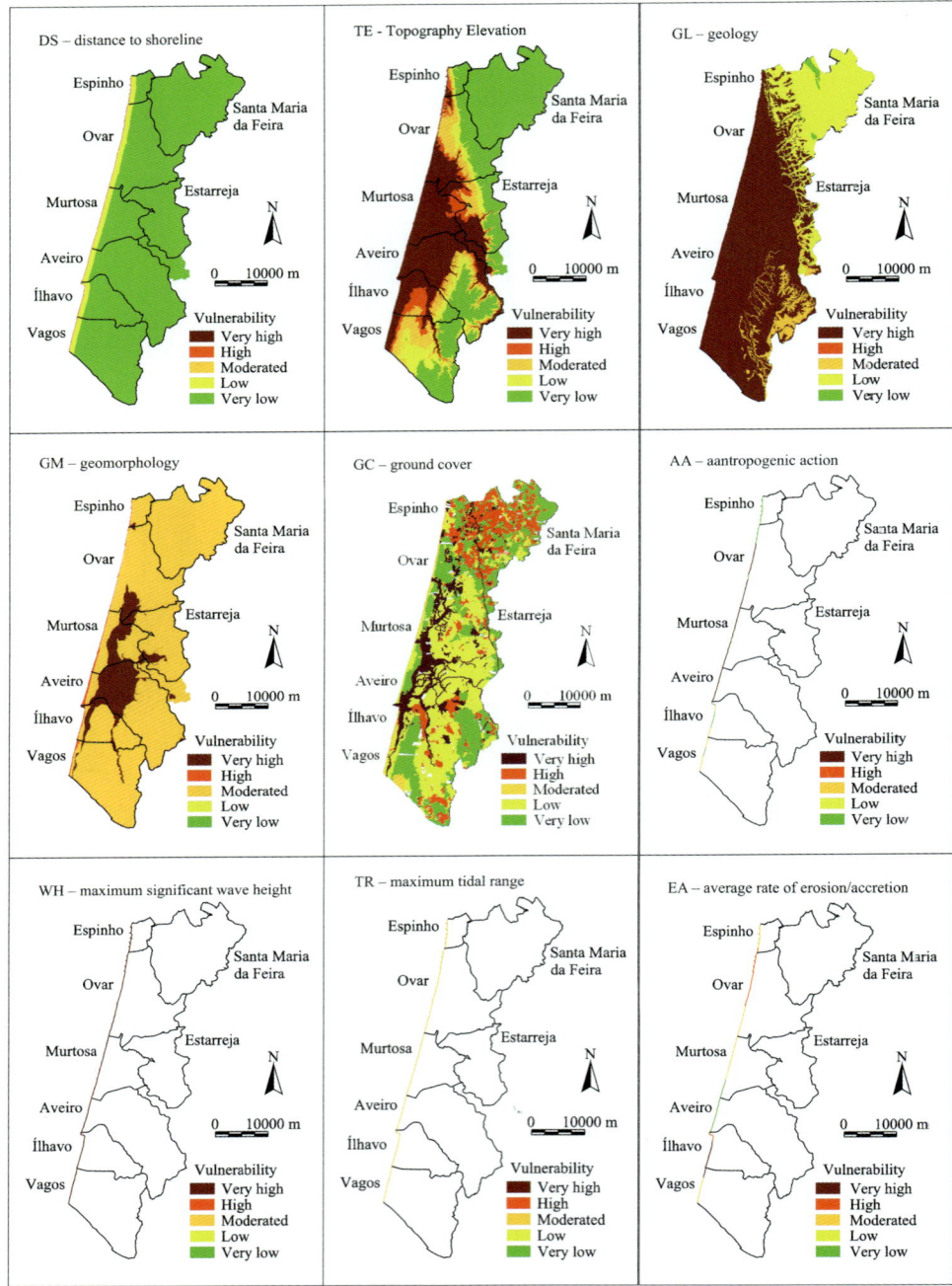

Plate 24 (Figure 24.2) Representation of the vulnerability parameters for the district of Aveiro (based on Segurado-Silva and Granjo, 2006).

Plate 25 (Table 6.2) Overview of test results conducted on the revised EU progress indicator.

Action	NE England (national)	Belgium (national)	NE England (national)	Belgium (national)	NW England (regional)	NW England (local)	NE England (local)
1	green	green	green	green	green	green	green
2	green	green	green	green	green	red	green
3	green	green	green	red	green	green	green
4	green	green	green	green	green	red	green
5	green	green	green	green	green	green	green
6	green	green	green	green	green	red	green
7	red	red	red	red	red	red	red
8	green	green	green	green	red	green	green
9	green	green	green	green	green	green	green
10	green	green	green	green	green	green	green
11	red	red	red	red	red	yellow	red
12	green	green	green	green	green	green	green
13	green	red	red	yellow	red	red	red
14	red	red	red	green	red	red	yellow
15	red	red	red	red	red	red	red
16	yellow	green	red	red	red	green	red
17	red	red	red	red	red	red	red
18	red	red	red	green	red	red	red
19	red	red	red	red	red	red	red
20	red	red	red	yellow	red	red	red
21	red	green	red	green	red	red	red
22	red	red	red	red	red	red	red
23	red	red	red	red	red	red	red
24	red	red	red	red	red	red	red
25	red	red	red	red	red	red	yellow
26	red	red	red	red	red	green	green
27	red	red	red	red	red	red	red
28	red	red	red	red	red	red	red
29	red	green	red	green	red	red	red
30	red	red	red	red	red	red	red
31	red	yellow	green	green	red	red	green

Green, affirmative response; red, negative response; yellow, undecided.

Plate 26 (Table 6.3) Overview of test results in three different administrative units of NE England.

Action	National	Regional	Local
1	green	green	green
2	green	green	green
3	green	green	green
4	green	green	green
5	green	green	green
6	green	green	green
7	red	red	red
8	green	green	green
9	green	green	green
10	green	green	green
11	red	red	red
12	green	green	green
13	green	red	red
14	green	green	yellow
15	red	red	red
16	yellow	green	red
17	red	red	red
18	red	red	red
19	red	red	red
20	red	red	red
21	red	red	red
22	red	red	red
23	red	red	red
24	red	red	red
25	red	red	yellow
26	red	red	green
27	red	red	red
28	red	red	red
29	red	red	red
30	red	red	red
31	red	green	green

Green, affirmative response; red, negative response; yellow, undecided.

Chapter 16
Exploring the Future of Seas and Coasts: Scenarios within the Joint Research Project 'Zukunft Küste – Coastal Futures'

Andreas Kannen, Kai Ahrendt, Antje Bruns, Benjamin Burkhard, Doris Diembeck, Kira Gee, Bernhard Glaeser, Katharina Licht-Eggert, Tanja Michler, Ophelia Meyer-Engelhard, Corinna Nunneri, Sebastian Stragies and Wilhelm Windhorst

Abstract

Driven by a range of global developments, marine and coastal areas are faced with profound change. In Germany, the emergence of offshore wind farms is a new permanent large-scale activity which is much debated as a symbol for the increasing 'industrialisation' of coastal and marine waters. Within the research project 'Zukunft Küste – Coastal Futures', the development of scenarios is used to describe different development pathways that could emerge over the next 50 years. The storylines of these scenarios are built around five different patterns of sea use, each describing a different mix of driver settings and therefore resulting in different priorities of spatial use. The five patterns of use describe sea and coast (1) as a natural area, (2) as leisure and tourism area, (3) as a source for renewable energies, (4) as an industrial area and (5) as a transport area. It is suggested that scenario-based assessment of potential 'futures' could be a useful tool in integrated coastal zone management (ICZM). Scenarios as used within 'Zukunft Küste – Coastal Futures' can serve as a tool for visualising complex relationships between system elements and variables, providing an analytical context for subsequent system analysis.

16.1 Introduction

Offshore wind farming is a relatively recent phenomenon that has captured the imagination of many coastal countries. Driven by energy and climate policy, it is representative of a new type of sea use which is spatially intense and relies on creating permanent structures in the sea. In Germany, offshore wind farming is not without controversy in terms of technological, economic, ecological and political feasibility (e.g. Byzio et al., 2005), but plans exist to significantly extend offshore capacities in the years to come (Bundesregierung, 2002).

But how much offshore wind is really feasible in terms of the system's capacity to adapt? It is obvious that offshore wind farming heralds significant change at many levels. Socioeconomic effects are often portrayed in a positive light, focusing on employment or a general boost to the economy, whilst ecological impacts on the marine environment are still controversial at times. Without a doubt, however, offshore wind farming will impact on a wide range of other maritime activities and uses, many of which are experiencing growth themselves. A recent stocktake of human activities and current trends in German coastal and marine areas (Gee et al., 2006a) concludes that growing pressure from offshore wind farms is paralleled by intensification of shipping and demands for new marine nature reserves, for example. Competition for marine space is therefore expected to increase (Gee and Kannen, 2004).

In order to assess the potential future for offshore wind farming in Germany and to determine opportunities and risks associated with offshore wind farm development, a comprehensive impact analysis of offshore wind farming is required. This needs to take into account cumulative rather than isolated impacts of offshore wind farming and consider the wider socio-ecological system including the political and administrative context. The federally funded joint research project 'Zukunft Küste – Coastal Futures' aims to provide such a comprehensive approach. Funded by the Federal Ministry of Education and Research (BMBF), it comprises 12 interlinked work packages that deal with different aspects of offshore wind farming and concepts of integrative impact assessment (Kannen, 2004). The overall aim of the project is to develop an interdisciplinary, integrative approach for assessing and evaluating change on the coast and in the sea.

Within the project, scenarios represent a central tool for integrating and structuring different aspects of change. Focusing on the entire North Sea area, scenarios were developed to describe different patterns of use that could feasibly emerge in the next 50 years. Pointing towards possible 'futures' makes them a useful tool for analysing the systems context in which offshore wind farming will need to take place. Here we show the core approach behind the development of the scenarios, which we present as a first step towards a more comprehensive impact assessment of offshore wind farming on the German North Sea coast.

16.2 Offshore wind farms: proposed developments and challenges

Table 16.1 is evidence of the dynamism currently experienced by German offshore wind farm development. A substantial amount of space has been set aside for offshore wind farms in the German North Sea (Figure 16.1), especially considering that Figure 16.1 only shows pilot projects and not the full extent of long-term plans of investors. In July 2006, 14 offshore wind parks had received official planning approval in the German EEZ (12 in the North Sea and 2 in the Baltic Sea).

In some areas on the German coast, wind power has already become a key economic driving force. In the town of Husum, for example, with its approximately 20 000 inhabitants, this sector accounts for about 30% of the local business tax revenue and provides more than

Table 16.1 Approved offshore wind farms in North Sea and Baltic Sea.

Name of wind farm	Approved number of piles	Total number of piles planned	Capacity/pile in MW	Total capacity in MW[a]	Water depths
North Sea (EEZ)					
Borkum West	12	208	5	60 (1040)	30 m
Butendiek	80	80	3	240	20 m
Borkum Riffgrund	77	180	3–5	231 (746)	23–29 m
Borkum Riffgrund West	80	458	3.5	280 (1800)	30 m
Amrumbank West	80	80	5	400	20–25 m
Nordsee Ost	80	250	4–5	400 (1250)	22 m
Sandbank 24	80	980	4–5	400 (4720)	20–35 m
ENOVA Offshore Northsea Windpower	48	251	4.5	202.5 (1255)	36–34 m
DanTysk	80	300	3.6–5	400 (1500)	Bis 30 m
Nördlicher Grund	80	402	Maximum 5	400 (2010)	23–40 m
Globatech I	80	240	5	400 (1200)	39–41 m
Hochsee Windpark Nordsee	80	508	5	400 (2286)	
Baltic Sea (EEZ)					
Kriegers Flak	80	80	5	400	20–40 m
Arkona Becken Südost	80		5	400	21–38 m

Source: As of July 2006, from Kannen et al. (2008).
[a]Figures in brackets indicate the capacity of the total number of piles planned.

1000 jobs (Volmari, 2002). Offshore wind farming is expected to add to this. This highlights a specific aspect of offshore wind farming. Although the actual wind harvesting in case of offshore farming takes place in the sea, often at considerable distance from the mainland, economic effects and related knock-on effects on demography and social infrastructure manifest themselves exclusively on land. Direct ecological impacts, on the other hand (discounting indirect impacts on the mainland through infrastructure development), are restricted to the sea. This highlights the fact that the impacts of offshore wind farming cannot be adequately assessed from a local perspective alone. Offshore wind farming is characterised by many complex land–sea interactions, many of which extend to areas that are at a considerable physical distance from the actual wind farms.

It follows that sustainable offshore wind farming has many facets. One is to achieve a balance between climate protection, implying large-scale expansion of offshore wind farming, and the need for marine nature conservation, a potential competitor for marine space. The needs of traditional and established activities in coastal and marine waters also need to be considered in terms of potential threats that could be posed by offshore wind farming. Last but not least, wind power can be a key local industry, with knock-on effects on demography, regional infrastructure and wider economic development. Regional development needs, impacts on the ecosystem and sociocultural structures are therefore key elements when it comes to assessing opportunities and risks associated with offshore wind farming in specific regions (Kannen et al., 2004).

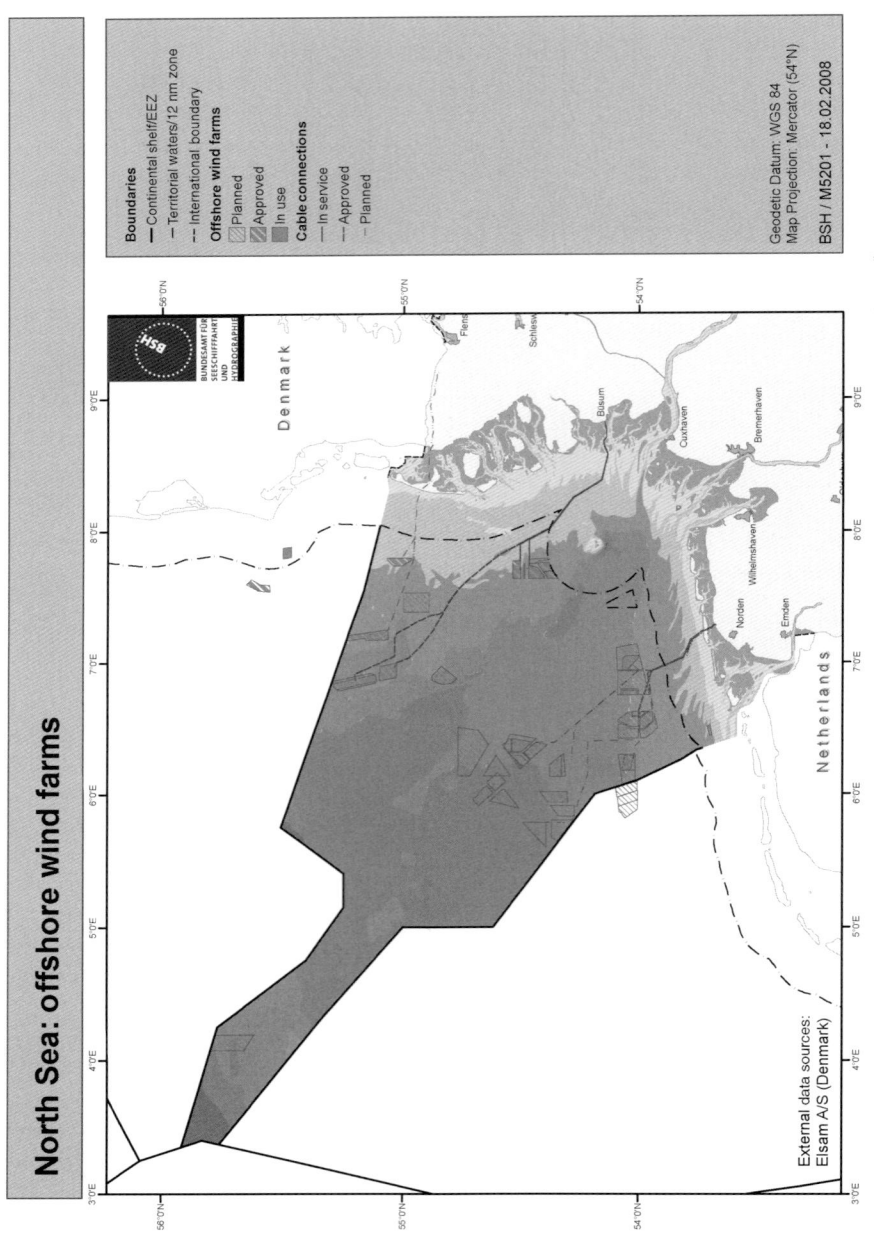

Figure 16.1 Planned and approved wind farms in the German North Sea. (*Source:* Federal Maritime and Hydrographic Agency CONTIS Information System 2008.) (To see this figure in colour, please see colour plate 13.)

From the perspective of spatial planning, polyculture has been proposed as a fundamental guiding principle for marine spatial use (Gee and Kannen, 2004; Gee et al., 2006b). Polyculture means the coexistence of several forms of use in a given spatial framework, with the aim of minimising total spatial use whilst maximising economic valorisation of that space. Polyculture is illustrated for instance by combining mariculture with offshore wind farms or offshore wind farms with nursery zones for commercial fish species.

The above makes clear that integrative assessment of offshore wind farming cannot focus on an actual wind farm or wind farming alone. Even if this specific activity is recognised as a major cause of change, focus must be on polycultural patterns of use in order to establish the thematic and spatial context for offshore wind farming. In Coastal Futures, scenarios have been designed to provide this context as a first step of analysis.

16.3 Development of scenarios within Coastal Futures

Scenarios can be applied in many different contexts for many different purposes, ranging from impact assessment of certain developments to providing a decision support system for politicians and decision-makers. In recent years, scenarios have become well established as a method of assessing potential future developments and are increasingly popular in various scientific contexts (WBCSD, 1997; Raskin et al., 1998; IPCC, 2000; Rotmans et al., 2000, 2001; Alcamo, 2001; UNEP, 2002; Alcamo and Bennet, 2003; Shell International Ltd., 2005). Three methodological elements always need to be specified in some detail because they affect the basic application of the scenarios. These are (a) spatial scales (Section 16.3.1), (b) time scales (Section 16.3.2) and (c) the information structure (Section 16.3.3).

The scenarios in Coastal Futures bring together two methodological approaches. On the one hand, they are explorative, explaining plausible future developments based on assessments of the current state. On the other hand, they are anticipatory in that they look back from a certain future situation and investigate why and how these situations could have come about. Because they were intended as an analytical tool and means of structuring complex interrelationships, the scenarios were developed by the project scientists in an interdisciplinary exercise. No stakeholders were involved.

In principle, the scenarios are based on the ecological, economic and social system at the west coast of Schleswig-Holstein and the German North Sea. Each scenario consists of a storyline, which describes a potential 'future' for this area: what it will be like in 2055 and what the corresponding political, social and economic climate would be. The counterpoint is a description of the current social, economic and ecological state, where current patterns of sea use are expressed in terms of pressures and these in turn in measurable ecological or socioeconomic impacts. Various interchanges between spatial and political scales, as well as driving forces, are taken into account both in the storyline and in the description of the status quo. Pressures and impacts are described by key parameters that are able to reflect change over time. A rigid structure of time scales, spatial scales and quality of information was imposed to ensure the scenarios are plausible and comparable.

Although the 'future worlds' set out in the storylines can seem quite convincing, it is important to point out that the scenarios are no prognoses. They simply reflect a range of assumptions based on expert knowledge and publicly available information: How might current trends develop in future, what role might be played by certain factors, and what new factors might become significant in future (e.g. WBCSD, 1997; UNEP, 2002)? Also, it should be pointed out that offshore wind farms only represent one of many system elements considered. Each scenario storyline assumes a different degree of offshore wind farm development in the patterns of use it describes. The next stage of the project will be to assess the ecological, socioeconomic and institutional impacts of offshore wind farms in the context of these scenarios and the respective constraints they impose.

Storylines were based on the premise that different forms of resource use and different societal, political and technological drivers can be given different weight. Based on the concept of polyculture, five patterns were developed where one form of sea use is always given priority without entirely discounting the others. The following options for regional development were used:

- Coast and sea (primarily) as natural space,
- Coast and sea (primarily) as tourism and recreation space,
- Coast and sea (primarily) as a source of renewable energies,
- Coast and sea (primarily) as industrial space,
- Coast and sea (primarily) as transport space.

Following a first workshop, 13 raw scenarios were drafted by individual project scientists. Five much more comprehensive scenarios were developed in a second stage by interdisciplinary teams, merging elements from the raw scenarios where appropriate and including additional information from external sources. Each of the five final storylines reflects one of the above thematic strands and incorporates a specific level of offshore wind farm development. For each scenario storyline, ecological and regional economic effects can thus be modelled (at least partially) in later project stages. Modelling will focus on different degrees of offshore wind farm development to illustrate the impacts this will have. On the other hand, feeding modelling results back into the storylines will ensure that the results of these impact assessments link back to the original context.

16.3.1 Spatial scales

The importance of scales (not only spatial) is shown by Cash et al. (2006), who distinguish between scales in terms of space, time, jurisdiction, institutions, management, networks and knowledge. Young (2006) analyses the vertical interplay of scales for governance. Kannen et al. (2008) point to effects of scale in the context of offshore wind farming in Germany, linking scales to different interests and perceptions on the coast and in the sea. Differences between local and national actors are particularly important in this context. The former often argue from a very local perspective, whilst the latter often exist in a political and strategic environment with conflicting targets and shifting patterns of power. Selecting spatial scales for developing scenarios is crucial because certain effects and processes are only visible at a very local level, although others can emerge on much larger scales such as the entire North Sea. Within 'Zukunft Küste – Coastal Futures', a multi-scale approach was chosen, with political scales ranging from European to national, regional and local decision-making and spatial scales ranging from the Southern North Sea down to individual offshore wind turbine piles or individual municipalities.

16.3.2 Time scales

The definition of an appropriate time frame and temporal steps for scenarios is one of the most crucial factors as far as modelling and simulation results are concerned. Here, the year 2005 was chosen as base year and the years 2010, 2030 and 2055 selected as subsequent steps for analysis. These steps correspond to German plans for a gradual expansion of offshore wind farming in the German North Sea. The temporal stages include the so-called first extension phase up to the year 2010 (up to 3.000 MW of installed offshore wind power at suitable sites), the currently last extension phase up to the year 2030 (between 20.000 and 25.000 MW of installed offshore capacity) and the year 2055 as a projected end point. However, several years of delay in the construction of offshore wind farms in Germany have meant that the stages assumed in the scenarios no longer correspond to real developments.

It is obvious that the accuracy of a scenario's foresight decreases as the simulated time span increases. Analysis therefore initially focuses on the period 2005–2010, continuing to the year 2030 after that. For these periods, statements can be made that have a certain degree of likelihood. For the period after 2030, only hypotheses can be given even though these are supported by model calculations and sound assumptions. Since the scenarios are not intended to be prognoses, but simply illustrate cause and effects, this is not considered to be a problem.

16.3.3 Information structure

The scenario storylines are narrative descriptions of developments which highlight the main features and processes. The storylines are linked to forces that drive this particular development in a plausible and illustrative manner (Alcamo, 2001). Information contained within the scenarios is structured along the Driver–Pressure–State–Impact–Response (DPSIR) approach (EEA, 1999; Bowen and Riley, 2003). Definitions of the DPSIR categories can vary. Here, the definitions applied are those from the Millennium Ecosystem Assessment (UNEP, 2006) as modified for spatial planning by Gee et al. (2006b):

- *Driver*: All natural or man-made factors which cause changes in a system, either directly or indirectly. 'System' here refers to coastal and marine systems in the context of spatial planning including social, economic and institutional as well as ecological components.
- *Pressure*: Environmental pressure or pressure of use acting on the coastal and marine system as a result of the drivers. In the context of land–sea interchanges, pressures for instance comprise material and energy flows between rivers and coastal waters. In the context of spatial planning, pressure includes all forms of marine and coastal resource use.
- *State*: A stocktake description of the system which is influenced by drivers and pressures. In the context of spatial planning, analysis includes ecosystems, the current state and trends in spatial use, social infrastructure, the institutional system governing coastal and marine resource use and the current state of particularly relevant sectors. A description of the state should also include societal values and standards, as well as attitudes of local communities to central developments (e.g. coastal defence, wind energy).
- *Impact*: Specifies the effects of drivers and pressures on the coastal and marine systems. Contrary to the previous step, which is purely descriptive, impacts already contain an element of evaluation. The classic definition of impact in environmental assessment

refers to the knock-on effects of ecosystem deterioration on resource use (ecosystem goods and services), using, for example, algal blooms as indicator of excess eutrophication with potentially negative impacts on tourism. The spatial planning context takes a wider perspective of impacts, including impacts of drivers and pressures on all elements listed in 'state'.

- *Response*: Describes societal forms of response to systems change. Central elements include management options, institutional responses and their framework conditions, as well as individual changes of behaviour (e.g. changes in holiday destinations). Response also includes changes to the existing legal framework, the introduction of monitoring systems or investment decisions affecting the location of businesses.

16.4 Comparing assumptions for two scenarios

In order to demonstrate the outcomes of the scenario development, this section compares two of the five scenarios. These are 'coast and sea predominantly used for energy production' (subsequently called Energy Scenario) and 'coast and sea predominantly used for transport' (subsequently called Transport Scenario). Drivers were identified and described for the presumed changes in spatial use, as were resulting pressures on natural and anthropogenic systems. Figure 16.2 illustrates differences in patterns of use, whilst Figure 16.3 highlights differences in the respective driving forces.

The Energy Scenario is based on the assumption that overall energy demands will continue to be high, on a par perhaps to what they are now (see the driver 'Natural Resources and Energy' in Figure 16.3). The global economy will continue to develop, but climate change and nuclear proliferation represent real fears. Within Germany at least, nuclear energy is not perceived as a solution to CO_2 reduction. National energy policy strongly favours the use of renewables even if this requires regulation or high economic incentives, and there is strong societal support for renewable energies too. Under these assumptions, the sea would be an ideal place for energy production because of several options for renewable energy generation. This in turn would stimulate marine technology

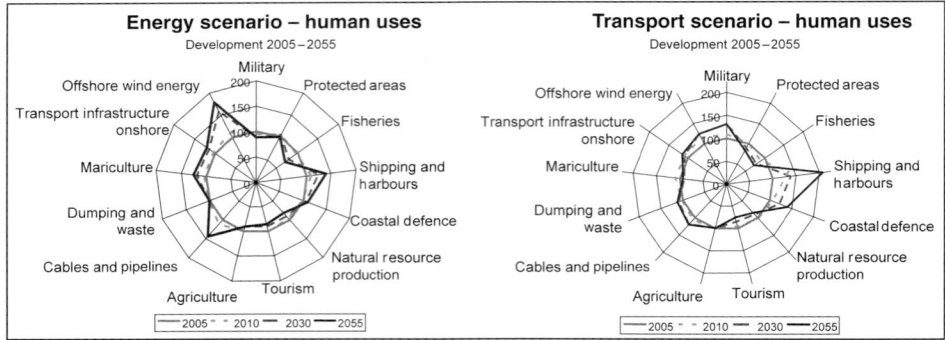

Figure 16.2 Use patterns in the scenarios 'coast and sea as energy production area' and 'coast and sea as a transport area'. (To see this figure in colour, please see colour plate 14.)

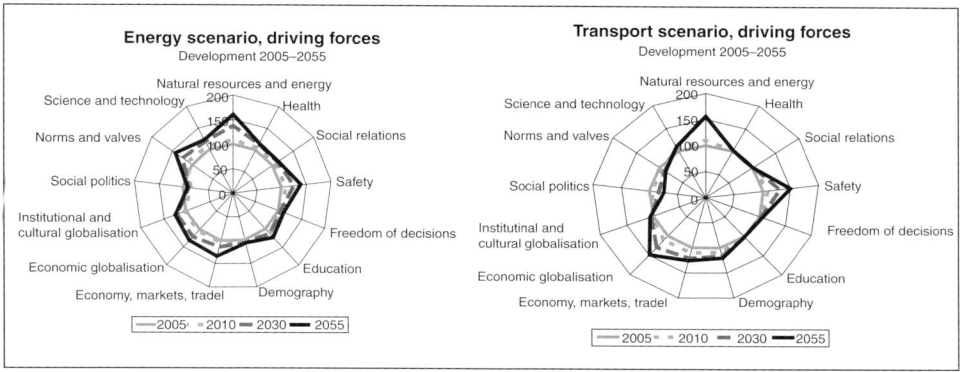

Figure 16.3 Driving forces in the scenarios 'coast and sea as energy production area' and 'coast and sea as a transport area'. (To see this figure in colour, please see colour plate 15.)

development, as transport or storage of electricity produced at sea will be very limited without technical innovation. It is therefore assumed that the sea will gradually develop into an intensive energy production site. Other forms of use will be forced to take second place through appropriate regulation.

In the Transport Scenario, it is assumed that globalisation only applies to global production and the resulting exchange of goods. This would increase the intensity of shipping and result in larger ships. Under these circumstances, keeping the sea free for shipping becomes a top priority. Other forms of use including energy production can only develop as long as they do not interfere with shipping routes and shipping safety. As a result, they are strongly limited in their development. At its core, this scenario is one of a neo-liberal world, where social and cultural drivers matter little except for safety from terrorist attacks and safety of investment (including the securing of infrastructure against natural hazards).

The two scenarios in Figures 16.2 and 16.3 make clear that strong development of offshore wind farming and increasing shipping intensity do not necessarily exclude each other. Both uses show an increase in both scenarios compared to their original status quo. The difference is in their relative development. While both scenarios are driven by economic globalisation, assumptions differ as far as other drivers are concerned. In the Energy Scenario, strong public concern is assumed for climate change, which is reflected in the driver 'norms and values'. Strong institutional globalisation acts as another driver. Here the assumption is that global rules have been put into place to facilitate the use of renewables, accompanied by global environmental policies that are not contradictory. The result is that the sea is not only available but also required as an energy-generating powerhouse.

Based on the above descriptions, impact assessments are now being developed that trace as a first case study the wind energy related consequences of these storylines in the ecological, regional economic and institutional domain. Once this next stage is completed, the scenarios will represent a methodology and tool that can be used to inform political debates and stakeholder dialogues on the future of marine areas and the role of wind energy within this future.

16.5 Discussion and outlook

Taking the current dynamics of change as a starting point and taking into consideration the multitude of demands on coastal areas, this chapter described the use of scenarios in the research project 'Zukunft Küste – Coastal Futures'.

Analysis of current trends in the German North Sea makes clear that impact assessments need to be placed in a context of varying temporal and spatial scales, levels of decision-making and existing or developing patterns of use. This significantly increases the complexity of analysis and means that many uncertainties need to be taken into account. This is particularly the case where future developments, normative aspects and non-linear, complex cause-and-effect relationships are concerned.

There is a danger that planning and management decisions are taken on the basis of incomplete information. This is the case when only individual uses are taken into account instead of patterns. Planning approaches such as spatial planning and Integrated Coastal Zone Management (ICZM) need to ensure their analysis includes spatial and temporal patterns of use, focusing also on political and administrative mechanisms that link local and national or even international interests. What is required is multi-scale and multi-temporal analysis of cause-and-effect chains, supported by an analysis of governance as part of the socio-ecological system.

Scenarios provide a useful tool for the inclusion of complexity and uncertainty in integrated assessments. This is illustrated by their widespread application in climate research. In 'Zukunft Küste – Coastal Futures', scenarios were developed to describe and analyse a specific socio-ecological system, namely offshore wind farm development in the German North Sea. Bringing together explorative and anticipatory methodologies, they create an instrument which is well placed to highlight and structure complex interrelationships.

In Coastal Futures, the five scenarios serve as background information for later modelling exercises and more detailed assessments of specific aspects of system behaviour. Their use as a visionary planning tool is limited because they have not been developed in a participatory manner through stakeholder dialogue. On the other hand, they successfully complete the first step in a highly complex analytical chain, thus serving the purpose for which they were designed. The scenarios were able to integrate approaches taken from the social and natural sciences and assisted interdisciplinary analyses by providing a common framework. A prerequisite was to rigorously structure the scenarios along the DPSIR chain and to agree on temporal as well as spatial scales before any assessments were made.

In times of far-reaching change, effective coastal and marine planning and management depend on appropriate processes of dialogue and decision-making. This in turn requires appropriate platforms for sharing information and communicating with all relevant stakeholders. In this context, the Coastal Futures scenarios can be used to stimulate more systems-oriented thinking among stakeholders and a comprehensive and long-term debate that goes beyond the immediate concerns of planning permissions or short-term controversies. An initial presentation of the scenarios to decision-makers has shown that they are useful for creating awareness for the complexity of the system, a factor which is not always taken into account in contemporary decision-making. A similar approach has been successfully used by the trilateral Wadden Sea Forum (Wadden Sea Forum, 2006) in order to develop a common vision and joint activity targets among a wide range of governmental and non-governmental stakeholders acting at the local level in the Dutch, German and Danish Wadden Sea.

The scenario approach in general can therefore aid the development of long-term visions for coastal and marine areas and contribute to consensus-oriented management (Glaeser et al., 2005). Scenarios can help to establish concepts such as multi-functional spaces (polyculture) and participatory assessment, and also increase awareness of various forms of sea–land, land–sea and sea–sea interactions. As such, they can help to implement innovative approaches, such as integrated maritime spatial planning recently outlined in the Blue Book for the EU Maritime Policy.

Acknowledgement

This chapter is based on the work of 'Zukunft Küste – Coastal Futures', a joint research project funded by the BMBF (FKZ 03F0404A-D).

References

Alcamo, J. (2001) *Scenarios as Tools for International Environmental Assessments*. European Environment Agency, Copenhagen, 31 pp.

Alcamo, J. and Bennet, E.M. (2003) *Ecosystems and Human Well-being: A Framework for Assessment*. Island Press, Washington, DC, 200 pp.

Bowen, R.E. and Riley, C. (2003) Socio-economic indicators and integrated coastal management. *Ocean and Coastal Management* **46**, 299–312.

Bundesregierung (2002) *Strategie der Bundesregierung zur Windenergienutzung auf See*, 26 pp. Available from www.offshore-wind.de

Byzio, A., Mautz, R. and Rosenbaum, W. (2005) *Energiewende in schwerer See? Konflikteum die Offshore-Windkraftnutzung*. Oekom-Verlag, München, 184 pp.

Cash, D.W., Adger, W., Berkes, F., Garden, P., Lebel, L., Olsson, P., Pritchard, L. and Young, O. (2006) Scale and cross-scale dynamics: governance and information in a multilevel world. *Ecology and Society* **11**, 8. Available from http://www.ecologyandsociety.org/vol11/iss2/art8/.

EEA (1999) Environmental indicators: typology and overview. Technical Report No. 25, European Environment Agency, Copenhagen, 19 pp.

Gee, K. and Kannen, A. (2004) Raumplanungsstrategien an der Küste und im Meer. Erste Ergebnisse der Bestandsaufnahme und Thesenpapier. In: *Bundesministerium für Verkehr, Bau- und Wohnungswesen (Hrsg.)*. Auf dem Weg zur nationalen Integrierten Küstenzonenmanagement-Strategie – Perspektiven der Raumordnung. Dokumentation des Workshops vom 23. und 24. October 2003, Berlin, pp. 10–21.

Gee, K., Kannen, A., Licht-Eggert, K., Glaeser, B. and Sterr, H. (2006a) *Bestandsaufnahme der Nutzungstrends, Planungsherausforderungen und des strategischen Umfelds in der deutschen Küstenzone*. Berichte aus dem Forschungs- und Technologiezentrum Westküste der Universität Kiel Nr. 38, Büsum 2006, 149 pp.

Gee, K., Kannen, A., Licht-Eggert, K., Glaeser, B. and Sterr, H. (2006b) *The Role of Spatial Planning and ICZM in the Sustainable Development of Coasts and Seas*. Research project of the Federal Ministry of Transport, Building and Urban Affairs (BMVBS) and Federal Office for Building and Spatial Planning (BBR): Integrated Coastal Zone Management (ICZM):

Strategies for coastal and marine spatial planning (Az: Z6 – 4.4 – 02.119). October 2006. Available from http://athos.ecology.uni-kiel.de/servlet/is/7122/Abschlussbericht_engl.pdf?command=downloadContent&filename=Abschlussbericht_engl.pdf.

Glaeser, B., Gee, K., Kannen, A. and Sterr, H. (2005) Vorschläge für eine nationale IKZM-Strategie aus Sicht der Raumordnung. In: *Bundesministerium für Verkehr, Bau- und Wohnungswesen u. Bundesamt für Bauwesen und Raumordnung (Hrsg.)*. Nationale IKZM-Strategien – Europäische Perspektiven und Entwicklungstrends. Konferenzbericht zur Nationalen Konferenz, 28 February to 1 March 2005, Berlin, pp. 47–52, 78.

IPCC (2000) *Emission Scenarios. A Special Report on Working Group III of the Intergovernmental Panel on Climate Change*. Cambridge University Press, Cambridge, MA.

Kannen, A. (2004) Holistic systems analysis for ICZM: the coastal futures approach. In: Schernewski, G. and Dolch, T. (eds). *Geographie der Meere und Küsten, AMK 2004 Conference Proceedings, Coastline Reports 1*. EUCC – The Coastal Union, EUCC – Die Küstenunion Deutschland e.V. Warnemünde, pp. 177–181.

Kannen, A., Gee, K. and Glaeser, B. (2004) Offshore wind farms, spatial planning and the German ICZM strategy. In: *Delivering Sustainable Coasts: Connecting Science and Policy*, Proceedings of Littoral. Cambridge Publications, Aberdeen, pp. 450–455.

Kannen, A., Gee, K. and Licht-Eggert, K. (2008) Managing changes in sea use across scales: North Sea and North Sea coast of Schleswig-Holstein. In: Krishnamurthy, R.R., Glavocic, B., Kannen, A., Green, D.R., Ramanathan, A.L., Han, Z., Tinti, S. and Agardy, T.S. (eds). *ICZM – The Global Challenge*. Research Publishing, Singapore, pp. 93–108.

Raskin, P., Gallopin, G., Gutman, P., Hammond, A. and Swart, R. (1998) *Bending the Curve: Toward Global Sustainability*. PoleStar Series Report, Environment Institute Stockholm.

Rotmans, J., Anastasi, C., van Asselt, M., Rothman, D., Mellors, J., Greeuw, S. and van Bers, C. (2001) *VISIONS – The European Scenario Methodology*. International Centre for Integrative Studies, Maastricht, 23 pp.

Rotmans, J., van Asselt, M., Anastasi, C., Greeuw, S., Mellors, J., Peters, S., Rothman, D. and Rijkens, N. (2000) Visions for a sustainable Europe. *Futures* **32**, 809–831.

Shell International Ltd (2005) Shell Global Scenarios to 2025. In: *The Future Business Environment: Trends, Trade-Offs and Choices*. Global Business Environment (PXG), London, 242 pp.

UNEP (2002) *Global Environment Outlook 3*. Earthscan, London, Sterling VA, 410 pp.

UNEP (2006) *Marine and Coastal Ecosystems and Human Wellbeing: A Synthesis Report Based on the Findings of the Millennium Ecosystem Assessment*. UNEP, 65 pp.

Volmari, M. (2002) Positionspapier zur Windkraftbranche in Nordfriesland, Husum, 12 p. Available from http://www.windcomm-sh.de/Downloads/positionspapier_windkraft_de.pdf.

Wadden Sea Forum (2006) *Breaking the Ice*. Wilhelmshaven, 76 pp.

WBCSD (1997) *Exploring Sustainable Development – Global Scenarios 2000–2050*. World Business Council for Sustainable Development, London, Summary Brochure, 29 pp.

Young, O. (2006) Vertical interplay among scale-dependent environmental and resource regimes. *Ecology and Society* **11**, 27. Available from http://www.ecologyandsociety.org/vol11/iss1/art27/.

Chapter 17
Integrated Coastal Zone Management (ICZM) in Guyana: Development Barriers, Opportunities and Recommendations

Robin McCall and Talia Choy

Abstract

Guyana is in urgent need of improving the present integrated coastal zone management (ICZM) programme, as the current one does not properly address the coastal zone issues faced by the country. The ICZM Plan drafted in 2000 by the Environmental Protection Agency (EPA) forms the foundation of the ICZM programme in Guyana. According to EPA representatives, there has been no meaningful follow-up undertaken to determine the status of the ICZM Action Plan. This chapter represents a fundamental effort in this direction, which is to provide a status report on the current ICZM programme in Guyana. Further, the chapter identifies critical barriers that impede its success and offers recommendations on how such barriers may be overcome. Related information was gathered from structured and unstructured interviews conducted with representatives from various governmental and non-governmental agencies and the ICZM Committee. A review of environmental reports, feasibility studies, action plans and relevant literature was also done. Additionally, a case study was done on six rural coastal communities from the North West District of Guyana, where the perceptions of coastal resource users were garnered regarding issues of coastal resource use and management. While there have been some accomplishments during the first 7 years of the creation of the ICZM Action Plan; real, sustained progress of the ICZM agenda in Guyana has been circumscribed by the lack of supporting legislation, competent and committed human resources, workable mechanisms for integrated enforcement/management and adequate financial and scientific support. Some of the recommendations for addressing these barriers include providing incentives to ICZM Committee members so as to educe their support and commitment to ICZM goals; expanding ICZM training programmes to include a focus on building participants' skills in interagency cooperation, negotiation and conflict resolution; setting up practical arrangements for sharing the costs of ICZM activities with private sector organisations, NGOs and financial institutions; and setting up administrative mechanisms for more meaningful public involvement in ICZM decisions.

17.1 Introduction

Notwithstanding its ecological value, the economic significance of the coastal zone has been the main incentive for the settlement and development of human society. Approximately 60% of the world's population resides in this relatively small area, which occupies about 18% of the surface of the globe (Field et al., 2002). The future of the earth's coastal areas is under continuous threat due to pollution, eutrophication, land reclamation, mining, over fishing and changing sediment loads; all of which are associated with intense urbanisation and industrialisation.

These problems are pronounced in developing countries where social demands have obligated the need to seek short-term economic growth without careful consideration of the adverse impacts on the environment (Tagliani et al., 2003). Paradoxically, environmental degradation of coastal areas has directly affected those economic sectors that depend on environmental quality, such as fisheries and eco-tourism, and may also be associated with other indirect costs, including reduced human health conditions (Tagliani et al., 2003). What occurs in many of these countries is a negative feedback cycle between environmental degradation and poverty of rural coastal communities (Tagliani et al., 2003). However, many coastal nations are becoming aware of the environmental costs of rapid or inappropriate development, and correspondingly cite environmentally related issues as reasons for initiating integrated coastal zone management (ICZM) programmes (Cicin-Saine and Knecht, 1998).

As a developing nation, Guyana is no exception. The country is subjected to the same negative feedback cycle, as a result of mismanagement of their coastal resources. To deal with these issues, affirmative action was taken in the form of an ICZM Plan that was developed by the Environmental Protection Agency (EPA) in 2000. The Plan laid the foundation for an ICZM programme in Guyana. It outlined a number of pressing coastal issues and proposed several related 'actions' that were to be implemented by and among the various sectors and stakeholders in the years that followed (EPA, 2000). However, the extent to which the proposed actions have been implemented is, to date, largely unknown. According to EPA representatives, there has been no meaningful follow-up undertaken to determine the status of the ICZM Action Plan. This chapter represents a fundamental effort in this direction, which is to provide a status report on the current ICZM programme in Guyana. Further, the chapter identifies critical barriers that impede its progress and offers recommendations on how such barriers may be overcome.

17.2 Objectives and methodology

In an attempt to determine the factors that serve to impede the successful adoption of the ICZM programme in Guyana, it was important for the researchers to ascertain its current successes and shortcomings. To achieve this objective, structured and unstructured interviews were conducted with representatives from various governmental and non-governmental agencies and the ICZM Committee. A review of environmental reports, feasibility studies, action plans and relevant literature was also done. Additionally, a case study was done on six rural coastal communities from the North West District of Guyana, where the perceptions of coastal resource users were garnered regarding issues of coastal resource use and management. The chapter sets the stage with an overview of the coastal resource issues in Guyana, followed by a historical background of ICZM. Findings from the interviews and literature review are incorporated in the discussion of the barriers that prevent or limit the successful adoption of the ICZM programme and recommendations for improving the ICZM approach.

17.3 The need for improving the ICZM programme in Guyana: background, issues and priorities

17.3.1 Background

Guyana is in urgent need of improving the present ICZM programme. Currently, the most serious environmental issues are found within 7% of Guyana's surface area (approximately 10 500 km^2), the coastal plain, where over 90% of the population (approximately 753 000 according to the 2004 Census) and the majority of economic activities are concentrated (Ramraj, 2003; FAO, 2005; Da Silva, 2007). This narrow stretch of land extends 470 km between the borders of Venezuela (estuary into Waini River) and Suriname (estuary into Corentyne River) (see Figure 17.1), and lies between 0.5 and 1.0 m below spring tide level, thus making it prone to flood and erosion (EU, 2005). Cultivation of Guyana's three main crops (sugar, rice and coconut) occurs almost exclusively within the coastal plain because of its rich alluvial soil. The capital city, Georgetown, and most of the major towns are located along the coastline. The following section offers a brief discussion of the most

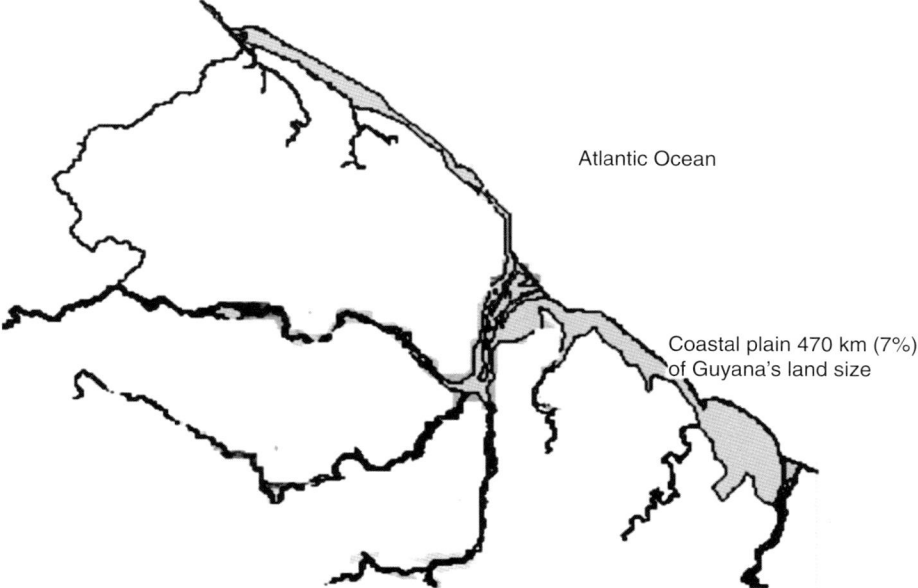

Figure 17.1 Coastal plain of Guyana. (Adapted from Da Silva, 2007.) (To see this figure in colour, please see colour plate 16.)

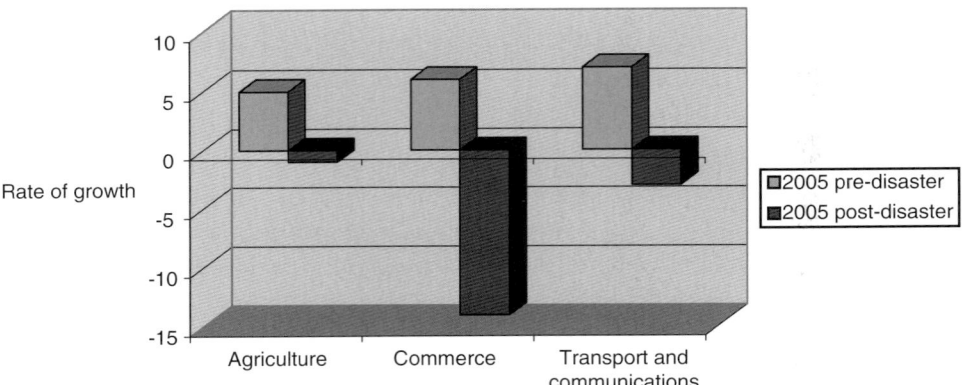

Figure 17.2 Growth of sectors in Guyana showing pre- and post-disaster scenarios. (Adapted from UNDP, 2006.) (To see this figure in colour, please see colour plate 17.)

pressing issues in the coastal zone, including poor drainage, irrigation and waste disposal systems; deteriorating sea defence infrastructure; mangrove degradation; unclear fisheries status; and rising aquaculture in an unprepared environment.

17.3.2 Poor drainage, irrigation and waste disposal systems

Both life and property are constantly under threat as a result of Guyana's topographic and geographic characteristics and the generally poor drainage infrastructure and waste disposal system (Pelling, 1997; Ramraj, 2003). The coastal area of Guyana lies below sea level, making effective water management and flood control systems crucial. For instance, inadequate maintenance of drainage works has significantly reduced agricultural yields and kept some land unproductive (Da Silva, 2007). Coastal flooding, which occurred in early 2005, severely affected 37% of the country's population, with a total estimated damage of 59.49% of the GDP (UNDP, 2006). The agricultural sector, which contributes more than 30% of the GDP and approximately 40% of foreign exchange earnings, was especially affected (UNDP, 2006). Although abnormally heavy rainfall initiated the disaster, the resulting damage can be significantly attributed to malfunctioning drainage structures and a poor waste disposal system. Figure 17.2 illustrates the redirection in the rate of growth of three major sectors in Guyana.

17.3.3 Sea defence infrastructure

Rising sea level, which has been projected to increase by up to 20–25 mm/year (between 0.8 and 1.0 m before 2050) in some regions due to weakening of the thermohaline circulation (Kont et al., 2002; Anders et al., 2005), will serve to further exacerbate flooding conditions in Guyana. To combat this threat, the coast is protected by natural sea defences (46.8%), as well as concrete structures (47.8%), as shown in Figure 17.3.

The majority of concrete sea wall is between 30 and 75 years old and has had little to no maintenance over the last 20 years. A recent study by the EU (2006) concluded that at least 35 km (or 18.2%) of defence is in a 'poor' or 'bad' condition, and approximately 20% of

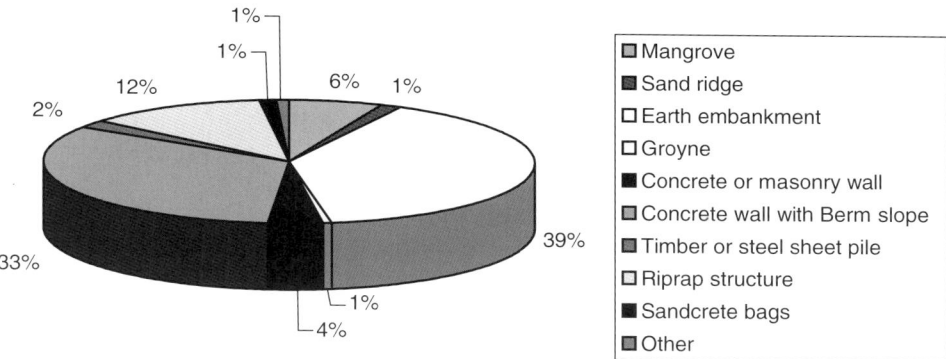

Figure 17.3 Lengths of Guyana's coastal sea defence in terms of percentages. (Adapted from EU, 2006.) (To see this figure in colour, please see colour plate 18.)

the existing sea defences, some 40 km, have a 'residual life' of less than 5 years, and up to 50% have less than 10 years.

17.3.4 Mangrove degradation

Mangroves are distributed along most of Guyana's coastline (see Figure 17.4), comprising an area estimated at 80 442 hectares (GFC, 2001).

Unfortunately, in Guyana, there seems to be a limited understanding of the values and functions of mangroves. The importance of mangroves is only realised when large investments have to be emptied into constructing coastal protection structures that attempt to re-establish the natural functions of mangrove forests (EU, 2006). Some major causes of mangrove degradation in Guyana include (adapted from National Development Strategy, 1996; EU, 2006; Ramsay and McCall, 2006) (1) population pressure and related waste generation associated with human settlement, (2) unregulated extraction to meet fuel needs by communities, (3) an increase in aquaculture projects via pond construction and clearing lanes to allow inflow of water, (4) agriculture activities that encroach on mangrove forests (e.g. cultivation of water melons) and add to the accumulation of fertilizers and chemical pesticides, (5) coastal pollution from increased industrial activities, (6) siltation of streams and rivers by mining activities and (7) urban and industrial development.

Currently, there is no specific legislation on mangroves in Guyana. There are, however, a Mangrove Action Plan (MAP) and clauses in four legal instruments, namely, the Environmental Protection Act, Sea Defence Act, Fisheries Act and the Forests Act, which address protection and management issues (EU, 2006). A review of existing policies and legislation regarding mangrove conservation and mangrove restoration are among the proposed actions of the MAP (Mangrove Action Plan 2001).

17.3.5 Fisheries and aquaculture

The Fisheries sector of Guyana comprises three main components, namely, Marine Fishery, Inland Fishery and Aquaculture. According to the revised Draft Fisheries Management Action Plan (FMP, 2006), Guyana's fishing activities occur primarily on the continental

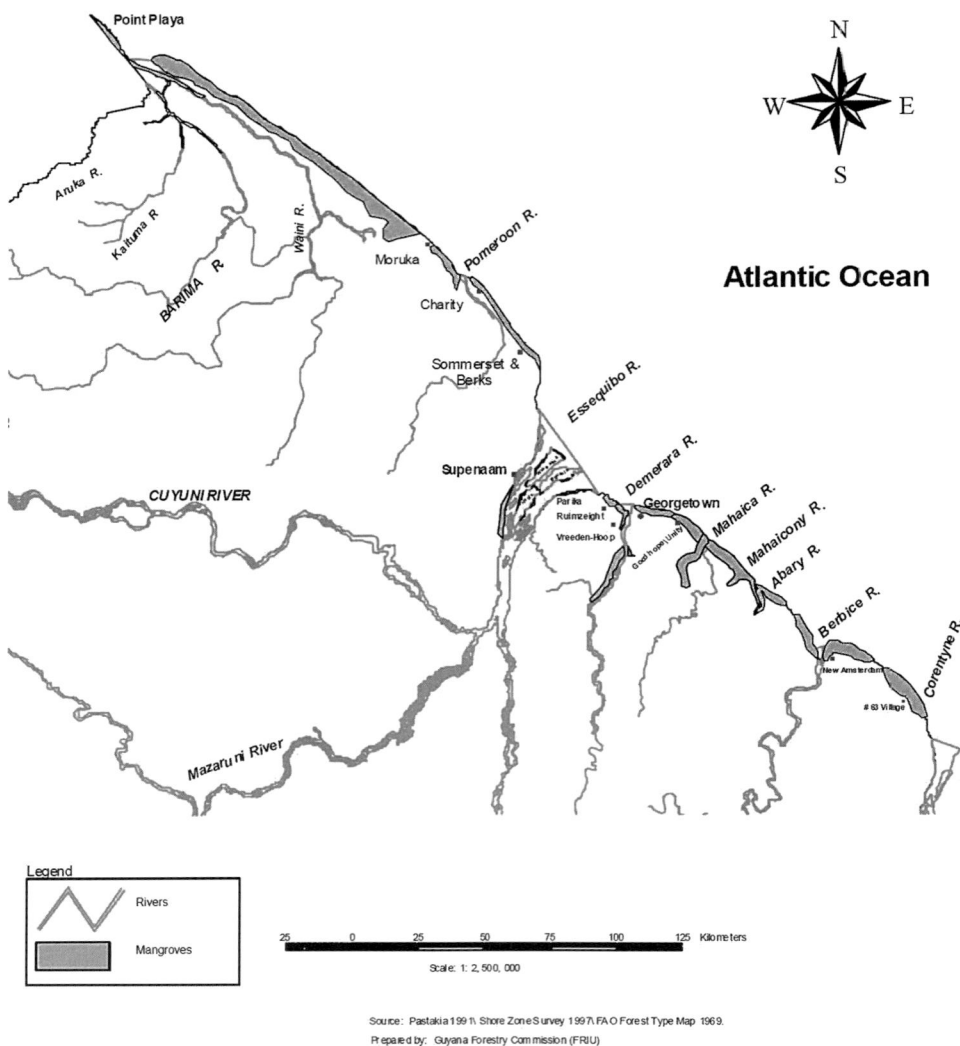

Figure 17.4 Distribution of Mangroves within the coastal zone of Guyana. (Adapted from Pastakia, 1991.) (To see this figure in colour, please see colour plate 19.)

shelf and to a minimal degree on the continental slope, but no activity occurs in the far offshore area where the potential of coastal and oceanic pelagic species is unknown. The fishing pressure on traditional inshore fisheries continues to increase, as reports of low abundances (Figures 17.5 and 17.6) and decreased recruitment rates of several commercially viable fish species have been made (FAO, 2005) (refer to Case Study).

While some of the fisheries in Figures 17.5 and 17.6 seem to show a steady level in fisheries catch (or production), the total effort on each fishery during the same time period is unknown, but is thought to have increased (FMP, 2006). The Case Study below,

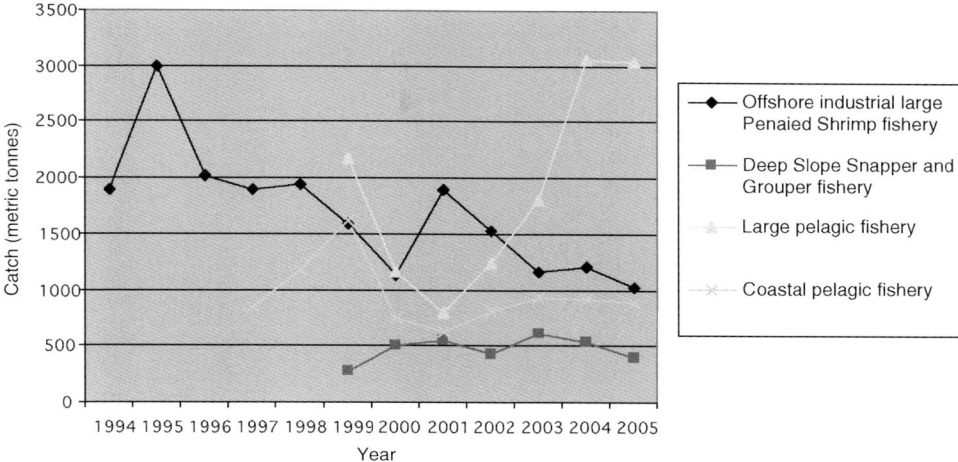

Figure 17.5 Fishery trends based on catch history of commercially viable species. (Adapted from FMP, 2006, and FAO, 2005.)

which examines coastal resource use issues relevant to North West coastal communities of Guyana, reveals that a majority of community fishers felt that while their fishing effort increased, the size and quantity of their catch diminished. According to the FMP (2006), these low abundances may be a result of the following: (1) full- or over-exploitation of certain fish species (e.g. Penaeid Shrimp, Seabob, Grouper and Snapper), (2) unsustainable fishing practices (e.g. trawling of ocean bottom, small mesh size) causing high levels of bycatch and (3) an overall uncontrolled, poorly regulated and minimally enforced fisheries management regime (FMP, 2006).

The Fisheries Department has the mandate to manage and develop all the fisheries resources. Specifically, the Fisheries Act 2002 mandates that the Chief Fisheries Officer

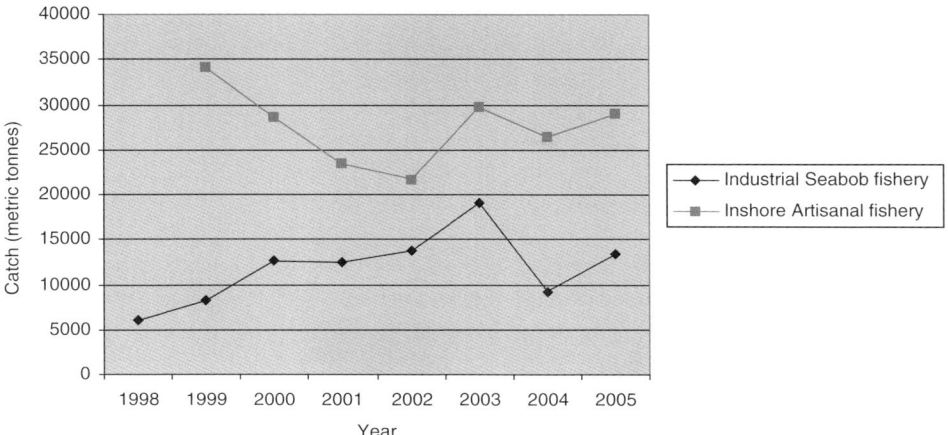

Figure 17.6 Trends based on catch history of two (2) commercially viable fisheries. (Adapted from FMP, 2006, and FAO, 2005.)

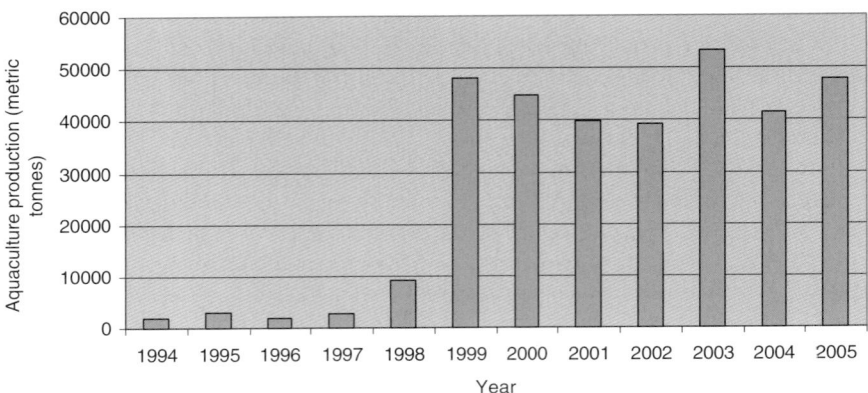

Figure 17.7 Production trend of cultured fish in Guyana between 1994 and 2005. (Adapted from FMP, 2006, and FAO, 2005.)

consults with fisher folk and other stakeholders. Currently, regulations do not exist for the small pelagic fishery; and while the other fisheries are regulated, mechanisms for ensuring that such regulations are complied with are weak. The Department of Fisheries liaises with the Coast Guard and the Marine Police on fisheries enforcement issues. However, the surveillance and monitoring capacities of these agencies have been severely limited by inadequate resources, maritime delimitation matters with Suriname,[1] and relatively expansive maritime zones (FMP, 2006). Furthermore, current fisheries regulations are broad and non-specific. Regulations do not appropriately take into account the nature of the fish species being managed. Additionally, comprehensive and reliable assessments of fishing effort and fish stocks have not been recently done for most commercially viable species, for which regulations have been developed. Recent assessments of fishing effort relative to the industrial Seabob fishery have been done, but the reliability of the findings from the assessments has been questioned, thereby prompting the need to adopt precautionary approaches for reducing fishing effort (FMP, 2006).

The aquaculture industry in Guyana has gained much attention, as the production of cultured fish has significantly increased over the last decade (Figure 17.7). In response to this increase, the Government of Guyana, with technical and financial assistance from the Food and Agriculture Organisation (FAO) and the Canadian International Development Agency, implemented several initiatives, including the establishment of an aquaculture demonstration farm and the National Aquaculture Association of Guyana (NAAG) (adapted from FMP, 2006).

While some initiatives are required to improve aquaculture development in Guyana, they seem to focus on production and economic gain, and inadequately address environmental repercussions. As with the active fisheries industry, appropriate regulations for sustainable

---

[1] A ruling on this maritime delimitation dispute between the two countries was made by the United Nations Tribunal in September 2007, on the basis of the United Nations Convention on Law of the Sea (UNCLOS).

management of the aquaculture industry are lacking, and the negative feedback cycle described by Tagliani et al. (2003) is imminent.

These threats to Guyana's coastal zone diversity and an overall failure by the Government to sustainably maximise economic benefits are among the principal challenges that urge the need to improve upon the current ICZM approach (National Development Strategy, 1996; EPA, 2005). An ICZM regime should not only create new regulations but should generate mechanisms to resolve conflicts between and among sectoral, coastal and ocean-based activities; provide adequate scientific and technical information for making informed (policy) decisions on ocean/coastal use issues; and establish simple and realistic mechanisms for promoting education, awareness and citizen participation.

17.4 Case study: resource use issues as reported by communities along the North West coastal area of Guyana

Focus group and individual interviews were undertaken with representatives (e.g. Village Captain and/or Vice Captain, Councillors, Treasurers and Woman's Group Leaders) from six (6) of twelve (12) coastal communities in Regions 1 and 2 (North West District of Guyana). The coastal communities studied were Assakata, Warapoka, Waini, Almond Beach, Santa Rosa and Mabaruma. With the exception of Mabaruma, these communities consisted mainly of indigenous peoples of various tribes depending on the location and were quite small in population (from 208 to just over 500 people). The interviews concentrated on the following topics: local resource use, developmental changes within the community and surrounding area, knowledge of fisheries and marine turtle species specific to the area, knowledge/understanding of existing governmental regulations and recommendations on improving management of local resources. Information was also gathered during a 'walk-around' of each community.

It was reported by 5/6 (83%) of the coastal communities visited that local fish species, historically abundant, have been greatly reduced in number and size. Representatives from these communities complained of having to 'travel further to catch smaller fish' within the last 10–15 years. The main reason for such a decline, suggested by 4/5 (80%) of these communities, was increased motorboat traffic that results in high gasoline pollution and physical disruption of habitat. A few of the communities (50%) also mentioned an increase in fishing in the area. Concerning the issue of 'poaching' of marine turtle meat and eggs, all six (6) of the communities admitted that 'some turtle harvesting still does take place', but due to the efforts of the Guyana Marine Turtle Conservation Society (GMTCS) and a general increase in environmental education in the region, it has been significantly reduced. According to Dr Peter Pritchard (interviewed at Almond Beach), co-founder of GMTCS who has researched marine turtle activities in the region for over 30 years, the combination of these and other efforts in the region (Suriname, French Guiana and Costa Rica) 'have led to stabilised populations of marine turtles, such as the leatherback, and improved turtle numbers'.

In an attempt to take pressure off the fisheries resource, several community-based projects related to the farming and/or processing of Pineapples, Crabwood oil, Heart-of-Palm and Cocoa were recently initiated in the region. However, community representatives perceived themselves as being 'unfairly treated' and received 'broken promises' from the supporting agencies. Such claims were maintained by a representative of the National Agricultural Research Institute (concerning the Cocoa project) and a recent report by Sullivan and O'Regan (2003) (concerning the Crabwood oil project), who stated that 'incomes are not always positive for primary producers, and... middle men make most of the profits to be made from Crabwood oil commercialisation'. Such projects seem to be mismanaged and unsustainable, and community members are now reverting to some of the more traditional and dependable methods of subsistence, primarily fishing. Overall, communities were unaware of the financial and regulatory logistics pertinent to the commercial sale of their products. This may be a result of a lack of transparency of and information sharing by the supporting agencies, which effectively allows them to take advantage of the primary producers.

Communication from governmental and non-governmental agencies, and overall involvement of community members in policy formation concerning sustainable resource use, is insufficient. Representatives from the communities held limited knowledge on existing regulations on the use of their resources and claimed to have minimal input into any regulations made. Furthermore, enforcement of fisheries and other formal environmental regulations was extremely limited. In contrast, the majority of the communities (67%) expressed the use of their own enforcement rules, premised on social pressure. The recent Amerindian Act 2005 was developed through consultations with Amerindians, who had a chance to inform the content of the Act, regarding rules governing resource use.

17.5 Overview of ICZM in Guyana

In Guyana, the coastal zone management regime was gradually centralised, beginning with the Sea Defence Ordinance in 1883 and culminating with the enactment of the Drainage Act in 1994 (Pelling, date unknown, as cited in EPA, 2000, p. 8). Just before then, in 1990, a subcommittee on coastal zone management was established and tasked with the responsibility of coordinating all interagency bodies and activities, with some relevance to coastal zone development. However several conditions, including the inability to source, attract and retain qualified and competent human resources to manage the committee and coastal zone development activities, as well as the lack of funding to support these efforts, contributed to weak institutional capacity. Some of these very weaknesses continue to challenge the achievement of ICZM goals in Guyana.

The mandate to coordinate ICZM activities was transferred to the EPA, following the drafting and adoption of the Environmental Protection Act in 1996, which formally empowered the Agency with the directive over the Integrated Coastal Zone Programme. The National Environmental Action Plan (NEAP) 1994 relates another support instrument for the ICZM programme in Guyana (EPA, 2005).

Still, the absence of appropriate legislation and resources limits local government's involvement and power in coastal resource management decisions; as such, power over coastal resource management decisions is shared among several centralised ministries and line agencies. In an attempt to strengthen the environmental protection and coastal resource management role of local government, capacity building efforts geared towards improving their ability to deal with related issues have been undertaken.

The ICZM programme in Guyana has, primarily, a horizontal dimension. Zhang et al. (2006) citing Guo (2001) suggests that out of two possible dimensions to ICZM institutional arrangements vertical and horizontal – the difficulty of coastal management lies more with how to deal with the horizontal issues. Indeed, the adoption of a horizontal model for ICZM in Guyana has been subjected to a number of disadvantages, which have undermined the process of adopting and executing more integrated approaches to coastal management. An ICZM Committee was established in 1999 by the EPA, and represents one of the three suggested horizontal institutional models for ICZM as proposed by Cicin-Sain and Knecht (1998). Essentially the national level ICZM Committee acts as an interagency coordinating body for the promotion and execution of ICZM activities in Guyana. To this end the Committee has several responsibilities, namely:

- Coordinating the activities of sectoral agencies with some mandate in coastal resource management.
- Recommending policies for the general management of coastal resources.
- Advising on issues of concern to coastal zone management.
- Advising the EPA on priorities for research, management and monitoring of activities within the coastal zone.
- Advising the EPA on programmes, strategies and plans relating to coastal zone management.
- Participating in training activities for coastal zone management.

The Committee's membership is drawn from several line agencies, environmental research institutions and ministries, including the EPA; Hydro-meteorological Service; Sea Defence Division; Lands and Surveys Commission; Hydrographic Office, Transport and Harbours Department; Ministry of Housing and Water; Fisheries Department; University of Guyana; Guyana Water Incorporated; Guyana Sewerage and Water Commission; Guyana Forestry Commission; Conservation International; and the National Drainage and Irrigation Board.

17.6   ICZM in Guyana: achievements

As stated before, the EPA has a very important role to play in leading ICZM efforts in Guyana. Apart from monitoring and coordinating the progress and activities of the ICZM Committee, the Agency, through its ICZM unit, facilitates and conducts ICZM-related training programmes. Table 17.1 presents a review of the critical activities set out in the ICZM Action Plan (EPA, 2000). The table also relates achievements made with respect to the ICZM activities (see EPA, 2000, and Da Silva, 2007).

17.7   Barriers to developing an effective ICZM programme in Guyana

This section examines some of the principal barriers to effective implementation of ICZM activities. This examination is guided by findings extrapolated from interviews conducted with various ICZM stakeholders, and document reviews.

1. There are overlapping jurisdictions/functions among sectoral agencies over coastal areas and resources, which make the adoption of an integrated coastal resource management

Table 17.1 Summary of actions proposed by the EPA, as part of its ICZM programme, and the related achievements to date.

Actions	Achievements
Delineation of the coastal zone	The proposed actions in this regard include the development and evaluation of criteria for delineation of the coastal zone. To date, no significant achievements have been realised along these lines, although there is general consensus among the relevant institutions and stakeholders on the importance of boundary delineations of the coast for improved coastal zone management
Identifying gaps among sectoral agencies and reviewing existing legislation so as to strengthen the institutional and legal framework for ICZM	EPA's work in this regard has been quite limited. A study was done by consultant Judy Daniels, but no other notable action has been taken by the EPA (Da Silva, 2007). An evaluation of the legislation on, and land use context of the coastal zone was done as part of an Inter-American Development (IADB) sponsored feasibility study of the Shore Zone Management Programme
Creating a database to facilitate improved ICZM	While data on coastal resources exist in some respects, information sharing among sectoral agencies is measured; which has been a major constraint to creating and managing a database. There are gaps in information as well, for example, current comprehensive assessments on commercial fish stocks have not been done recently. While this is not a direct responsibility of the EPA, the related data are very relevant to developing and executing ICZM programme objectives. A National Resource Management Project (NRMP) funded by the German Government was implemented by the now disbarred Guyana National Resource Agency during 1994–2002. The establishment of a National Database (Guyana Integrated Natural Resources Information System – GINRIS) on natural resources for Guyana, and the proposal of a framework for future maintenance and upgrade of the database by governmental agencies were among the objectives of the NRMP
Monitoring the shore zone and rehabilitating of shore zone structures	Some work has been completed. For example, the Land and Surveys Department, which is one of the key agencies involved in the implementation of this ICZM activity, has developed digital maps of the coast. However, much work still needs to be done, as the ability of the Agency to utilise GIS technology is quite limited due to a shortage of the needed technical and financial resources. Furthermore, existing maps are relatively 'basic' and contain minimal information on coastal systems and biodiversity. The ICZM Committee has directed the preparation of a proposal for funding, which was submitted to the Caribbean Development Bank (CDB). The objectives of the project are to create digital map of the coast that contains information on coastal systems, and to fill relevant data gaps (Da Silva, 2007). The Sea and River Defence Division (SRDD), another Agency involved in the fulfilment of this ICZM objective, has been actively involved in sea defence maintenance and rehabilitation, and also embarked on institutional capacity building activities in this regard, for example, it has developed a GIS-linked system for capturing and storing data relevant to sea defence, which can be used for making informed decisions (EU, 2006)

Table 17.1 (*Continued*)

Actions	Achievements
Raising awareness on coastal vulnerability and climate change	To this end, EPA has worked with the Organisation of American States (OAS) in the execution of a project known as the Caribbean Planning for Adaptation to Climate Change (CPACC). Component 3 of the Global Environmental Facility (GEF) funded CPACC project, involves the development of an inventory of coastal resources and uses, to provide baseline data for the execution of other project activities (EPA, 2000)
Assessing the status of coastal and marine resources	Component 6 of the CPACC Project entails an assessment of the vulnerability of the Country's coast, and the risks contained therein (EPA, 2000: 11). According to Da Silva (2007), actions taken to assess the status of coastal and marine resources have been quite limited. On the other hand, achievements have been made in regards to the preparation of a Draft Fisheries Management Plan, and a Mangrove Action Plan. Additionally, there has been ongoing research led by the Guyana Marine Turtle Conservation Society (GMTCS) in partnership with the World Wildlife Fund (WWF) on turtle bycatch and harvesting. An ecologically important coastal area for turtle populations, referred to as Shell Beach, has been identified as a potential Protected Area (PA), as part of the National Protected Area System project
Conducting public awareness	Some work has been done by the EPA to spread awareness on environmental issues, particularly as these relate to solid waste management, biodiversity sustainability and coastal zone protection (EU, 2005). Regional sensitisation workshops and public awareness programmes have been undertaken by the Agency (Da Silva, 2007). The SRDD is also leading a public awareness programme (as part of its capacity building effort), which is funded by the European Union (EU). The aims of the programme are to promote sustainable use of the coastal environment, improve public awareness on related issues and public participation. One of the possible reasons for the generally low level of environmental awareness in Guyana is the absence of administrative mechanisms for facilitating meaningful public involvement. Additionally, public awareness endeavours sometimes exclude essential stakeholders, or involve the exchange of excess, often technical information

framework problematic. Furthermore, agencies are sometimes confused about their role in coastal resource management, which can in effect reduce their commitment to ICZM activities. In fact, several agency representatives who were interviewed concurred that this is a primary factor that has thwarted ICZM efforts. Many cite the need for developing appropriate legislation to clarify jurisdictional and functional powers, as the first step towards any real solution. A legal framework is vital for legitimising policy objectives and establishing a structure for implementing policy initiatives related to ICZM. Yet in Guyana, related legislation is either not there to support the ICZM agenda, or obsolete.

The absence of legislation that spells out the delineation of the coastal zone is another related barrier.

2. There is the problem of a lack of political will and agency commitment, which is demonstrated by poor information sharing among agencies and interagency conflicts, arising from competing objectives and resource scarcity. Perhaps on account of the fact that ICZM in Guyana is still in its developmental stage, the ICZM Committee falls short on the required commitment and enthusiasm needed to direct and progress ICZM efforts. While views might differ among Committee members on this issue, observations and consultations with a few of them have led to the researchers' conclusion that members are first and foremost loyal to those organisations that they are employed with full-time, and perceive their membership on the ICZM Committee as a secondary priority. The resultant subordination of ICZM objectives to Agency mission is a problem that needs to be addressed.

3. In instances where the political will is present (as demonstrated by the government's policies on the construction, rehabilitation and maintenance of the country's sea defence), other constraints such as the inability of the economy to support and sustain advanced levels of public investment can impede ICZM (European Union, 2006). By extension, it is a 'strain' on the economy that often motivates the prioritisation of policies of economic interests over environmental and conservation interests.

4. The approach to coastal resources management is still largely fragmented and sector-based, even with the establishment of the ICZM Committee as an interagency coordinating body. One reason why the transition by agencies to an integrated management regime has been slothful relates the fact that the 'sectoral mentality' is deeply imbedded in the culture and tradition of the participant agencies (Cicin-Sain and Knecht, 1998). Poor information sharing and communication among agencies further perpetuate fragmentation among agencies.

5. The ICZM programme is implemented predominantly through the national government. To date, efforts to decentralise or devolve coastal resources management responsibility to lower levels of government, such as the municipal level, have been minimal. Institutional reformation of local government, vis-à-vis decentralisation, as an option for improved coastal resource management has been approached with caution. One of the reasons accounting for this relates to the state of current local government capacity, which constrains their ability to assume additional responsibilities in regards to coastal zone management (personal communication with the Advisor to the President of the Cooperative Republic of Guyana on Sustainable Development, May 2007). Efforts to initiate and cultivate responsible community-based coastal resource management practices have been few and far between (information provided by the Research Coordinator the GMTCS through email correspondence, April 2007). While the Environmental Protection Act 1996 mandates the involvement of members of the public in the process of integrating their environmental concerns in sustainable development plans, mechanisms for public involvement are not specific (personal communication with the Chairman of the ICZM Committee, April 2007). In particular, no known procedures exist to facilitate community-based involvement in the development of ICZM-related environmental regulations (information provided by the Director of the EPA, through formal written correspondence, April 2007). As such, these regulations, which are characteristically issue specific and sectoral, seldom reflect the socioeconomic realties faced by local

coastal resource users (personal communication with the Chairman of the ICZM Committee, April 2007). This presents a problem, as more often than not, the legitimacy of these regulations is grounded in the active support of those affected (Fanning, 2000). Therefore, the intentional or unintentional exclusion of local coastal resource users from the process of developing regulations can lead to them questioning the legitimacy and fairness of such regulations, as the case study illustrates.

6. Limited human resources trained in coastal zone management, as well as the related issues of low salary levels for and brain drain of skilled personnel are major institutional barriers to the success of the ICZM programme.
7. Vertical agencies, such as the Coast Guard, which as the point Agency for monitoring the Exclusive Economic Zone and territorial seas are not appropriately represented on the ICZM Committee. This exclusion could create enforcement shortcomings in relation to some of the EPA's planned activities that pertain to ICZM (personal communications with the Head of the Coast Guard, Guyana Defence Force, April 2007).

17.8 The way forward: recommendations and opportunities

1. The ICZM Committee needs to play a more active role in sourcing and creating opportunities for developing innovative solutions in response to coastal resource use, jurisdictional, interagency conflict and legislative issues. Interviews with several representatives of the Committee revealed that funding remains a fundamental constraint to meeting related responsibilities. The Committee's main weakness is that its membership is not being appropriately utilised where sourcing the needed funds is concerned. The Committee should elect a head to lead the identification of funding opportunities, and delegate duties to other Committee members related to preparing proposals for funding. This level of commitment can only be realised if the ICZM Unit of the EPA plays a more pronounced and complementary management role in the proceedings and developments of the Committee. Additionally, the Committee could explore more multifaceted revenue generation measures to financially sustain ICZM programmes and projects beyond the initial funding period, for example, they can set up practical arrangements for sharing the costs of ICZM activities with private sector organisations, NGOs and financial institutions (Inter-American Development Bank, 1998).
2. To effectively deal with the problems concerning the lack of political will and agency commitment, the ICZM Unit within the EPA needs to embark on efforts devoted to assuring member agencies that their participation is vital for the success of the ICZM programme. Representatives from the various sectoral agencies who sit on the ICZM Committee possess the expertise and have access to vital information on ICZM management issues. The importance of their role in ICZM cannot be overemphasised. It is also worthwhile to create incentives for participation among agencies. Efforts should be geared towards ensuring that Committee members do not view the ICZM programme as separate and antagonistic to their home agencies' mission and mandate, but rather as complementary and integrated.
3. Given the lack of basic knowledge on the status of resources (e.g. fisheries), and the limited means to find out, involvement at the community level in the development of

relevant and appropriate environmental regulations is crucial if successful management of coastal resources is to be realised. Mechanisms for public involvement must be made specific to the socioeconomic and cultural realities of the coastal resource users. For example, frequent meaningful visits (both formal and informal) to communities should be made to establish a sense of familiarity and trust between groups, and the mode(s) of communication, during such visits, should be relevant and understandable by both groups. As suggested by Tagliani et al. (2003), the ICZM programme should be strongly oriented towards building capacity of such 'social actors' and developing economic alternatives to rural communities that can be sustained by the users.

4. The ICZM Committee, with the approval of the EPA, should engage the attention of vertical agencies, such as the Coast Guard, in the planning, execution and overall management of ICZM activities. These agencies can provide valuable information that can be used to inform ICZM-related decisions and action plans.
5. Legislation is not enough to prevent the unsustainable use of coastal resources; the general population has to be made to understand how inappropriate coastal resource uses affect them. They have to be made understand the importance of minimising detrimental actions. This can only be done if efforts are expanded to promote their participation in monitoring, reporting and maintenance decisions and actions (EU, 2005).
6. Training programmes should be expanded, with a focus on building participants' skills in interagency cooperation, negotiation and conflict resolution. Such training should be aimed at selected agencies and stakeholder groups, such as the Department of Fisheries, the Forestry Commission, Lands and Surveys Division, University of Guyana, fisheries cooperatives and municipalities (and other levels of local government). Training has to be complemented by the installation of mechanisms for retaining skilled personnel.
7. Base line data generated from EIA studies and other project-based research should be better managed and documented for optimal utilisation. Access to this information is often limited. Much of this data is very valuable and relevant to coastal zone management issues.

17.9 Conclusion

While there have been some accomplishments during the first 7 years of the creation of the ICZM Action Plan, real, sustained progress of the ICZM agenda in Guyana has been circumscribed by the existence of several deficiencies and barriers. The ICZM programme, set up by the EPA, is wanting in several respects. This programme lacks supporting legislation, competent and committed human resources, workable mechanisms for integrated enforcement/management and adequate financial and scientific support. Additionally, the system suffers from poor information sharing among the relevant agencies. Some of the recommendations for addressing some of the stated barriers include providing incentives to ICZM Committee members so as to educe their support and commitment to ICZM goals; expanding ICZM training programmes to include a focus on building participants' skills in interagency cooperation, negotiation and conflict resolution; setting up practical arrangements for sharing the costs of ICZM activities with private sector organisations, NGOs and financial institutions; exploring more multifaceted revenue generation measures to financially sustain ICZM programmes and projects; setting up administrative mechanisms for more meaningful public involvement in ICZM decisions; and better management and documentation of EIA studies and project-based research to improve their utilisation.

Acknowledgements

The authors wish to extend their gratitude to those persons who were interviewed, who shared their views and expertise willingly and graciously. We thank the following organisations and persons for their invaluable contributions to the study: University of Guyana, European Union – Guyana Delegation, Regional Chairman for Regions 1 and 2 and Jermaine Clarke (Junior Researcher, Centre for the Study of Biological Diversity, University of Guyana).

Interviews (structured and unstructured) were conducted with the following persons: Chairman of the ICZM Committee, Chairman of the National Climate Committee (NCC), Advisor to the President of Guyana on Sustainable Development, co-founder of and Research Coordinator for the GMTCS, ICZM Specialist of the Sea Defence Division/ICZM Committee member, Director of the EPA, the Head of the Coast Guard Division of the Guyana Defence Force (GDF) and Lands and Surveys Commission.

Although a number of other agencies and ICZM Committee members were approached (sometimes on several occasions), they were reluctant to participate in the study for unspecified reasons.

References

Anders, L., Griesel, A., Hofmann, M., Montoya, M. and Rahmstorf, S. (2005) Dynamic sea level changes following changes in thermohaline circulation. *Climate Dynamics* **24** (4), 347–354.
Cicin-Saine, B. and Knecht, R.W. (1998) *Integrated Coastal and Ocean Management: Concepts and Practices*. Island Press, Washington, DC.
Da Silva, P. (2007) Situational Analysis: Thematic Area Coastal Biodiversity. National Biodiversity Action Plan, Georgetown, Guyana.
EPA (2000) *Integrated Coastal Zone Management Action Plan*. Environmental Protection Agency, Greater Georgetown, Guyana.
EPA (2005) *National Environmental Action Plan*. Environmental Protection Agency, Greater Georgetown, Guyana.
EU (2005) *Institutional Capacity Building on Guyana Sea Defences: Socio-Economic Survey and Public Awareness Programme*. Delegation of the European Commission to Guyana, Georgetown, Guyana.
EU (2006) Feasibility Study for the Sea Defences Programme Financed under the 9th EDF for 'The Co-operative Republic of Guyana': Feasibility Report. European Union. A project implemented by Mott MacDonald.
Fanning, L.M. (2000) *The Comanagement Paradigm: Examining Criteria for Meaningful Public Involvement in Sustainable Marine Resource Management*. University of Chicago, Chicago.
FAO (2005) Status of Fisheries and Aquaculture Production in Guyana, Georgetown, Guyana.
Field, J.G., Hempel, G. and Summerhayes, C.P. (2002) *Oceans 2020: Science, Trends and the Challenge of Sustainability*. Island Press, Washington, DC.

FMP (2006) *Revised Fisheries Management Action Plan*. Fisheries Department, Government of Guyana, Georgetown, Guyana.
GFC (2001) *Guyana National Mangrove Management Action Plan*. Guyana Forestry Commission and the Integrated Coastal Zone Management Committee, Georgetown, Guyana.
Inter-American Development Bank (1998) *Strategy for Coastal and Marine Resources Management in Latin America and the Caribbean*. Bank Strategy Paper, Washington, DC, December 1998.
Kont, A., Jaagus, J. and Aunap, R. (2002) Climate change scenarios and the effect of sea-level rise for Estonia. *Global and Planetary Change* **36** (1–2), 1–15.
National Development Strategy (1996) National Development Strategy for Guyana, Georgetown, Guyana.
Pastakia, C.M.R. (1991) Preliminary Study of the Mangroves of Guyana. (Article B 946/89 Contract No: 8912) Final Report. Aquatic Biological Consultancy Services Limited. European Community, Georgetown, Guyana.
Pelling, M. (1997) What determines vulnerability to floods; a case study in Georgetown, Guyana. *Environment and Urbanization* **9** (1), 203–206.
Ramraj, R. (2003) *Guyana: Population, Environments and Economic Activities*. Battleground Printing and Publishing, Greensboro, NC.
Ramsay, S. and McCall, R. (2006) *Coastal Communities and Mangrove Management; an Investigation of Use Patterns and Awareness of Mangrove Ecosystems in Guyana*. University of Guyana.
Sullivan, C.A. and O'Regan, D.P. (2003) Winners and Losers in Forest Product Commercialization. Document prepared for the International Development and Natural Environment Research Council. CEH, Wallingford.
Tagliani, P.R.A., Landazuri, H., Reis, E.G., Tagliani, C.R., Asmus, M.L. and Sanchez-Arcilla, A. (2003) Integrated coastal zone management in the Patos Lagoon estuary: perspectives in context of developing country. *Ocean and Coastal Management* **46**, 807–822.
UNDP (United Nations Development Programme) (2006) *Guyana Macro-Socio Economic Assessment of the Damage and Losses Caused by the January–February 2005 Flooding*. United Nations Development Programme, ECLAC Subregional Headquarters for the Caribbean, Georgetown, Guyana.
Zhang, L., Xue, X., Fang, Q. and Shen, S.S. (2006) A proposal for the application of integrated coastal management for institutional arrangements in Xiamen. *Aquatic Ecosystem Health and Management* **9** (1), 111–115.

Further Reading

Acioly, C., Jr. (2000) Reviewing urban revitalisation strategies in Rio De Janeiro: from urban project to urban management approaches. *Geoforum* **32** (4), 509–520.
Chopyak, J. and Levesque, P. (2002) Public participation in science and technology decision making: trends for the future. *Technology in Society* **24** (1–2), 155–166.
Christie, P. (2005) Is integrated coastal management sustainable? *Ocean and Coastal Management* **38** (3–6), 208–232.
Eisma, R.V., Christie, P. and Hershman, M. (2005) Legal issues affecting sustainability of integrated coastal management in the Philippines. *Ocean and Coastal Management* **48** (3–6), 336–359.

Chapter 18
Strategies for the Beneficial Use of Dredged Material in Japan

Ryoji Naito and Yoshiyuki Nakamura

Abstract

In this chapter, strategies for the beneficial use of dredged material in Japan are summarized. Whereas the total amount of dredged material per annum is predicted to remain almost constant over the next 10–20 years, marine disposal will be restricted due to ratification of the London Dumping Convention, in particular its 1996 Protocol. To date, most material dredged from ports and harbors in Japan has been used for coastal landfill projects, airport construction and other industrial or commercial purposes. Because the availability of possible landfill sites will be restricted in ports and harbor areas due to the convention named above, the beneficial use of dredged material needs to be promoted urgently. For about 20 years, dredged material in Japan has been used for marine environmental restoration projects such as the construction of shallows, tidal flats and eelgrass beds. More than 1000 ha of tidal flats have been constructed in many coastal regions of Japan so far. Very recently, the re-contouring of borrow pits has received wider attention because most subaqueous borrow pits cause severe deterioration of water quality due to oxygen depletion and the formation of blue tides. Thus, filling borrow pits with dredged material may be an alternative way of putting dredged material to good use. Further research and development themes for promoting the beneficial use of dredged material are also discussed.

18.1 Introduction

As waterways, ports and harbors usually have an enclosed typography, suspended sediment tends to be deposited on the seabed in a cumulative manner. In order to maintain waterborne commercial activities, dredging is essential in most ports and harbors. Additionally, as the draught and capacity of vessels such as container ships have increased over the years, they need deeper channels to navigate safely. Thus, new dredging sites are being continually developed. The management of dredged material thus forms an important task for harbor engineering services. This material is basically a valuable resource which can be used, although much of it is currently disposed because of economic, logistic or environmental constraints. PIANC working group 19 (PIANC, 1992) provided guidance on the procedure to assess the possibilities of beneficial use of dredged material. Beneficial use is not yet common practice in all parts of the world, although some countries already make extensive use of dredged material and value it as a resource. In Japan, for example, more than 60%

of dredged material is used beneficially. In other countries, a number of constraints have prevented more extensive use (CEDA, 2005). However, public concern about conservation of the coastal ecosystem is increasing and recent national and international regulations on oceanic disposal, such as the 1996 Protocol of the London Convention, have been strictly observed. However, more attention must be paid to the handling of dredged material and to promote its beneficial use.

PIANC working group 14 (PIANC, 2009) has recently identified a range of beneficial uses and have classified these into two categories. One is engineering uses such as construction materials, isolation and flood defense. Construction of Chubu International Airport in Japan (Sato et al., 2004) and brownfields site development, Fasiver, Belgium (PIANC, 2009) are the examples. The other category is environmental enhancement including habitat creation and enhancement, maintenance of sediment supply, aquaculture and recreation. Habitat creation in Galveston Bay, environmental restoration project in Poplar Island in Chesapeake Bay, USA (US Army Corps of Engineers and Maryland Port Administration, 2004), and wetland creation in Wallasea Island, UK (DEFRA, 2006) are the examples. However, monitoring results of biology and ecology at the construction sites are less frequently documented. For environmental enhancement, research to promote better understanding of ecological effects in the application of dredged material for habitat creation is needed.

In Japan, not only habitat creation such as tidal flat construction but also relatively unique projects such as filling depressions have recently been performed in the context of environmental enhancement. Filling depressions of borrow pits with dredged material is potentially an alternative and promising way of good use of dredged material in other countries (e.g. DEC, Department of Environmental Conservation, New York; Findings of the technical review panel: Jamaica Bay borrow pit evaluation project; http://www.dec.state.ny.us/website/reg2/jbborrow/findings.html), but few reports on this kind of project are documented so far.

The purpose of the study is therefore to review the unique characteristics of the Japanese activities and to provide strategies for the beneficial use of dredged material in Japan. The information would be of benefit to the readers in other countries where promotion of the beneficial use is an urgent task. In the present report, we first review recent trends in dredging activities in Japan. We follow this by introducing two typical examples of the beneficial use of dredged material in Japan, both are categorized in environmental enhancement: one is a traditional approach and addresses restoration of the coastal environment, such as the construction of tidal flats and shallows; the other is a rather new approach to re-contour or fill depressions, called subaqueous borrow pits, with dredged material. Further research and development themes for promoting the beneficial use of dredged material are also discussed.

18.2 Dredging activity in Japan

Dredged material from ports and harbors, except fishing ports, in FY 2002 amounted to 26.65 million cubic meters in Japan. In terms of the total amount of material dredged, 51% and 39% originated from waterways and anchorage areas, respectively. The total annual amount of dredged material in Japan generally fluctuates between 20 and 30 million cubic meters, although in the 3 years encompassing 1998–2000, up to 40 million cubic meters

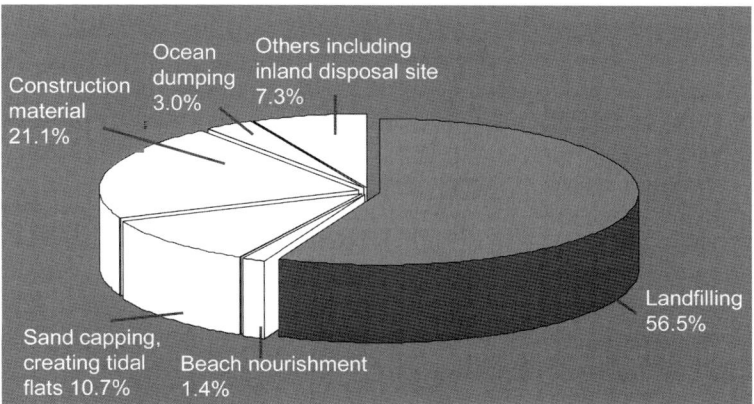

Figure 18.1 Proportional representation of the use of dredged material in FY 2002. (To see this figure in color, please see color plate 20.)

were dredged. The amount of dredged material is predicted to remain relatively constant for the next 10–20 years (Tanabe, 2005). Historically, dredged material has been deposited in the ocean or used for a range of other purposes. The different uses of dredged material as a percentage of total for FY 2002 are shown in Figure 18.1. More than half of dredged material was used as landfill. Use for restoration, including beach nourishment, sand capping and the construction of tidal flats, was 12%. Use of dredged material as construction material in ports and airports accounted for 21%. Only 3% of the total volume was disposed in the ocean. The total amount of dredged material deposited in the ocean compared to that deposited by several other countries is shown in Figure 18.2 (Tanabe, 2005).

In anticipation of the 1996 Protocol of the London Convention, Japan's Marine Pollution Prevention Law was revised in 2004 and enacted in 2007. In the Japanese law, dredged material is defined as waste, and the oceanic disposal of waste must be authorized by the

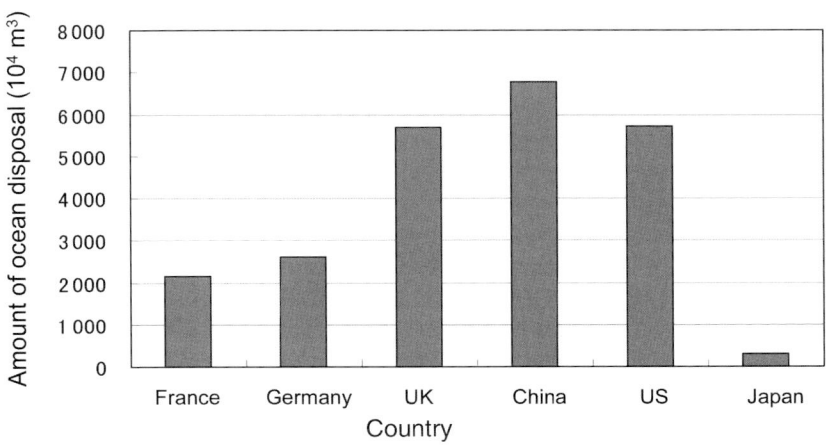

Figure 18.2 Country-by-country data showing the volume of dredged material disposed in oceans in 1999. (*Source*: Data from the International Maritime Organization (IMO).)

Environment Minister. In the assessment procedure for issuing a permit, an environmental impact assessment will be necessary to assess any adverse effects on the environment. Following such regulatory changes, the Ports and Harbors Bureau of the Ministry of Land, Infrastructure and Transport (MLIT) recently published the 'Technical guideline for ocean disposal and beneficial use of dredged material' (Ports and Harbors Bureau, MLIT, 2005). Traditionally, most dredged material is put to good use in Japan. However, the percentage destined for landfill use has decreased slightly because the number of landfill sites has also decreased steadily. The nationwide promotion of the reuse, reduction or recycling of various waste materials in many sectors is another factor that has diminished the number of available landfill sites. This tendency is backed by wider public concern about degradation of the coastal ecosystem. Therefore, it is essential to encourage beneficial use of dredged material, other than as landfill.

18.3 Construction and restoration of habitats

18.3.1 *National activity*

Tidal flats are recognized as crucial coastal areas for both wildlife and humankind, because they provide several benefits, such as biological production, a habitat for wildlife, water purification and recreational amenities. The important aspect is that many of these functions are supported by the organisms living there. For example, bivalves have the ability to purify seawater effectively by filtering large amounts of it. Some species of bivalves have economic value and as such are targeted by commercial fisheries. Gathering the bivalve Asari (the short-necked clam), on the other hand, is a recreational activity in Japan's coastal regions. A wide variety of other organisms also play key roles in providing food for fish and birds. Given their production of these organisms, some tidal flats have been designated as important sanctuaries for wild birds. Since 1945, however, 300 km^2 of Japan's natural tidal flats have vanished due to reclamation and subsidence. From the 1960s to around the 1980s in particular, not only an abrupt increase in anthropogenic nutrient input but also a loss of tidal flats and shallow areas due to reclamation caused water quality deterioration and ecological degradation of Japan's coastal waters.

The recovery and reconstruction of tidal flats is therefore critical in restoring damaged coastal ecosystems. The construction of tidal flats in Japan thus commenced in the 1980s along with active studies on creating tidal flats. The recent enactment of the 'Law for the Promotion of Nature Restoration' has encouraged such activities (Ministry of Environment, 2004). Several documents have been published that summarize numerous nature restoration projects in the coastal waters of Japan in relation to tidal flat construction and monitoring projects, such as the experimental cooperative project at the Han-nan second section in Osaka Bay, the cooperative research project to restore Ago Bay where the pearl oyster has been cultivated for more than a century, tidal flat and shallows construction at Mikawa Bay using dredged sediment from the navigation channel, and long-term monitoring results at Kasai Constructed Beach in Kasai Marine Park, located in the inner part of Tokyo Bay (PARI and KORDI, 2004; Working Group of Coastal Area Restoration, 2004; Nakamura, 2006a). Among these, we will report on two examples of constructed tidal flats and the monitoring activities in Mikawa Bay and the Seto Inland Sea. Additional projects for the restoration of coastal waters will also be discussed.

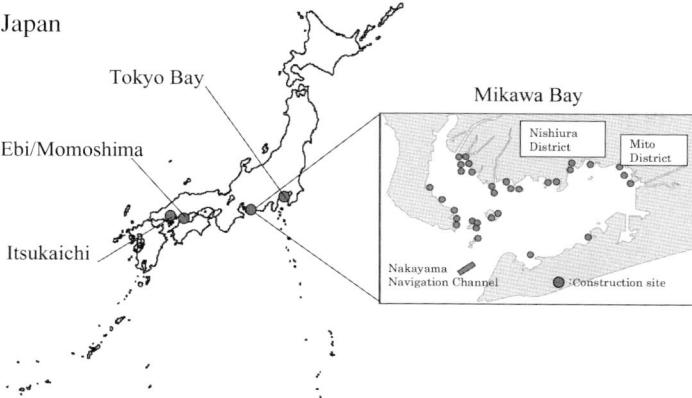

Figure 18.3 Areas in which tidal flats and shallows construction, as well as sand capping, were carried out in Mikawa Bay. The material used was fine sand dredged from widening projects in Nakayama Navigation Channel. The project started in 1998, and 32 habitats (construction site area approximately 620 ha) had been constructed in the bay by FY 2004. (To see this figure in color, please see color plate 21.)

18.3.2 Examples in Mikawa Bay

The MLIT, together with the local government of Aichi Prefecture, drew up plans for constructing tidal flats/shallows as well as sand capping for the environmental restoration of Mikawa Bay by making use of the large quantities of dredged material from the Nakayama Navigation Channel at the mouth of Mikawa Bay. The project was started in 1998 and had been implemented in 32 locations (construction site area approximately 620 ha) by FY 2004 (Figure 18.3). Following construction, the physical properties of the sediment and its quality and geomorphology, as well as water quality and benthic organisms, were continuously monitored in and around each tidal flat site. Recently, Nakata et al. (2005) summarized the monitoring results for two of the tidal flats constructed in the Mito and Nishiura districts. Both construction areas retained sediment with which they were initially filled; this was of uniform particle size, of which fine sand content accounted for 90% or more. However, particle size at the Mito site began to resemble that of the neighboring sea area where the ground level was low (D.L. −2.0 to −1.0 m). Immediately after construction, the terrain of the constructed tidal flats, particularly the landward parts, changed dramatically to form higher terrace due to wave actions. However, the degree of change diminished with the passage of time, and while the initial sinking continued, it seems that the topography has recently stabilized.

The Mito site is close to the mouth of the Toyo River and tends to be of lower salinity than the Nishiura site. The dissolved oxygen (DO) concentration tends to be high at both constructed tidal flat sites throughout the year, but it was confirmed that an anoxic water mass had developed in the neighboring sea areas (see Figure 18.5). Figure 18.4 compares the succession of benthic organisms in the Mito and Nishiura sites. Rapid recruitment of benthic organisms was observed in the subtidal areas of the constructed tidal flats (D.L. −1.0 to ±0.0 m) for approximately 6 months following construction. This was due to the recruitment of adult bivalves such as short-necked clams, which were not present

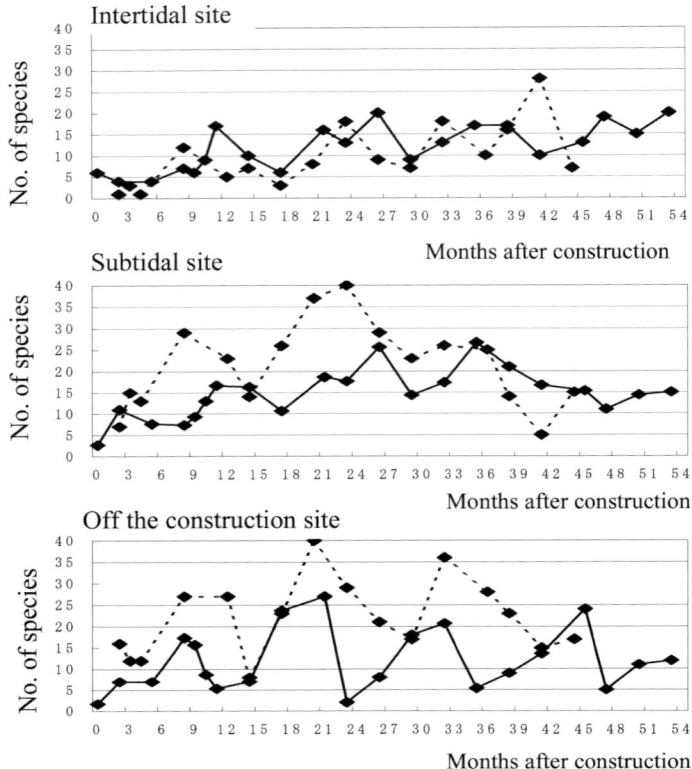

Figure 18.4 Number of species of benthic organisms in Mikawa Bay following construction of artificial tidal flats. Solid lines: Mito district; broken lines: Nishiura district (Kuwae and Nakamura, 2005).

immediately after construction. Approximately 2 years after construction, the number of emerging species began to decline somewhat. The recruitment of benthic organisms was less rapid in the intertidal flats (D.L. ±0.0 to +1.0 m) than in the subtidal areas, but the number of species of benthic organisms continued to increase gradually for 3–4 years after construction was completed. There was also a large fluctuation in the number of species of benthic organisms in the neighboring sea areas: the number of organisms declined only in the summer period and appeared to have been affected by anoxic conditions. Apparently, anoxic events of this kind are not commonplace for the most part in the tidal flats that were constructed (see Figure 18.5). In addition to simply surveying organisms by monitoring, we also conducted studies to clarify forms of recruitment of these organisms to the constructed areas (Kuwae and Nakamura, 2005). There appeared to be various processes by which organisms were recruited onto the constructed tidal flat after the dredged material had been placed over the original ground: recruitment of larvae to a newly created habitat; apparent recruitment of organisms previously contained within the placed material; organisms moving upward from the original ground, over which covering sandy materials were placed; and organisms moving or drifting in from neighboring areas and not being placed there in the material. Monitoring results with special devices to identify the various

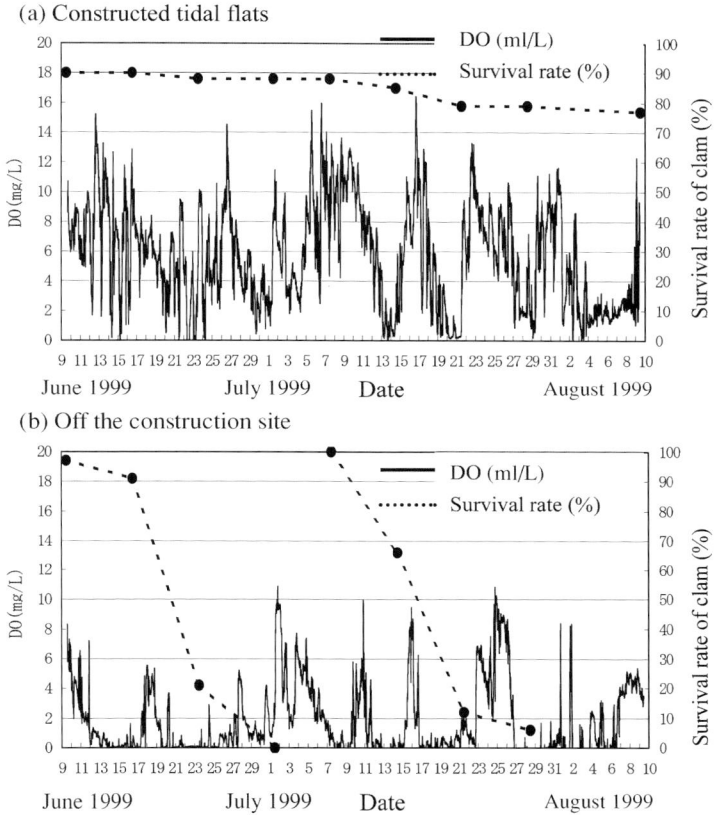

Figure 18.5 Variations in DO observed in the constructed tidal flat at Mito district and its adjoining deeper area as well as the survival rate of caged short-necked clams (Nakata et al., 2005).

ways of recruitment showed that larval recruitment was minimal and that the contribution of apparent recruitment of adult animals that moved to or drifted in from surrounding areas was larger.

The results of studies on the constructed tidal flats in Mikawa Bay showed that organisms were recruited somewhat earlier than those that had been observed during the 1-year period after the start of the experiment with the tidal flat mesocosm (Kuwae and Hosokawa, 2000). The order in which the organisms arrived was also found to be different. In a mesocosm of the tidal flat, benthic microalgae and bacteria first recruited and grew, and this was followed after about 10 months by growth of the meiobenthos and macrobenthos that fed on them. In contrast, the constructed tidal flat in Mikawa Bay was marked by the arrival of numerous adult macrobenthic organisms at an even earlier stage (after about 6 months). It was expected that organisms would enter in a relatively shorter time scale if adult organisms could enter from neighboring natural or formerly constructed tidal flats.

Providing habitats for birds may be regarded as one of the major objectives in constructing tidal flats. Kuwae and Nakamura (2005) monitored for 7 months, from summer to winter, three constructed tidal flats and three natural tidal flats situated in Mikawa Bay to determine the species and numbers of birds flying to the tidal flats. The populations of birds occupying the natural tidal flats were yet to be observed flying to the constructed sites. We studied the numbers of snipe and plover (two types of birds that are observed feeding actively on tidal flats) and the numbers of individual organisms serving as prey (polycheate and crabs), and found a mutual relationship between the two. It is therefore possible that the scarcity of birds flying may be due to lack of food in the constructed tidal flat compared to natural tidal flats. It should be noted that the results obtained represented in the initial stages (around 3 years) after construction. Follow-up studies will be needed to determine whether the number of benthic organisms and the number of birds visiting tidal flats changes over time.

18.3.3 Examples in the Seto Inland Sea

In the central region of the Seto Inland Sea, two constructed tidal flats, the Itsukaichi tidal flats and the Ebi/Momoshima tidal flats, are located in Hiroshima Prefecture. Tidal flats constructed in the Ebi and Momoshima districts have typically been considered as successful examples of such construction and were honored by an environmental prize from the Japan Society of Civil Engineers. The objective of the construction was to create a habitat for the short-necked clam by using dredged sediments taken from the Onomichi-Itozaki Ports. Dredged sediments comprising mud and sandy materials were capped with dredged materials. The construction period was 1984–1987 and the cost of the project was 1700 million yen. The total area of the redeveloped site was 36.9 ha. Geomorphological changes, due mainly to land subsidence, took place immediately after construction; however, long-term biological monitoring results revealed that the sustainable natural reproduction of targeted shellfish had been achieved. Eelgrass beds also became abruptly widespread 10 years after construction.

The Itsukaichi tidal flat, the construction of which was completed in 1991, was designed for mitigation purposes. An estuary-type tidal flat existed at the mouth of the Yawata River, which runs into the western part of Hiroshima Bay. This location served as a feeding and resting place for 84 species of migrating birds. Due to a landfill project, an alternative artificial tidal flat was planned in the adjacent area as a mitigating factor. The designed tidal flat had an area of 24 ha and an intertidal zone width of 250 m, which was about the same size and width as the disappearing natural tidal flat. Ground elevation was designed to be C.D.L. -2.4 to $+1.0$ m, which is suitable for wild birds to feed on benthic fauna and for resting. The ground slope was 1.65%, exactly the same gradient as the natural tidal flat (Working Group of Coastal Area Restoration, 2004). Immediately after construction, the number of wild bird species quickly recovered to former levels prior to construction. However, the horizontal area of the intertidal zone, which is suitable as a feeding place for migrating birds, has been gradually lost according to land subsidence due to compaction. In the 4 years after construction, the ground level descended by 2 m. Geomorphological characteristics of the tidal flat describe a so-called two-layer structure: the main body of the constructed tidal flats consists of dredged sediments with a muddy component. Sandy materials are capped with dredged sediments. Reconstruction action against subsidence has now been conducted to ensure that the tidal flat characteristics meet the original goal of the project.

18.4 Re-contouring subaqueous borrow pits

Mining for commercial volumes of aggregate for use in construction and for landfill projects has left borrow pits on the seafloor of several coastal regions of Japan. A nationwide survey revealed that these borrow pits can be classified geomorphologically into two types: flattened-out and local depressions. Most of the sand mining pits in the Seto Inland Sea are classified as the flattened-out type. The sand mining there is considered to have caused the direct loss of a habitat for economically valuable fish species and indirect loss of eelgrass beds (Nakamura, 2006b). Despite the loss of eelgrass beds, flattened-out type pits appear to have less evidence of oxygen depletion compared to local depressions. Even so, borrow pits classified as local depressions often cause severe deterioration of water quality, which can manifest as anoxia and blue tides. The local depression type of borrow pits are widespread in Tokyo Bay, Osaka Bay, Mikawa Bay and other eutrophic embayments in Japan. Re-contouring (raising the bottom) is expected to be an effective method of restoring damaged ecosystems. Although only a few geomorphological restoration projects have been implemented in Japan to date, actions to re-contour borrow pits and to assess the effectiveness of this approach, including monitoring activities, have started in Mikawa Bay. In order to promote effective restoration and to minimize negative effects, we conducted several research initiatives, including an evaluation of the procedures for assessing the effect of restoration on water quality and coastal ecosystems, and an analysis of the technical methods available to prevent the dispersion of turbid water masses produced by placing material in borrow pits. We also assessed the adverse effects of chemical compounds contained or accumulated in the borrow pits on benthic fauna and coastal ecosystems. Considering that the amount of material necessary for complete filling is greater than the annual supply from dredging (e.g. in Tokyo Bay), the countrywide systematic management of dredged sediment or the use of materials other than dredged sediment, such as slug, would appear necessary.

18.4.1 *Example of Tokyo Bay*

Huge borrow pits are located in the inner part of Tokyo Bay along the northeastern coast, off Chiba Prefecture. In this area, which has been dredged for sand for landfill projects, all of the borrow pits are classified as local depressions. The total sand mining volume in the entire Tokyo Bay is estimated to be 123 million cubic meters, for which some of the borrow pits have been partly filled with dredged material. However, about 100 million cubic meters of borrow pits remain in this area. The still remained maximum borrow pit has volume of 55.3 million cubic meters. Such borrow pits provide an enormous potential for filling given that the annual amount of dredging in Tokyo Bay, where maintenance dredging is ongoing, accounts for only 2.9 million cubic meters.

All of the borrow pits in Tokyo Bay are geometrically steep-sided, with the bottoms of the pits being 20–30 m deeper than the surrounding terrain. Therefore, they are stagnant in nature and accumulate organic materials. Water quality deterioration, such as oxygen depletion, inside the borrow pits is severe. Moreover, blue tide formation because of upwelling events of the bottom anoxic water mass is frequently observed in the inner part of the bay. Sasaki et al. (1996) estimated from field observations and by numerical modeling which borrow pits are responsible for the formation of large-scale blue tides, such as the one observed in August 2002. Because a blue tide consists of an anoxic water mass, its influence on adjoining coastal areas, including the biomass-rich area of Sanbanze, often leads to severe damage to fisheries.

Re-contouring (raising the bottom) by placing dredged material in one of the borrow pits off Kemigawa, Chiba Prefecture has been ongoing since FY 1997 as an interlocal governmental project. The filling project is ongoing as the 'Shallow water fishing ground restoration project' in Chiba Prefecture; the dredging project that provides the dredged material is being carried out in Tokyo Port by the Tokyo Metropolitan Government. Every year about 1 million cubic meters of sediment is placed in the borrow pit, with the project expected to continue until around FY 2010.

18.4.2 Example of Mikawa Bay

Two large borrow pits are located in the eastern part of Mikawa Bay. One is the Mito District pit (East pit), the volume of which is 1.4 million cubic meters, with a surface area of 46.8 ha. The other is the Otsuka District pit (West pit) with a surface area of 69.4 ha and a volume of 1.8 million cubic meters (see Figure 18.6) of each pit, which is lower than the surrounding area by 3–4 m. While the pits have relatively gently sloping sides compared to those in Tokyo Bay, these borrow pits have been associated with frequent cases of oxygen depletion (see Figure 18.7).

The Rokujo tidal flat is close to the borrow pits and is considered to be a natural reproduction site for the short-necked clam, which is valuable to commercial fisheries. Large-scale blue tide events in the eastern part of Mikawa Bay took place in the summers of 2001 and 2002, and subsequent severe damage to the clam population in the tidal flat

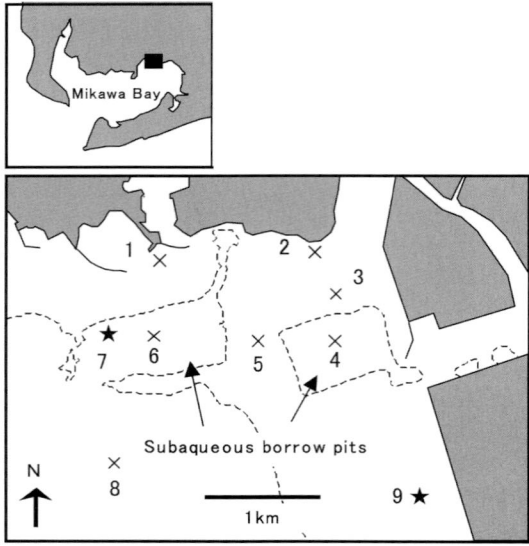

Figure 18.6 Location of subaqueous borrow pits in Mikawa Bay. Numerals denote stations for continuous observations of sediment and water quality and benthic organisms.

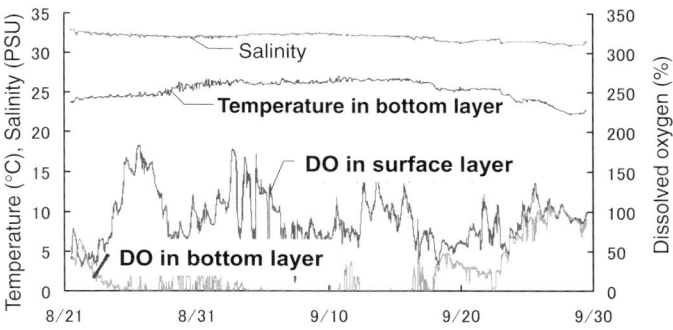

Figure 18.7 Typical results of continuous monitoring of salinity, temperature and DO in Stn. 4 of a borrow pit before filling. (To see this figure in color, please see color plate 22.)

area occurred as a result. Continuous monitoring of the water quality inside and outside the borrow pits by the Aichi Fisheries Experimental Station supported the view that the borrow pits were the main cause of blue tides in both seasons. Local port authorities and the local branch of MLIT considered the use of dredged material from Mikawa Port and conducted a preliminary field survey of the impact of placing dredged material in the borrow pits. The results of the survey revealed that dispersion of turbidity would be minimal and that depositing dredged material in the pit would potentially benefit restoration of the area. Such activities promoted negotiation among several stakeholders and encouraged agreement on restoration action. Thus, a re-contouring project using dredged material from Mikawa Port was started in May 2003. By the end of FY 2006, the East pit was completely filled with dredged material and sand capping is now ongoing to restore the habitat for macrofauna. Re-contouring is also taking place in the western pit.

Besides the practical re-contouring projects in Mikawa Bay, a research project was started in 2005 to assess the effectiveness of re-contouring. The research project aims not only to evaluate the effectiveness of the re-contouring undertaken in Mikawa Bay, but also to draft a standard manual for the promotion of re-contouring. To evaluate the effectiveness of this approach, frequent monitoring of sediment and water quality as well as bacterial and benthic faunal activities inside and outside the borrow pits have been conducted. Three-dimensional numerical modeling is also being conducted to accurately reproduce stratification and destratification processes and to estimate water quality, such as the DO content in the study area. The predicted results of DO and water temperature in several restoration scenarios will be used to create a semi-empirical scheme to estimate macrofaunal biomass and thus to evaluate natural purification activities in terms of organic carbon, nitrogen and phosphorus. Furthermore, in order to promote effective restoration and to minimize the negative effects of placement activities, the project is conducting several research activities, including the evaluation of procedures for assessing the effect of restoration on water quality and coastal ecosystems, and an analysis of technical methods available to prevent the dispersion of turbid water masses produced by placing material in borrow pits. The project also aims to assess the effects of chemical compounds contained or accumulated in the borrow pits (see Figure 18.8) and the methods available to manage dredged sediment and/or identify suitable material other than dredged sediment that could be used as a substitute placement material (e.g. slug).

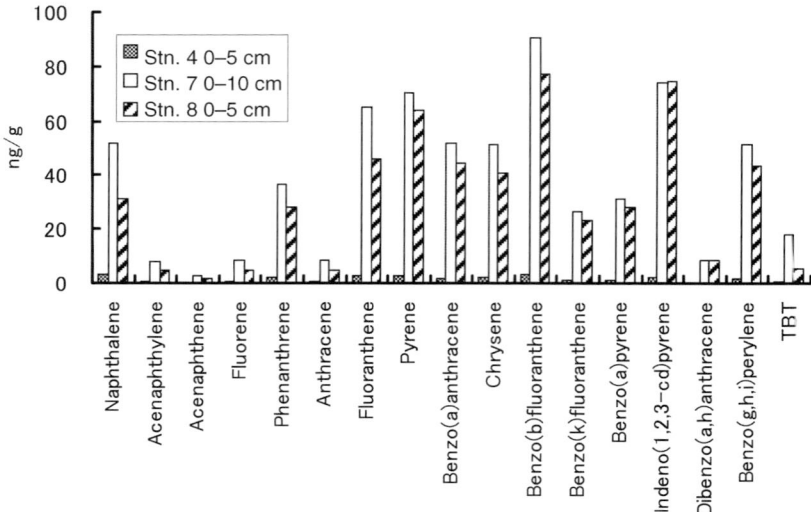

Figure 18.8 Example of sediment quality surveys. PAHs (polycyclic aromatic hydrocarbons) in the surface layer of sediment in and out of the borrow pits in Mikawa Bay. Note that results of Stn. 4 are those after re-contouring with sand capping.

18.5 Further research needed to promote the beneficial use of dredged material

Considering the above-mentioned state-of-the-art uses of dredged material, other aspects of research and development that are necessary to promote the beneficial use of dredged material in Japan are described as follows:

1. Countrywide and long-term analysis for the supply and demand of dredged material.
2. Research on placement materials other than dredged material: a combination or mixture of slug or deposited material in reservoirs.
3. Development of estimation methods to evaluate the effectiveness of ecosystem recovery/habitat creation in constructed areas, and so on.
4. Criteria for prioritizing the different ways for using dredged material beneficially.
5. Methods of minimizing the negative impact on surrounding waters during dredging or placement actions.

Japan is a small country with few natural resources. Dredged material is a good natural resource and has potential for various applications. Therefore, it is a national strategy to further develop and promote the beneficial use of this material.

18.6 Conclusions

Habitat construction or environmental restoration, such as the construction of tidal flats and shallows, has been conducted throughout Japan's coastal regions. Additionally, restoration activities such as filling borrow pits with dredged material is another viable option for the beneficial use of this material. In order to promote such use of dredged material, several guidelines including the 'Technical guideline for ocean disposal and beneficial use of dredged material' edited by Ports and Harbors Bureau, MLIT and the 'Guideline for filling subaqueous borrow pits (provisional)' edited by the Ministry of Environment (2006) have been published. Related research projects are also being pursued in order to successfully restore coastal areas as examples of the beneficial use of dredged material.

References

CEDA (2005) First results of a questionnaire 'Dredged material and sediment management in Europe', Unpublished report.

Department for Environment Food and Rural Affairs (DEFRA) (2006) Wallasea wetland creation project. *Newsletter* **2**, 6.

Kuwae, T. and Hosokawa, Y. (2000) Mesocosm experiments for the restoration and creation of intertidal flat ecosystems. *Environmental Sciences* **7**, 129–137.

Kuwae, T. and Nakamura, Y. (2005) Recruitment of benthic fauna and birds to constructed tidal flats. *Confirmation of Constructed Tidal Flats and Shallows in Mikawa Bay as a Beneficial Use of Dredged Sediment*. Edited by Mikawa Port and Harbor Office, Chubu Regional Branch, Ministry of Land, Infrastructure and Transport, pp. 44–51.

Ministry of Environment (2004) Outline of Law for the Promotion of Nature Restoration. Available from http://www.env.go.jp/nature/saisei/law-saisei/gaiyo.html.

Ministry of Environment (2006) Guideline for filling subaqueous borrow pits (provisional).

Nakamura, Y. (2006a) Restoration activities of tidal flats in Japan. *Proceedings of the second International Conference on Estuaries and Coasts (ICES 2006)*, Vol. **1**, 60–67.

Nakamura, Y. (2006b) Current status of subaqueous borrow pits and their restoration in Japan. *Journal of Advanced Marine Science and Technology Society* **12**, 43–50.

Nakata, K., Nakamura, Y., Suzuki, T., Ishida, M., Tanabe, Y., Hasegawa, M., Nagakura, T., Oshima, I. and Kazama, T. (2005) Effects of tidal flats construction by monitoring water quality, sediment quality and benthic fauna in and around constructed tidal flats. *Confirmation of Constructed Tidal Flats and Shallows in Mikawa Bay as a Beneficial Use of Dredged Sediment*. Edited by Mikawa Port and Harbor Office, Chubu Regional Branch, Ministry of Land, Infrastructure and Transport, pp. 15–30.

PIANC (1992) *Beneficial Uses of Dredged Material: A Practical Guide*. Report of PIANC working group 19.

PIANC (2009) *Dredged Material as a Resource Options and Constraints*. Report of PIANC working group 14, 49 pp.

Port and Airport Research Institute (PARI) and Korea Ocean Research and Development (KORDI) (2004) *Proceedings of 3rd and 4th Japan–Korea Joint Workshop on Tidal Flats*, 68 pp.

Ports and Harbors Bureau, Ministry of Land Infrastructure (MLIT) (2005) *Technical Guideline for Ocean Disposal and Beneficial Use of Dredged Material*. Edited by Ministry of Land, Infrastructure and Transport, 86 pp.

Sasaki, J., Isobe, M., Watanabe, A. and Gomyo, M. (1996) A note on the scale of blue tide in Tokyo Bay. *Proceedings of Coastal Engineering, JSCE* **43**, 1111–1115.

Sato, T., Kitazume, M. and Suganuma, F. (2004) Application of pneumatic flow mixing method. *PIANC Bulletin* **116**, 45–50.

Tanabe, T. (2005) Current status on the disposal of dredged material in Japanese Ports and Harbors. *Journal of Japan Society of Waste Management Experts* **16**, 61–65.

US Army Corps and Engineers Baltimore District and Maryland Port Administration (2004) Poplar Island environmental restoration project, adaptive management plan, draft final, 13 pp.

Working Group of Coastal Area Restoration (2004) *Handbook for Restoration of Coastal Area: Its Design, Technology and Practice*. Edited by Ports and Harbors Bureau, Ministry of Land, Infrastructure and Transport, Gyosei, Tokyo.

Section 4
Coastal Governance

Chapter 19
A Practitioner's Perspective on Coastal Ecosystem Governance

Stephen Bloye Olsen

Abstract

This chapter explores emerging conceptual frameworks that help analyze and address the issues brought by accelerating change in coastal ecosystems. It briefly reviews the evolution of practices and programs that work to modulate human activity in coastal regions areas where human activity is the dominant driver of change and where needs for both conservation and development must be met simultaneously. These issues are considered from the perspective of a practitioner engaged in the design and administration of coastal management projects and programs that must be tailored to the unique contexts present in each locale. The chapter looks ahead to how coastal governance initiatives designed to address the challenges of social and environmental change in the twenty-first century will need to be adapted and strengthened in the face of mounting pressures and accelerating human-induced change along coastlines.

19.1 The evolution of contemporary coastal management

Contemporary coastal management reflects a growing awareness of the difficulties that are inherent in modulating human activity in the ecosystems where our species is most concentrated. According to the Millennium Ecosystem Assessment (2005a), 40% of the human population is concentrated in coastal regions on only 5% of the inhabited land space. Today 12 of 15 largest cities of the planet are coastal. This implies that coastal regions contain the lion's share of the infrastructure that supports industry, transportation and trade, energy processing, tourism as well as several important sources of food production (National Research Council, 2008). Coastal regions therefore generate a disproportionate share of the global consumption of man-made and natural resources and the resulting wastes that such consumption generates (Crossland et al., 2005). How humanity manages its activities and their impacts in coastal ecosystems is therefore a major challenge for the twenty-first century.

Coastal management has foreshadowed the challenges and the development of integrating approaches to the management of the sectors (agriculture, fisheries, urban development and public health) by which planning and management have traditionally been organized. As described in a recent report issued by the National Research Council (2008), the evolution of coastal management can be divided in three periods. The first extends over the 22 years from the release of the Stratton Commission report (Commission on Marine Science, Engineering, and Resources, 1969) in the US in 1969 to the United Nations Conference on the Environment and Development (UNCED) in 1992. The second extends over the decade between UNCED and the World Summit on Sustainable Development in 2002 (United Nations, 2002). The third period begins with the World Summit and continues today.

Contemporary coastal management has its roots in the seminal Stratton Commission report to the US Congress, *Our Nation and the Sea*, in 1969. This promoted a forward-looking and comprehensive approach to ocean and coastal management. It gave rise in the following decade to federal legislation in the US that set new priorities and policies for research, planning and decision making for coastal development and restoration, the management of fisheries and the allocation of ocean space and resources. The US Coastal Zone Management (CZM) Act of 1972, as one response to *Our Nation and the Sea*, launched an innovative program featuring incentives for state–federal partnerships designed to address priority issues raised by the intensification of human activity along the coastlines of the US and the degraded condition of many estuaries and the Great Lakes. These CZM programs and plans relied primarily on regulations, construction standards, zoning and the designation of protected areas to make operational good practices that address the need for both coastal conservation and development. The CZM programs developed by individual states in the US assumed that the process of coastal development will and should continue and that improved local–state–federal coordination and a holistic approach will avoid or minimize undesirable social and environmental side effects. However, a comprehensive evaluation (Hershman et al, 1999) of the impacts of CZM at the state and national scales concluded that there is insufficient evidence to demonstrate that CZM programs have been a significant cause of improvements in the condition and appropriate use of coastlines. The study found that interagency coordination had in many instances improved and the processes of change had become more predictable. However, linking the impacts of such improved coordination to the condition of coastal environments proved difficult in good measure because the necessary data had not been collected.

Beginning in the early 1980s, efforts to address similar issues produced by the misuse and overuse of coastal resources in developing nations (Chua and Scura, 1992; Chua, 1998; Olsen et al, 1998; Olsen and Christie, 2000) began by applying the CZM model, but soon recognized that major adaptations are necessary where governments are weak and regulatory approaches may have little impact on human activity. In countries where governmental authorities are weak, illegal activity is common, and poverty prevails, these early efforts in coastal management suggested that the approaches that proved most successful were those that stress (1) strong involvement by those affected by coastal change, (2) incentive-based methods and (3) the willing compliance with plans of action.

The second period in the evolution of contemporary coastal management was prompted by the UNCED in 1992. UNCED put forward integrated coastal management (ICM) as the recommended approach for managing the world's coastal regions. Agenda 21 of Chapter 17 of the UNCED agenda details the ICM approach and its defining features. The text was influenced by the US experience with CZM and by the initial experience from coastal management programs in a number of developing nations including Sri Lanka, Ecuador and the Philippines. As subsequently defined by GESAMP (1996), ICM is:

> ... a continuous and dynamic process that unites government and the community, science and management, sectoral and public interests in preparing and implementing an integrated plan for the protection and development of coastal ecosystems and resources. The overall goal of ICM is to improve the quality of life of human communities who depend on coastal resources while maintaining the biological diversity and productivity of coastal ecosystems .... Central to success in achieving this goal is the need for ICM to provide an equitable, transparent and dynamic governance process that is acceptable to the community.

ICM, like CZM, focuses on coasts as interconnected systems where natural resources and natural processes are intimately linked to complex combinations of human activities. It also develops coordinating institutional mechanisms to overcome vertical and horizontal fragmentation and reconcile the interests of coastal stakeholders (Glavovic, 2008). In contrast to CZM, however, ICM defines the goal as sustainable forms of coastal development. The concept of sustainable development is absent in the CZM act and in the state programs that it catalyzed. The emphasis on sustainable forms of development in ICM is important since it suggests that the trajectory of change – unless corrected – may lead to unsustainable outcomes that will make it more difficult, or impossible, for future generations to meet their needs. The warning inherent in adopting sustainable development as the ultimate goal contrasts sharply with the confident optimism of the Stratton Commission and its belief that better coordination, combined with major investments in research and the application of new technologies, will generate a positive future for all.

UNCED triggered major investments in ICM projects by bilateral and multilateral donors. With few exceptions, these were designed as short-term investments by multilateral and bilateral international organizations designed to help nations progress through the initial planning phase of an ICM program. The assumption was that securing the funds to implement a formally adopted ICM program would be arranged by the individual country's government. However, where promising beginnings have been made, the available but scanty evidence suggests (Cicin-Sain et al., 2006; National Research Council, 2008) that national ICM programs in most developing nations are not taking root and prospering.

In the period between UNCED and the Johannesburg Conference, the magnitude of the impacts of human activity on the global ecosystem became a mounting concern. These led Crutzen and Stoermer (2000) to coin the term *Anthropocene* to describe the epoch in which human activity equals or surpasses nature in modulating the behavior of the planet as an ecosystem. The changes underway in ecosystems at all scales are unprecedented and they are happening at speeds measured in decades. As summarized by the International Geosphere–Biosphere Program (IGBP, 2001):

- Planet Earth is a system that life itself helps control.
- Human enterprise drives multiple, interacting effects that cascade through the earth system.
- Earth's dynamics are characterized by critical thresholds and abrupt changes.
- Human activities can trigger changes with catastrophic consequences.
- Earth is currently operating in no-analogue state.

Global change is already having major impacts on coastal ecosystems. Current estimates are that sea level may be expected to rise 0.6–1.5 m by the end of the century (Rahmstorf, 2007). Storms and hurricanes are already more powerful and more frequent. Human activities are fixing nitrogen, and have thereby made it available for primary production at rates that match or succeed all natural nitrogen-fixing processes. This nutrient loading is threatening many estuaries and nearshore waters with eutrophication and producing larger dead zones off coasts (Crossland et al, 2005). All these expressions of global change have been linked to human activity and are expected to become more intense along coastlines with each decade. The result is that the ecosystem services that provide renewable resources maintain biodiversity and sustain and fulfill human life are declining (Millennium Ecosystem Assessment, 2005b). According to the Millennium Assessment, 60% of global ecosystem services (classified as provisioning, regulating, supporting and cultural services) are degraded and only 4 out of 24 services are increasing. Those increasing are provisioning services.

The World Summit on Sustainable Development (WSSD) held in Johannesburg, South Africa (United Nations, 2002), marks the beginning of the third period in the evolution of coastal management. The conference noted only modest progress toward the goal adopted at UNCED that had called for operational ICM programs in all coastal nations by 2000. Millennium Development Goals were adopted that set targets for global priorities within a timetable. Several millennium goals address coastal and ocean issues. Yet a review published 4 years later by the Global Forum on Oceans, Coasts and Small Islands (Cicin-Sain et al., 2006) pointed out that accepted methods and indicators for assessing progress have yet to be developed. While Sorensen (2002) and others (Cicin-Sain et al., 2006) have made inventories of the many coastal management initiatives, it has become increasingly obvious that the integrating approaches of both CZM and ICM are difficult to put into practice. Particularly in the tropics, examples of positive progress toward the goals of ICM, as defined by GESAMP, are principally in the form of small-scale 'pilot' programs (Olsen and Christie, 2000). The few national programs that have made the transition from issue analysis and planning to the implementation of a plan of action have focused primarily on a single coastal issue. The national coastal management programs of Sri Lanka (Coast Conservation Department, 1990), for example, have addressed coastal erosion and shorefront construction. Glavovic (2008), in a detailed review of the efforts to launch a national ICM program in South Africa, details the many challenges of bridging what he terms 'the chasm between rhetoric and reality'. In the US, a detailed assessment conducted by the US Government Accounting Office (GAO, 2005) of the results of massive investments over nearly four decades in the restoration and management of Chesapeake Bay, the nation's largest estuary, concluded that the results were far below expectations. The report documents the difficulties of controlling human activities in a densely populated watershed in which nonpoint sources of nutrients and other sources of ecosystem stress must be reduced if the qualities of an estuary are to be restored and sustained. Indeed, along most coastlines, the combined forces generated by population growth, the intensification of human activities and, in many countries, widespread poverty are creating conditions that perpetuate trajectories in the direction of less sustainable, rather than more sustainable, forms of development. These realities make it necessary to reconsider how the challenges and possible solutions are defined in a manner that takes into account the magnitude and complexity of the forces that make it so difficult to translate the principles of ICM into an operational reality.

19.2 How can integrating approaches be made more effective?

Preparations for the Johannesburg Conference and subsequent activity have produced a multitude of papers and meetings that have discussed the need for fresh integrating approaches to the multiple, interconnected forces and challenges of the Anthropocene. For example, Kates et al. (2001) called for 'sustainability science' as a new integrative field of study that draws together many disciplines and communities in the search for a transition to more sustainable forms of development. Biermann (2007) has suggested the term 'earth system governance' as the contribution of the social sciences to the analysis of human responses to earth system transformation to complement the quantitative, often computer-based methods, provided by the natural sciences. In this 'two pillar model' for global change research, Biermann calls for the development of an earth system governance theory. This would unite the social sciences that analyze organized human responses to earth system transformation to the causes of global environmental change and direct societal behavior to a 'safe coevolution with natural processes'. These ideas complement learning-based or adaptive approaches to ecosystem management developed by Holling (1978) and Lee (1993).

The author, as an early practitioner of CZM in the US in the 1970s and early 1980s and more than two decades of involvement in ICM programs in a diversity of settings in both high- and low-income nations, concludes that the accumulating experience in CZM and ICM suggests that future efforts need to place a much greater focus on the governance dimensions of the practice. Great attention to the dynamics of governance in turn requires new approaches to assessing progress toward the goal of sustainable forms of development.

Greater attention to the governance dimensions of responses to coastal change is consistent with the emergence of the ecosystem paradigm as the dominant approach to managing natural resources and the environment. The terms 'ecosystem-based management' and 'the ecosystem approach' have been endorsed by a number of studies and expert commissions (for example, the Pew Commission on the Oceans, 2003; the US Commission on Ocean Policy, 2004; the National Research Council, 2008) and are prominent in the work of such international organizations as the International Oceanographic Commission, the Food and Agriculture Organization, the United Nations Environmental Program and the Global Environmental Facility. In simple terms, ecosystem-based management recognizes that human communities, like plant and animal communities, are interdependent and interact with their physical environment to form distinct ecological units called ecosystems. These units provide the basis for all life including humanity, are transboundary in character, typically cutting across political and jurisdictional boundaries and thus subject to multiple management systems. In ecosystem-based management, the associated human population and economic/social systems are seen as integral parts of the ecosystem. Ecosystem-based management is therefore designed and executed as an adaptive, learning-based process that applies the principles of the scientific method to the processes of management.

Redefining responses to the changes underway in coastal regions as challenges of *governance* rather than *management* requires differentiating between the two. As suggested in an earlier paper (Olsen, 2003), governance probes the fundamental goals and the institutional processes and structures that are the basis for planning and decision making. Governance addresses the values, policies, laws and institutions by which a set of issues is addressed. Management, in contrast, is the process by which human and material resources are harnessed to achieve a known goal within a known institutional structure. We therefore speak of business management, park management, personnel management or disaster management. In these instances, the goals and the mechanisms of administration are well known and widely accepted. Governance, in contrast, questions the fundamental goals of human society and sets the stage within which management occurs. In the Anthropocene, the transformations underway in our planet's coastal regions must therefore be defined as challenges of governance rather than the more tractable and familiar challenges of management. In this view, the limited results of CZM and ICM suggests that challenges we face require rethinking the fundamental goals of contemporary culture – not least of which are the national and international goals that proclaim that continuing economic growth, with its attendant increases in goods, services and their attendant wastes define the path that leads to a positive future. It would, therefore, be more appropriate to use the term *ecosystem-based governance* rather than ecosystem-based management.

The dimensions and dynamics of governance as they can be applied to the analysis of the issues posed by ecosystem change have been examined in a series of papers and reports generated to support the application of the ecosystem approach to large marine ecosystems (LMEs). A series of papers (Sutinen, 2000) and more recently a handbook (Olsen et al., 2006) have focused on developing the governance dimensions of LMEs. Here, governance is defined (Juda, 1999; Juda and Hennessey, 2001) to encompass the formal and informal arrangements, institutions and values that structure and influence:

- How a resource or an environment is utilized
- How problems and opportunities are evaluated and analyzed
- What behavior is deemed acceptable or forbidden
- What rules and sanctions are applied to affect how natural resources are distributed and used

The processes of governance are expressed through the institutions and arrangements of markets, government and civil society. The manner in which these three mechanisms of governance interact with one another is complex and dynamic. The marketplace, in which goods and services are exchanged by profit-seeking producers, traders and consumers, affects how the environment is utilized, what resources are extracted and the manner in which these resources are utilized. Government policy and regulation is well recognized as a mechanism that can affect human behavior. Tax policies can provide incentives for some forms of behavior, and government spending policies apply a substantial portion of society's resources to achieve specific objectives. Social norms and networks – sometimes referred to as 'social capital' – shape individual and collective behavior and facilitate cooperation among individuals and between groups. By encouraging trust, civic engagement and social networks, social capital can enhance effective governance while reducing its costs (Grafton, 2005).

19.3 Assessing progress toward sustainable forms of development

As coastal management and ecosystem-based management have taken shape and matured, much energy has been expended on how to define, organize and sequence the *processes* of such management. There are many descriptions of the phases by which ocean and coastal governance initiatives evolve (Chua and Scura, 1992; GESAMP, 1996; Olsen et al., 1997; Cicin-Sain and Knecht, 1998; Olsen, 2003; United Nations Environment Program, 2006; Chua, 2007). The Joint Group of Experts on Scientific Aspects of Marine Environmental Protection (1996) selected the most essential steps, emphasizing that the process is a 'cycle of learning' that proceeds from an awareness of a set of problems and opportunities to their analysis, formulation of a course of action and then to the implementation and evaluation of a plan of action. In the case of the LME programs sponsored by the Global Environmental Facility, the same fundamental sequence of steps required to analyze issues and assemble a plan of action is detailed in a complex and lengthy sequence of actions and decision points organized as a transboundary diagnostic analysis (TDA) and strategic action program (SAP).

As applications of ecosystem-based management mature, the need to complement methods of organizing the *processes* of management with methods for assessing the *outcomes* of management has become apparent. The unifying framework developed by the author (Olsen et al., 1997, 1998; Olsen, 2003) – shown in its most recent form in Figure 19.1 – is useful for this purpose, since it segregates the ultimate goal of sustainable development into a sequence of more tangible thresholds of achievement. This framework suggests the sets of indicators (United Nations Environment Program, 2006) that can be used to trace the evolution of an ecosystem governance initiative as it progresses from an initial analysis

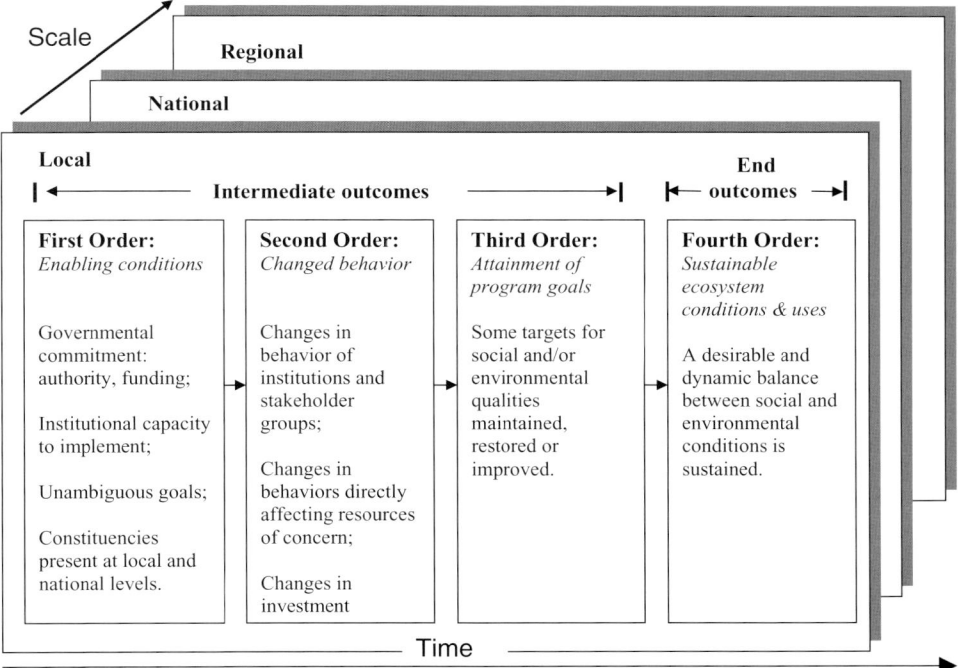

Figure 19.1 The four orders of outcomes in ecosystem-based governance. (Adapted from Olsen, 2003.)

of issues and trends in an ecosystem to the implementation of plans of action that produce progressively more sustainable conditions and patterns of use.

The framework defines the First Order as the creation of the enabling conditions that make it feasible to implement a sustained program of action. First Order outcomes are the institutional and societal conditions that must be present if an ecosystem-based initiative is to succeed in executing a sustained plan of action designed to influence the course of events in a coastal ecosystem. Experience in a wide diversity of settings suggests that the transition to implementation can be anticipated only when *all four* of the following outcomes are in place:

1. unambiguous *goals* have been adopted against which the efforts of the program can be measured,
2. sufficient initial *capacity* is present within the institutions responsible for the program to implement its policies and plan of action,
3. a core of well-informed and supportive *constituencies* composed of stakeholders in both the private sector and government agencies actively support the program, and
4. governmental *commitment* to the policies of a program has been expressed by the delegation of the necessary authorities and the allocation of the financial resources required for sustained program implementation (Olsen, 2003; United Nations Environment Program, 2006).

Building on methods developed by Canada's International Development Research Center (IDRC), the implementation of a program is defined in the Second Order as changes in behavior in the institutions and human population within and/or affecting the ecosystem in question (Earl et al., 2001). Second Order outcomes are evidence of successful implementation of an ecosystem-based management program. This includes evidence of new forms of collaborative action among institutions, the actions of state–civil society partnerships, the behavioral changes of resource users and changes in patterns of investment. An obvious example of a Second Order outcome is the cessation of such destructive practices as dynamite fishing or the release of untreated wastes into the environment. The successful implementation of pollution reduction practices signals a behavioral change that may be important to achieve water quality targets. Changes in investments may signal the provision of sustained financing for strengthening the capacity of institutions to practice ecosystem-based management, and the construction of necessary physical infrastructure supportive of a program's policies and programs is the third category of behavioral change.

Only after the requisite changes in behavior have been practiced for a sufficient period can improvements be expected in the environment and in the social benefits that constitute the Third Order achievement of the environmental and societal goals selected in the earlier phase of program design. Third Order outcomes mark the achievement of the program's goals as were defined during the issue selection and planning phase (the First Order) and may have been modified during implementation (the Second Order). These are the rewards for sustained achievements in institutional and behavioral change. Water quality improves, there are more fish, income levels rise and target communities' engagement in supplemental livelihoods stabilizes or improves. Third Order outcomes (Olsen et al., 2006; United Nations Environment Program, 2006) can be allocated to the two categories of ecosystem-based management goals.

- Targets for sustained or restored qualities of the biophysical environment
- Targets in human quality of life may be expressed as greater equity and diversified livelihoods

Examples of Third Order outcomes are:

- Measurable improvements in chemical, physical and biological parameters
- Improved recruitment of priority fish species
- Demonstrable reduction of persistent organic pollutants in the food chain
- Changes in local community income and social conditions as a result of improved environmental conditions
- Reductions in the loading of nutrients and the associated evidence of eutrophic conditions

The difference between Third and Fourth Order outcomes is that sustainable development requires achieving a dynamic equilibrium among *both* social and environmental qualities. Sustainable development has not been achieved if, for example, the condition of the coral reefs of a place is sustained or improved, but the people associated with them continue to live in unacceptable poverty. Similarly, sustainable development has not been achieved if some measures of quality of life are high, but such achievements are eroding the resource base or require the exploitation of other social groups. The challenge is vastly complicated by the imperative of defining an acceptable balance in terms of both intergenerational equity and a planetary perspective on both societal and environmental conditions and trends. Recognizing that all living systems are in a constant process of change, sustainable forms of development will be dynamic, not static, and must be capable of responding to the surprises that Mother Nature delivers.

It is important to recognize that some expressions of First, Second and Third Order outcomes will accumulate concurrently within a given time period. While there are causal relationships between the three orders, they are not, and should not, be achieved in a strictly sequential order. For example, once some progress has been made in assembling First Order outcomes, programs should work to achieve some evidence of Second and Third Order outcomes in a learning-by-doing mode. This can be accomplished, for example, by management activities at a pilot scale. Experience has repeatedly confirmed that the most successful initiatives focus their efforts on one or two issues and then expand the scope of the program as experience, capacity and constituencies are built. Particularly in developing country contexts, it is usually a mistake to launch a fully integrated program directed at multiple issues and goals.

19.4 Revised strategies for advancing coastal ecosystem governance

The orders of outcomes framework suggest a number of significant changes to how coastal management and coastal governance initiatives are designed, implemented and evaluated. Where international organizations are making investments in the application of the ecosystem approach in coastal and marine regions, the current practice is to require evidence of Third Order social and/or environmental improvements with little consideration of whether the full suite of First Order conditions are present. Often scant attention is paid to the social, economic and governmental challenges posed by achieving and then sustaining the necessary Second Order changes in behavior. Despite the obvious weakness of governments in many regions, current strategies typically place heavy reliance on government-imposed regulations to achieve desired changes in behavior. The importance of willing compliance and of harnessing market and civil society mechanisms to achieve desired goals is often discounted as impractical. Monitoring and evaluation tend to be directed at the performance of the organizations that are being funded rather than the prospects and achievements in terms of First, Second and Third Order outcomes that need to be achieved in the area of concern.

Adoption of the orders framework would change the expectations for many short-term projects and could redirect the available funds as contributions to a common effort directed at assembling the preconditions for implementation of a plan of action. Once the First Order threshold has been reached, additional funding – including specifically international funds – will often be required to sustain a coastal governance initiative long enough for it to achieve a harvest of Third Order goals at significant spatial scales.

An underlying problem in coastal governance initiatives generally is a chronic misjudgment of the time required to achieve a harvest of Third Order social and environmental outcomes in heavily stressed ecosystems. Expectations for significant improvements in the condition of a coastal ecosystem continue to be set within the time frame of the typical 5–10-year 'project', even though global experience demonstrates repeatedly that this is unrealistic. Duda (2002) points out that experience in the management of such large ecosystems as the North American Great Lakes, the Baltic Sea, the Rhine basin and the Mediterranean Sea shows that 15–20 years were required before meaningful commitments to joint management improvements can be secured from the several countries involved. Additional time is then needed before the transboundary water bodies respond to reductions in the stresses from pollution, overfishing, eutrophication and habitat alteration that are brought about by the implementation of a program of action. Attaining environmental and societal goals for desired ecosystem goals at this large scale may require 20–30 years.

Despite widespread acknowledgment in national and international funding institutions of the need for adaptive management, the prevailing rules make its practice difficult or impossible. The selection and monitoring of indicators that will be used to gauge progress in the First and Second Orders could provide an objective basis for the practice of adaptive management. Parsimonious sets of indicators for the First and Second Orders that could be used as a tool to promote adaptive management practices have been developed for linked watershed and coastal management programs (United Nations Environment Program, 2006).

The next phase in the refinement of these methods is being undertaken in collaboration with the international Land Ocean Interactions in the Coastal Zone (LOICZ) program. LOICZ has selected the comparative analysis of coastal governance practices as one of its priority topics. The LOICZ governance working group is applying the orders of outcomes framework to develop region-by-region assessments of current coastal governance practices that draw upon portfolios of case studies that document and analyze the processes and outcomes of past and current governance of human activities in specific locales. The case studies will be the basis for posing and testing hypotheses that probe the interactions of markets, government and civil society in coastal governance systems in diverse settings and the practices that enable progress toward desired outcomes. This will suggest how the contributions of the sciences can be more effectively incorporated into each step in governance processes that address coastal ecosystem change. The first application of this research is currently underway at selected sites in Latin America.

19.5 Conclusions

The emerging practices of coastal ecosystem governance need to be directed primarily at the management of human activities as these affect and change the goods and services that flow from coastal ecosystems and an advance to more sustainable forms of human activity. This requires a holistic approach that integrates the best available scientific knowledge with the processes by which a society sets goals, selects among options and adopts a plan of action. The knowledge, skills and attitudes to combine the best science with the processes by which societal choices are made lie at the heart of ecosystem governance. Today, the individual and institutional capacity to make this integration is weak. As a result, there is a major implementation gap in coastal management initiatives, the management of marine protected areas and associated endeavors. Growing the capacity to practice coastal ecosystem governance requires a greater appreciation of the need to harness not only governments but also markets and civil society in the practice of coastal ecosystem stewardship. Greater attention must be directed to understand the traditions and existing capacities for ecosystem governance in a given locale and to tailoring the responses to coastal change to these realities. The application of the orders of outcomes framework can help diagnose the needs in a given place and then to tailor good practices appropriately to the design and implementation of a sustained plan of action.

References

Biermann, F. (2007) 'Earth system governance' a crosscutting theme of global change research. *Global Environmental Change* **17**, 326–337.

Chua, T.E. (1998) Lessons from practicing integrated coastal management in Southeast Asia. *Ambio* **27**, 599–610.

Chua, T.E. (2007) *The Dynamics of Integrated Coastal Management: Practical Application in the Sustainable Coastal Developments in East Asia*. Partnerships in Environmental Management for the Seas of East Asia, Quezon City, Philippines.

Chua, T.E. and Scura, L.F. (eds) (1992) Integrative Framework and Methods for Coastal Area Management. *International Center for Living Aquatic Resources Management Conference Proceedings 37*, Manila, Philippines.

Cicin-Sain, B. and Knecht, R.W. (1998) *Integrated Coastal and Ocean Management Concepts and Practices*. Island Press, Washington, DC, 517 pp.

Cicin-Sain, B., Vandeweerd, V., Bernal, P.A., William, L.C. and Balgos, M.C. (2006) *Meeting the Commitments on Oceans, Coasts, and Small Island Developing States Made at the 2002 World Summit on Sustainable Development: How Well Are We Doing?* Global Forum on Oceans, Coasts and Islands, University of Delaware, Delaware, 63 pp.

Coast Conservation Department (1990) *Coastal Zone Management Plan*. Colombo, Sri Lanka, 35 pp.

Commission on Marine Science, Engineering, and Resources (1969) *Our Nation and the Sea: A Plan for National Action.* United States Government Printing Office, Washington, DC, 305 pp.

Crossland, C.J., Kremer, H.H., Lindeboom, H.J., Marshall Crossland, J.I. and Le Tissier, M.D. (eds) (2005) *Coastal Fluxes in the Anthropocene.* Springer-Varlag, Berlin, Heidelberg, 231 pp.

Crutzen, P.J. and Stoermer, E.F. (2000) The 'Anthropocene'. *Global Change Newsletter* **41**, 12–13.

Duda, A.M. (2002) *Monitoring and Evaluation Indicators for GEF International Waters Projects.* GEF Secretariate, The World Bank, Washington, DC, 11 pp.

Earl, S., Carden, F. and Smutylo, T.S. (2001) *Outcome Mapping: Building Learning and Reflection into Development Progress.* International Development Research Centre, Ottawa, 139 pp.

GESAMP (Group of Experts on the Scientific Aspects of Marine Protection) (1996) *The Contributions of Science to Integrated Coastal Management.* GESAMP Reports and Studies No. 61. Food and Agriculture Organization, Rome, 66 pp.

Glavovic, B. (2008) Sustainable coastal development in South Africa: bridging the chasm between rethoric and reality. In: Krishnamurthy, R.R., Glavovic, B.C., Kannen, A., Green, D.R., Ramanathan, A.L., Han, Z., Tinti, S. and Agardy, T. (eds). *Integrated Coastal Zone Management – The Global Challenge.* Research Publishing, Singapore, Chennai, pp. 129–153.

Grafton, R.Q. (2005) Social capital and fisheries governance. *Ocean and Coastal Management* **48**, 753–766.

Hershman, M.J., Good, J.W., Bernd-Cohen, T., Goodwin, R.F., Lee, V. and Pogue, P. (1999) The effectiveness of coastal zone management in the United States. *Coastal Management* **27**, 113–138.

Holling, C.S. (ed.) (1978) *Adaptive Environmental Assessment and Management.* John Wiley and Sons, New York.

International Geosphere Biosphere Program (IGBP) (2001) *Global Change and the Earth System: A Planet Under Pressure.* IGBP Science Series #4, Swedish Academy of Sciences, Stockholm, 332 pp.

Juda, L. (1999) Considerations in developing a functional approach to the governance of large marine ecosystems. *Ocean Development and International Law* **30**, 89–125.

Juda, L. and Hennessey, T. (2001) Governance profiles and the management and use of large marine ecosystems. *Ocean Development and International Law* **32**, 43–69.

Kates, R.W., Clark, W.C., Corell, R., Hall, J.M., Jaeger, C.C., Lowe, I., McCarthy, J.J., Joachim Schellnhuber, H., Bolin, B., Dickson, N.M., Faucheux, S., Gallopin, G.C., Grübler, A., Huntley, B., Jäger, J., Jodha, N.S., Kasperson, R.E., Mabogunje, A., Matson, P., Mooney, H., Moore III, B., O'Riordan, T. and Svedin, U. (2001) Sustainability science. *Science* **292**, 641–642.

Lee, K.N. (1993) *Compass and Gyroscope: Integrating Science and Politics for the Environment.* Island Press, Washington, DC, 255 pp.

Millennium Ecosystem Assessment (2005a) *Ecosystems and Human Well-Being: Scenarios.* World Resources Institute, Washington, DC, 596 pp.

Millennium Ecosystem Assessment (2005b) *Ecosystems and Human Well-Being: Synthesis.* World Resources Institute, Washington, DC, 160 pp.

National Research Council (2008) *Increasing Capacity for Stewardship of Oceans and Coasts.* The National Academies Press, Washington, DC.

Olsen, S., Tobey, J. and Kerr, M. (1997) A common framework for learning from ICM experience. *Ocean and Coastal Management* **37**, 155–174.

Olsen, S.B. (2003) Frameworks and indicators for assessing progress in integrated coastal management initiatives. *Ocean and Coastal Management* **46**, 347–361.

Olsen, S.B. and Christie, P. (2000) What are we learning from tropical coastal management experiences? *Coastal Zone Management Journal* **28**, 5–18.

Olsen, S.B., Tobey, J. and Hale, L. (1998) A learning-based approach to coastal management. *Ambio* **27**, 611–619.

Olsen, S.B., Sutinen, J.G., Juda, L., Hennessey, T.M. and Grigalunas, T.A. (2006) *A Handbook on Governance and Socioeconomics of Large Marine Ecosystems*. Coastal Resources Center, University of Rhode Island, Narragansett.

Pew Oceans Commission (2003) *America's Living Oceans: Charting a Course for Sea Change*. Pew Oceans Commission, Arlington, 144 pp.

Rahmstorf, S. (2007) A semi-empirical approach to projecting future sea-level rise. *Science* **315**, 368–370.

Sorensen, J. (2002) Baseline 2000 background report: the status of integrated coastal management as international practice. Second iteration – 26 August 2002. Available at www.uhi.umb.edu/b2k/baseline2000.pdf.

Sutinen, J. (ed.) (2000) A Framework for Monitoring and Assessing Socio-Economic and Governance of Large Marine Ecosystems. National Oceanic and Atmospheric Administration Technical Memorandum NMFS-NE-158.

U.S. Commission on Ocean Policy (2004) *An Ocean Blueprint for the 21st Century*. U.S. Commission on Ocean Policy, Washington, DC.

US Government Accounting Office (GAO) (2005) Chesapeake Bay program: improved strategies are needed to better assess, report, and manage restoration progress. Report to the Congressional Requesters. U.S. Government Accountability Office, Washington, DC, 87 pp.

United Nations (2002) *Johannesburg Plan of Implementation of the World Summit on Sustainable Development*. United Nations, New York, 88 pp.

United Nations Conference on Environment and Development (1992) *Agenda 21: A Programme for Action for Sustainable Development*. United Nations, New York, 294 pp.

United Nations Environment Programme/Global Programme of Action for the Protection of the Marine Environment from Land-Based Activities (2006) *Ecosystem-Based Management: Markers for Assessing Progress*. The Hague, 49 pp.

Chapter 20
Problem Perceptions and Use of Technical Knowledge in Decision making for the Extension of Mainport Rotterdam

Saskia Hommes, Suzanne J.M.H. Hulscher, Jan P.M. Mulder and Henriëtte S. Otter

Abstract

The Dutch government wants to expand Mainport Rotterdam, one of the largest ports in the world, by land reclamation in the North Sea. This may affect the Wadden Sea, a unique wetlands area protected by the European Bird and Habitat Directives. To assess the impact of the port extension on the Wadden Sea, an Appropriate Assessment procedure was carried out. We investigated how stakeholders' perceptions were dealt with and how knowledge was used in this decision-making process. Our findings form an argument for practitioners in coastal management to choose a process-oriented approach to deal with complex issues.

20.1 Introduction

Rotterdam was established as a fishing village at the western coast of the Netherlands in the second half of the thirteenth century. The village gradually developed into a prosperous merchant port. The first harbors were constructed at the beginning of the seventeenth century. The area began to fill up with warehouses, breweries, sugar refineries, gin distilleries, shipyards and ropeyards. Following the setting up of the United East India Company (VOC), trade and shipping rapidly increased. Nowadays, Mainport Rotterdam is one of the largest ports in the world. The most important activities of the port take place approximately 40 km from the center of Rotterdam on reclaimed land in the North Sea. In the 1990s, the Dutch government started a project to strengthen the port's position on the world market by expanding the reclaimed land area further into the North Sea. However, growth of the Mainport is not the only issue at stake, when the economic function fulfilled by Mainport Rotterdam puts pressure on the environment. Additionally, the port and industries with their associated living, working and infrastructure requirements make a large demand on the available space in the Rotterdam region. Furthermore, the expansion of the port, by means of land reclamation and associated sand dredging, affects other marine and coastal users such as the recreational and fishing sectors.

The extension of Mainport Rotterdam is an example of a complex coastal management issue. Decision makers involved in these complex issues have to deal with ecological effects, physical effects, economic costs and benefits and technical feasibility. On top of that, they operate within a complex web of policy, regulations, social and political processes. This illustrates that complex societal problems are not purely technical problems. Therefore, the classical[1] decision-making approach, in which problems are perceived to be of a technical nature, is not sufficient to deal with these kinds of problems. Several Dutch authors state that for these kind of complex issues a process-oriented approach should be adopted, which pays attention to differences in stakeholders' perceptions, and to interaction and communication (e.g. De Bruijn et al., 2002; Edelenbos et al., 2003). Many authors support this view; however, very few of their papers (e.g. Denters et al., 2003; Edelenbos and Klijn, 2005) are based on empiricism. Empiricism is therefore what we aim for in this chapter, thus providing a stronger theoretical basis for management practice in complex coastal management issues.

In this chapter, we aim to answer the following research questions:

- How can the decision-making process of the extension of Mainport Rotterdam be characterized?
- How is problem perceptions dealt with and how is technical knowledge used in this case study?

This chapter is organized as follows. In Section 20.2, the theoretical framework is presented, focusing on the following aspects: classical decision making, types of problems, stakeholders' perceptions and use of knowledge. Section 20.3 describes the case study on the decision-making process for the extension of Mainport Rotterdam. We thoroughly investigated this project, using project documents, news reports, in-depth interviews with project participants and observations from attended project meetings. In Section 20.4, the case study results are analyzed by using the theoretical framework from Section 20.2. Finally, conclusions on stakeholders' perceptions and use of knowledge are drawn in Section 20.5.

20.2 Theoretical framework

20.2.1 *Classical decision making*

Classical decision making refers to the approach that originates from system analysis. This approach arose after World War II when policy makers embraced the analytical approach of operations research, as a result of the successes it achieved in military issues during the war (Quade, 1989). The classical decision-making approach aims to support decision making on the basis of (scientific) knowledge. Therefore, the solution is sought in acquiring more knowledge and data on the basis of (policy analytical) research (Twaalfhoven, 1999; Arentsen et al., 2000; Koppenjan and Klijn, 2004). A key concept of the classical approach is *rationality*. Decision making should be rational and be supported by rational analysis.

---

[1] Other naming found in literature: analytical, traditional, technocratic, intellectual, (neo-)positivism.

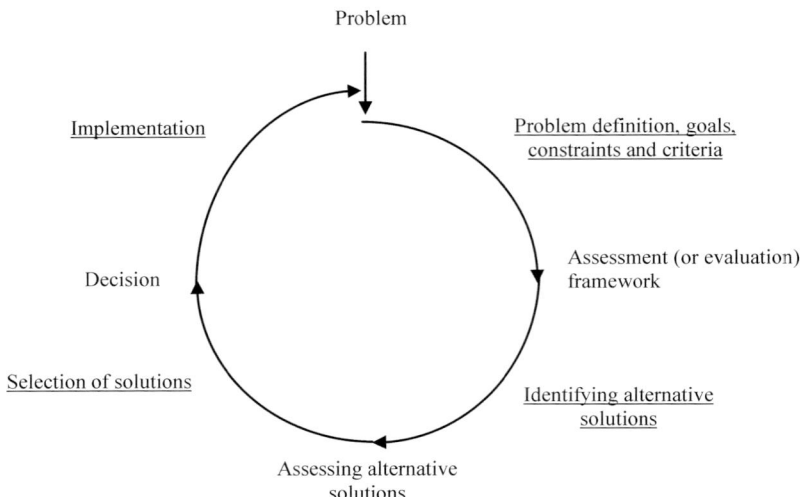

Figure 20.1 Phase model (underlined terms represent processes). (Adapted after Parsons, 1995; Koppenjan and Klijn, 2004.)

Crucial to the achievement of rational outcomes is *objectivity*. This means that knowledge should be gathered about reality as it is and that facts should be separated from subjective and normative insights, theories and prejudices: the 'fact–value dichotomy' (Hawkesworth, 1988). Thus, in this approach, a policy problem is perceived to be of a technical nature (Hoppe, 1999).

The process of the classical decision-making approach can be conceptualized by the phase model (Figure 20.1). The phase model describes a decision-making process as subsequent phases with a clear beginning and end (Koppenjan and Klijn, 2004). A process starts with the definition of problems and goals, and the identification of constraints and criteria. Based on this, an assessment (or evaluation) framework is constructed. Next, alternatives are designed and assessed using the assessment framework, and compared with each other. A decision is made and the solution is implemented (Miser and Quade, 1985; Dunn, 1994; Parsons, 1995).

Some authors criticize the phase model. Sabatier (1991, 1999) states that it is not a causal theory since it never identifies a set of causal drivers that govern the process within and across phases. Instead, work within each phase has tended to develop on its own, almost totally without reference to research in the other phases. In addition, without causal drivers there can be no coherent set of hypotheses within and across stages. Furthermore, according to Sabatier (1999), the phase model oversimplifies the usual process of multiple, interacting cycles. Also, Deleon (1999), who in principle defends the phase model as being a useful heuristic tool, criticizes the normative usage of the phase model. He states that each specific phase of the process is executed by a different set of actors and that thereby the process in its entirety is neglected. Besides that, the phase model implies a certain linearity as opposed to a series of feedback actions or recursive loops (e.g. the assessment of alternative solutions can lead back to problem definition rather than to the next phase) that characterize the policy process (Deleon, 1999). Finally, the phase-model rests on the assumption that a

Knowledge base → Values and norms ↓	Certain	Uncertain
Consensus	**1. Well structured** Knowledge use: instrumental & data	**2. Moderately structured** Knowledge use: strategic & argument
Disagreement	**3. Moderately structured** Knowledge use: conceptual	**4. Unstructured** Knowledge use: enlightment

Figure 20.2 Classification of policy problems and use of knowledge. (Adapted after Hisschemöller, 1993; Van de Graaf and Hoppe, 1996.)

problem can be solved analytically and that a central actor formulates objectives and solves problems in relative autonomy (Rhodes, 1997; Scharpf, 1997; Teisman, 2000; Van de Riet, 2003; Koppenjan and Klijn, 2004). Therefore, this approach is not adequate for dealing with complex unstructured problems, for which the available knowledge is uncertain and the stakeholders' perceptions diverge. In the following sections, we explain these aspects.

20.2.2 Types of problems

A problem occurs when a factual situation is in discrepancy with a desired situation. This implies that a problem always consists of normative and factual/empirical elements. Therefore, problems cannot be regarded as objective givens, but should be regarded as highly subjective social constructs (Dery, 1984; Hisschemöller, 1993; Van de Graaf and Hoppe, 1996). When this subjectivity is taken into account, two dimensions can be used to distinguish between different policy problems (Figure 20.2). The first dimension is consensus about values and norms (normative standards). The other dimension relates to the certainty of the knowledge base or content. By using these two dimensions, four types of policy problems emerge. Well-structured problems (type 1) are problems for which a certain knowledge base and consensus about values and norms (normative standards) exist. Some problems are moderately structured because knowledge is uncertain (type 2) or because of disagreement about values and norms (type 3). When objectives are at stake and knowledge is uncertain, a problem is unstructured (type 4) (Douglas and Wildavsky, 1982; Van de Graaf and Hoppe, 1996).

Figure 20.2 shows that the way knowledge is used differs by problem type. Experts play a dominant role in well-structured problems and take on the role of problem solver. In this case, policy is highly expert-driven (Turnhout, 2003). In moderately structured problems where the knowledge base is uncertain (type 2), the use of knowledge is strategic in that it will be used or rejected depending on the interests at stake. Willingly or unwillingly, knowledge becomes part of the debate, as the different sides tend to strengthen their position by the use of scientific arguments. In moderately structured problems with disagreement on values and norms (type 3), knowledge can accommodate the policy process. In the case of unstructured problems, knowledge can play a role as problem signaler. In doing that, knowledge takes the shape of ideas and can be used as enlightenment (Hisschemöller et al., 1998; Turnhout, 2003; Boogerd, 2005).

20.2.3 Stakeholders' perceptions

Dealing with complex societal problems takes place in an arena in which mutually dependent actors mold and shape problem definitions and solutions. If parties do not sufficiently consider the fact that they have different problem frames, typical results are knowledge conflicts and asymmetrical debates ('dialogues of the deaf') (Van Eeten, 1999; Koppenjan and Klijn, 2004). Information gathering, the use of experts and conducting research will then prove to be counterproductive: the variety of interpretations and the impasse between the parties is strengthened rather than reduced (Koppenjan and Klijn, 2004). Thus, solving complex societal problems is not only an intellectual design activity aimed at taming substantive uncertainties, but also a strategic game in a multi-actor and multipurpose setting. Only through engaging in interaction will parties gain information and only then will positions and standpoints become clear (Koppenjan and Klijn, 2004). In the classical decision-making approach, this interaction turns out to be insufficient, as the problem is handled as a technical problem.

Finally, one can distinguish single-actor and multi-actor complexity. Single-actor complexity arises on the one hand from fuzzy objectives and on the other hand from system complexity and uncertainty about the effects of solutions (Koppenjan, 1990; Van Heffen, 1993; Van de Riet, 2003). Multi-actor complexity is the result of the diversity of problem perceptions among the actors involved (Bennett et al., 1989; Rosenhead, 1989; Van de Riet, 2003). Van de Riet (2003) claims that the classical approach focuses on the requirements that single-actor complexity puts on decision making. While this generates knowledge that can probably withstand the test of scientific validity, it is unlikely to be useful in a multi-actor setting, because no attention is paid to the multi-actor complexity. As a consequence, the classical approach risks producing so-called *superfluous knowledge*.

20.2.4 Use of knowledge

In dealing with complex problems, actors are confronted with *substantive uncertainty* (Koppenjan and Klijn, 2004): uncertainty of problem definition and policy response (Arentsen et al., 2000). One major source of uncertainty has to do with the difficulty the actors experience in determining the nature of the problem, due to lack of information about the input (problem definition), output (policy response) and causal relations. Often, crucial information or knowledge is lacking or not instantly available. However, the problem is not only the lack of information and knowledge, but that it is ambiguous and contested as well (Edelenbos et al., 2003; Koppenjan and Klijn, 2004). Koppenjan and Klijn (2004) and Arentsen et al. (2000) describe two types of responses to substantive uncertainty in classical decision making. The first response is information gathering, using expert knowledge and (scientific) research, and sorting things out before taking steps.

... The assumption behind this reaction to uncertainty is that of neo-positivism: that scientific research into causal relations will lead to objective knowledge about the nature of the problem, the background causes, the possible interventions and their consequences [...] In this neo-positivist vision, science is separate from other societal domains such as government and the market, and is presumed to produce true, objective and universal knowledge.... (Koppenjan and Klijn, 2004)

However, 'facts' are the product of definitions and principles, influenced by people's view of the world, their so-called conceptual orientations, perceptions, mental maps and frames (see e.g. Schön, 1983; Edelenbos et al., 2003). The second response is that of counter-expertise. This results in a debate in which the various parties provide support for their own claims of truth. Research and policy analysis then acquire the function of policy advocacy (Koppenjan and Klijn, 2004).

Classical decision making assumes that the problem is exclusively of a technical nature (Hoppe, 1999). Therefore, technical experts play a dominant role; the problem is handled as a well-structured problem. However, problems are often more complex, because stakeholders have different norms and values (there is no consensus) and knowledge is uncertain or ambiguous (Koppenjan and Klijn, 2004). Thus, problems are more often moderately structured or unstructured. The latter types of problems imply a different usage of knowledge than is common in classical decision making.

In the next sections, a comparison is made between the theoretical framework and observations in the practice of Mainport Rotterdam.

20.3 Case study: decision-making process for the extension of Mainport Rotterdam, the Netherlands

20.3.1 Background

Mainport Rotterdam is located in the southwest of the Netherlands (Figure 20.3). The Dutch government wants to extend the port by 2000 ha (approximately 3 × 8 km) land reclamation

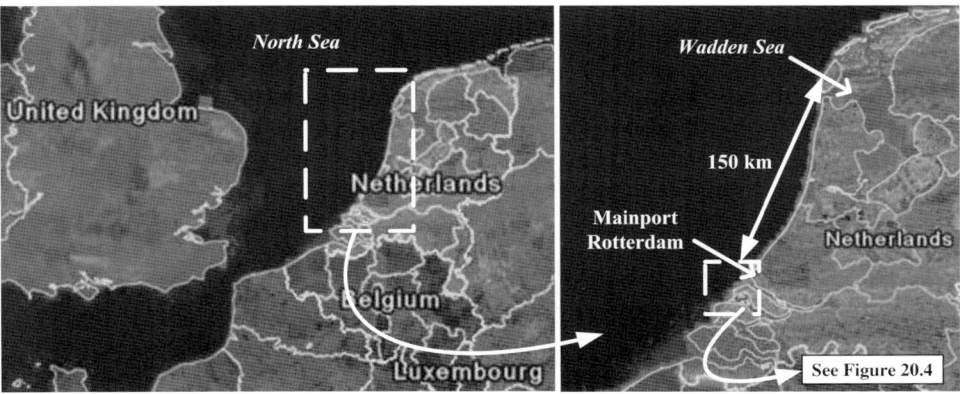

Figure 20.3 Location of Mainport Rotterdam and the Wadden Sea in the Netherlands (Google Earth).

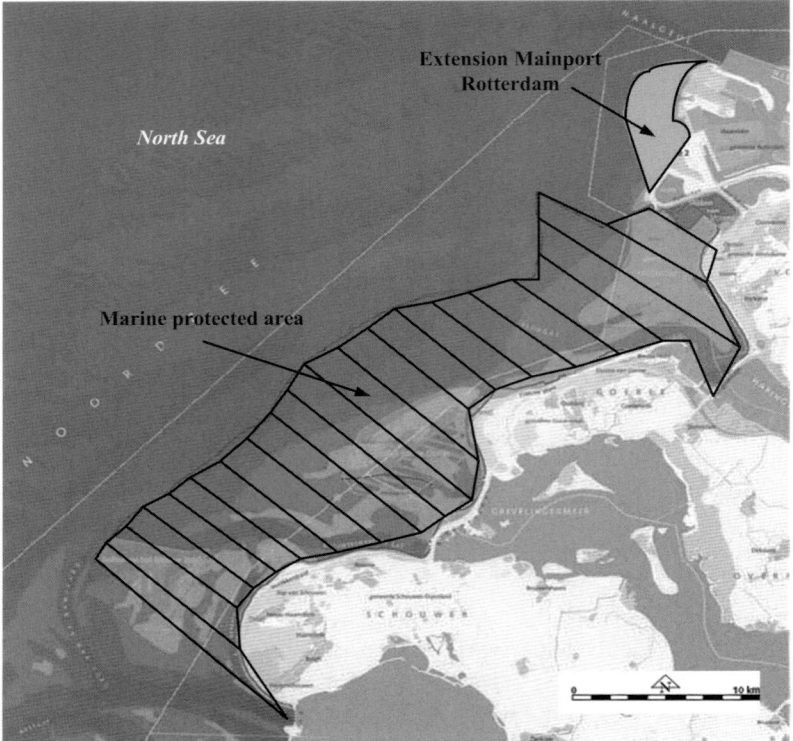

Figure 20.4 Extension of Mainport Rotterdam and location of marine protected area (Ministry of Transport Public Works and Water Management, http://www.rijkswaterstaat.nl, accessed on 31 August 2007).

in the North Sea, aiming to strengthen the port's position. Simultaneously, the government wishes to increase the quality of the region's living environment. This is achieved by more intensive utilization of space and the existing port and industrial area, improvement of the quality of the living environment in the Rotterdam area and by creating a 750-ha wildlife and recreational area near Rotterdam.

In our case study, we focus only on the seaward extension of Mainport Rotterdam. This land reclamation is located in the Voordelta, which is a protected (Natura 2000) site. To compensate for the negative effects on the local ecology, a marine protected area of 31 250 ha will be established in the vicinity of the new reclaimed land. Figure 20.4 shows the extension of Mainport Rotterdam and the location of the marine protected area (Ministry of Transport Public Works and Water Management, http://www.rijkswaterstaat.nl, accessed on 31 August 2007). The key measures to be taken in the protected area are termination of fishing that disturbs the seabed, fishing with stationary nets and snares, and mineral extraction, along with regulation and zoning of recreational activities (Project Mainport Rotterdam, 2001).

The decision process for the extension of Mainport Rotterdam took place by means of a Core Planning Decision-plus (CPD+). The CPD+ is a spatial plan formulated by the government. It contains so-called specific policy decisions, which are binding for lower authorities (i.e. provinces, municipalities). The CPD+ procedure for Mainport Rotterdam started in 1998 and ended with the government's position on the project in 2006. The aspects included in the CPD+ were the following: the location and scope of the land reclamation, the area within which sand dredging for the land reclamation may take place, possible locations for compensating natural resources lost through land reclamation (marine protected area), the location(s) and layout of a 750-ha wildlife and recreational area. Furthermore, for several of the activities from the CPD+, an Environmental Impact Assessment (EIA) procedure was followed. The EIA describes the various projects' effects on wildlife, recreation and the natural environment (Ministerie van V&W, 2001).

The CPD+ for the extension of Mainport Rotterdam was open for public participation. In 2004, a number of objections to the CPD+ were filed at the Council of State. The Department of Administrative Law of the Council of State forms the highest general administrative judge of the Netherlands. This department administers justice in matters of dispute between citizens and government. One of the objectors was the Dutch Fish Product Board. They claimed that, in the CPD+, the effects on fish larvae and mud transport to the Wadden Sea were not sufficiently investigated, as should have been done according to the Habitats Directive. The Wadden Sea is a wetlands area along the North Sea coasts of the Netherlands, Germany and Denmark. The Dutch part of the Wadden Sea consists of an area of approximately 2500 km^2. It is situated between the North Sea and the mainland, approximately 150 km north of Mainport Rotterdam (Figure 20.3), and is sheltered by barrier islands. The Wadden Sea is a highly dynamic ecosystem with tidal channels, sands, mud flats, salt marshes, beaches and dunes. An important aspect of the Wadden Sea is the tidal flats that emerge during low tide and cover about two thirds of the tidal area. The variety in transitional zones between land, sea and freshwater environment is the basis for high species richness. The Wadden Sea has a rich bottom-living fauna and is therefore an important nursery for many species of fish and crustaceans. Furthermore, it is an important feeding and resting area for many species of coastal and migrant birds (Van Berkel and Revier, 1991; De Jonge et al., 1993; Turnhout, 2003; Imeson and van den Bergh, 2006). For these reasons, a large part of the Dutch, German and Danish Wadden Sea has been designated as a wetland of international importance (Van Berkel and Revier, 1991; Verbeeten, 1999) and is protected by the European Bird and Habitat Directives. Furthermore, the responsible ministries of the Netherlands, Denmark and Germany have been working together on the protection and conservation of the Wadden Sea within the Trilateral Wadden Sea Cooperation (http://www.waddensea-secretariat.org/, accessed on August 2008) since 1978.

In January 2005, the Council of State judged that the objections by the Dutch Fish Product Board were valid. Furthermore, the Council of State judged that a so-called Appropriate Assessment procedure had to be carried out, to investigate the impact of the extension of Mainport Rotterdam on the integrity of the Wadden Sea area. This procedure had not been carried out before. The Council of State stated that it was possible that further research on changes in mud and fish larvae transport could give more insight into the impact of the extension of Mainport Rotterdam on the protected Wadden Sea. Due to the Council's decision, the extension of Mainport Rotterdam has become less certain. Therefore, part of the revision of the CPD+ was to perform an Appropriate Assessment procedure and the related investigations (Raad van State, 2005).

The objective of an Appropriate Assessment procedure is to assess whether there will be adverse effects on the integrity of the relevant Natura 2000 area, in this case the Wadden Sea, as defined by the so-called conservation objectives (European Commission, 2001). For the Wadden Sea area, the (provisional) conservation objectives are as follows:

> The policy and management [...] are focused on the sustainable protection and development of the Wadden Sea as a nature area, in which human influence is minimized, and on maintaining or restoring a favorable state of preservation for the structures, species, plants and animals that are designated for protection under the Bird and Habitat Directives for the Wadden Sea. To achieve this, the policy and management are focused on carrying out as naturally as possible the sustainable protection and development of [...] hydrological processes, water quality, soil and air, and also of the (soil) flora and fauna, including the foraging, breeding and resting areas of birds. (Harte et al., 2005)

The Appropriate Assessment Wadden Sea forms the basis for the case study in this chapter.

20.3.2 Process description: Appropriate Assessment Wadden Sea

The government (represented by the National Institute for Coastal and Marine Management) and the Port of Rotterdam worked together on the Appropriate Assessment Wadden Sea because they both were 'problem owner'. The Port of Rotterdam is responsible for the EIA and the government is responsible for the CPD+. Both are necessary to ensure the possibility of extension of Mainport Rotterdam. The Port of Rotterdam and the government commissioned a consortium for the investigations in the context of the Appropriate Assessment. The investigations took place in three parallel tracks. The government chose to run these tracks simultaneously, because the Dutch Parliament set a time limit of 7 months for the Appropriate Assessment. The tracks were as follows:

- *Model calculations (track 1)*: effects (of the extension of Mainport Rotterdam) on mud, nutrients and fish larvae transport to the Wadden Sea were predicted by using models.
- *Expert judgment (track 2)*: possible effects on species and habitats were determined by expert judgment.
- *Final assessment (track 3)*: effects were assessed using the results of track 1, track 2, historical data analysis and audit and stakeholder meetings.

20.3.2.1 Model calculations (track 1)

In track 1, the effects of the extension of Mainport Rotterdam on the Wadden Sea were predicted by using computational models. The objectives of the model calculations were

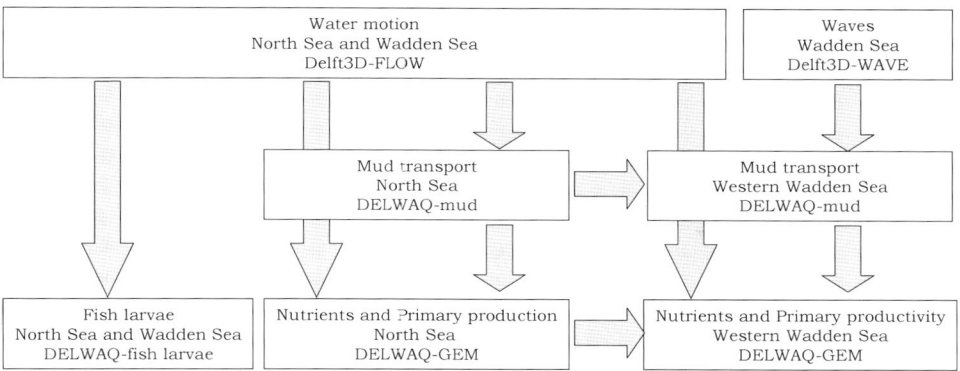

Figure 20.5 Models and relationships: each block lists subject, location and model.

to quantify changes in mud concentrations and fluxes; transport of nutrients and primary production; and larvae transport of herring, plaice and sole (De Jong et al., 2005). These effects were predicted starting at the intervention in the water system (the extension of Mainport Rotterdam). In this approach, several models are used; Figure 20.5 shows the relationships between these models.

20.3.2.2 Expert judgment (track 2)

In track 2, the effects on species and habitats were estimated by judgment of ecological experts. Starting from protected species and habitats, a reflection was made to nonbiological parameters that might be influenced by the extension of Mainport Rotterdam. Based on assumptions of changes in these parameters (nutrients and mud), attention was focused on species and habitats for which effects were expected. The ecological experts were consulted in three workshops. Furthermore, after a request from some of the ecological experts, an ecosystem model (EcoWasp, developed by Alterra, research institute of the Wageningen University and Research Centre concern) for the Wadden Sea was used (Minutes track 2 workshop, 19 May 2005). With this model, several scenarios for changes in mud and nutrients based on model results from the earlier performed Flyland study[2] were run to calculate the effect on ecological parameters (Interview track 2 leader (consortium), 21 June 2006). These parameters were used as input for the expert workshops. In track 1, another model (DELWAQ-GEM, developed by WL | Delft Hydraulics) was used to calculate the ecological parameters, but the results from this model were not yet available at the time the ecological expert workshop took place (Interview track 2 leader (consortium), 21 June 2006).

---

[2] Research program (1998–2003), initiated by the Dutch government, on the possibilities of placing Schiphol Airport on an artificial island in the North Sea. For more information on the Flyland studies, see also Van der Kleij et al. (2003).

The assessment framework of track 2 consisted of the following objective and criteria (Heinis et al., 2005). The objective was to determine the *significance of the impact* of the extension of Mainport Rotterdam on the favorable state of conservation. This was a sum of the following three criteria:

- *Criterion 1*: *extent of impact*, determined by the assessment of intervention–impact chains by expert judgment.
- *Criterion 2*: *conservation status of the species*, a reference value based on the Wadden Sea Area Birds Directive Assessment Framework (LNV DRZ-Noord 2005, in Heinis et al., 2005).
- *Criterion 3*: *proportion of the population that is affected*, is determined by the proportion of the biogeographical population that stays in the Wadden Sea.

In the expert judgment track, the significance of the impact on species and habitats was assessed by using this assessment framework. The assessment showed that '... under the 'basic scenario' [based on Flyland results] four [bird] species are subject to possibly significant negative impacts from the reduction in nutrient contents associated with the presence of the extension of Mainport Rotterdam... (Heinis et al., 2005)'.

20.3.2.3 Final assessment (track 3)

Besides these two research tracks, several other activities were initiated by the government to support the final assessment. This is what we call the final assessment track. Two analyses were performed by government experts using historical data. One analysis focused on the eutrophication state of the Wadden Sea and the other analysis investigated sedimentation in the Wadden Sea. Long-term records were used to investigate whether changes in eutrophication state and sedimentation in the Wadden Sea could be related to human interventions along the Dutch coast, such as the Delta Works and the first seaward extension of Mainport Rotterdam. Furthermore, the government requested an international audit panel to check the scientific underpinning and the consistent use of scientific results in the Appropriate Assessment Wadden Sea. The communication with the stakeholders was organized separately from the research investigations in three stakeholder meetings.

The assessment framework for the final assessment by the government consisted of two linked objectives, which were derived from the conservation objectives (see Section 20.3.2.1) and defined as follows (Harte et al., 2005; Project Mainport Rotterdam, 2006):

- *Objective 1*: Boundary conditions for dynamic processes that guarantee the existence of the natural relationship between species and habitats are not limited.
- *Objective 2*: Continued existence of protected species and habitats is guaranteed.

The final assessment was as follows: with regard to objective 1, the conclusion was that '... the land reclamation [the extension of Mainport Rotterdam] will have a very limited impact on the dynamic processes, which are responsible for the existence of the natural relationship between species and habitats... (Project Mainport Rotterdam, 2006)'. With regard to objective 2, the conclusion was that '... the land reclamation could have an effect on four bird species [as concluded by expert judgment]; however this effect is not significant. Therefore, the conservation status of protected species and habitats will not be influenced negatively by the land reclamation... (Project Mainport Rotterdam, 2006)'. Thus, the government concluded that the extension of Mainport Rotterdam will not have a significant effect on the integrity of the Wadden Sea.

20.4 Analysis

20.4.1 Decision-making process

The Appropriate Assessment Wadden Sea was a large research project (budget approximately $2.5 million) in which many investigations were carried out and much information was gathered. The used methods were model calculations (track 1), expert judgment (track 2), historical analyses (track 3) and audit and stakeholder meetings (track 3). Also, counter-expertise was used in the Appropriate Assessment Wadden Sea. First, the ecological model EcoWasp (Alterra) was used as counter-expertise for the ecological model DELWAQ-GEM (WL | Delft Hydraulics). Second, the historical analyses on nutrients and sediment transport were used as counter-expertise for the model calculations in track 1 on nutrients and mud transport. Thus, in the Appropriate Assessment Wadden Sea both standard responses, use of expert knowledge/research and counter-expertise, to deal with substantive uncertainty in classical decision making (Arentsen et al., 2000; Koppenjan and Klijn, 2004) can be recognized. This shows that in this case the solution of the problem is sought in acquiring more knowledge, indicating that the problem is handled as a purely technical problem, which can be solved by rational analysis. There is no discussion on, for example, the underlying model assumptions. These assumptions are based on values rather than on facts (this might not be recognized by the model experts). By not discussing the values, the model outcomes are presented as facts. This is a typical characteristic of classical decision making.

The process of the Appropriate Assessment Wadden Sea is roughly divided into the following phases:

- The problem and goal were defined in, respectively, the judgment of the Council of State and the conservation objectives.
- The effects of the extension of Mainport Rotterdam on the Wadden Sea were investigated in three research tracks.
- An assessment framework was formulated by the government.
- The effects were assessed by the government, using the results of the research tracks.
- The results of the Appropriate Assessment Wadden Sea formed the input for the revision process of the CPD+.

These are all sequential phases with a clear beginning and end. Furthermore, the government defined and solved the problem, a central steering actor. This is similar to the phase model.

In the literature (Sabatier, 1991, 1999), it is criticized that the phase model does not identify a set of causal drivers that govern the process within and across phases. In the case study, the communication between track 1 (model calculations) and track 2 (expert judgment) took place via the track leaders of the consortium. However, it was not possible to make adjustments in the model calculations (track 1). A project member of the consortium remarked that there was little room for changes due to the combination of short time period and a heavy methodology: '... a tanker that went in a certain direction...' was the metaphor he used to explain this (Interview project member WL | Delft Hydraulics, 14 February 2006). For example, when in the track 2 workshop the remark was made that from an ecological point of view, the year that was used for validation of the models in track 1 was inappropriate, it was not possible to adjust this because the model calculations had already been done (Interview track 1 leader (consortium), 31 July 2006). We observed that there is no causality between different phases as the work within each research track tended to develop on its own and no repeating, interacting cycles were incorporated. This is also illustrated by the fact that some knowledge, e.g. on mud transport to the Wadden Sea, that was brought up in the stakeholder meetings was not included in the investigations (Interview Dutch Fish Product Board, 7 August 2007; Minutes stakeholder meeting 15 July 2005 and 26 September 2005; Heinis et al., 2005; Van Ledden, 2005; Project Mainport Rotterdam, 2006).

Next, it is criticized in the literature (Deleon, 1999) that the phase model implies a certain linearity as opposed to a series of feedback actions or recursive loops that characterize the policy process. In the Appropriate Assessment Wadden Sea, we observed that the track 2 leader claimed that at the time, the effects calculated in track 1 proved to be less strong than those assumed in track 2, the whole expert judgment reasoning should have been repeated, but was not (Interview track 2 leader (consortium), 21 June 2006). Thus, linearity in the process was applied, whereas the outcomes of the model calculations (track 1) in fact lead to a change in the input of the expert judgment (track 2). Finally, it is claimed in the literature (Deleon, 1999) that each specific phase of the process is executed by a different set of actors and that thereby the process in its entirety is neglected. In the Appropriate Assessment Wadden Sea, the consortium had the task of determining the effects of the extension of Mainport Rotterdam on the Wadden Sea, whereas the government carried out the assessment of the effects (Interview project leader RIKZ, 13 March 2006). This separation of responsibilities was seen positively by the government participants (Interview project member RIKZ, 22 November 2005). However, the track 2 leader of the consortium claimed that

> ...the scientists should have the feeling that they would have formulated the assessment in the same way; they must agree on the conclusions and this was not the case.' Also, the track 1 leader remarked that '... as the government has a stake in the extension of Mainport Rotterdam [different caps], one could argue that the assessment should have been done by an independent party; e.g. the consortium. (Interview track 1 leader (consortium), 31 July 2006)

20.4.2 Stakeholders' perceptions

The formal problem perception of the objector, the Dutch Fish Product Board was made explicit in its claim to the CPD+. The Board claimed that the effects on fish larvae and mud transport to the Wadden Sea had not been sufficiently investigated, as should have been done according to the Habitats Directive. However, during an interview with the Dutch

Fish Product Board, we observed that its (implicit) problem perception is actually wider than that. First, the Dutch Fish Product Board was not satisfied with the decision-making process, because it was not allowed to participate in the negotiations with the minister at the time the project had just started, in the 1990s (Interview Dutch Fish Product Board, 7 August 2007). Second, the Dutch Fish Product Board was concerned about income losses due to the loss of fishing area where the extension of the Mainport Rotterdam (3000 ha) is envisaged and especially where the marine protected area (31 250 ha) is planned. Finally, as it was not actively involved in the decision-making process, no (financial) compensation was arranged for the fishing industry until its objection to the CPD+ in 2005 (Interview project leader RIKZ, 13 March 2006). From these remarks, we concluded that the implicit problem perception of the Dutch Fish Product Board is *socioeconomic*.

The problem perception by the government and of the Port of Rotterdam is that, by the invalidation of the CPD+, the extension of Mainport Rotterdam is no longer ensured by legal means. Thus, they perceive it as a *procedural* problem. Furthermore, project members from the government and the parties that executed the Appropriate Assessment Wadden Sea had the feeling that they must 'work on the edge of what is possible' to investigate the effects of the extension of Mainport Rotterdam on the Wadden Sea. According to them, the problem was that there was not sufficient knowledge on the effects; they perceived the problem as a *technical* problem. Therefore, the investigations did not focus on the implicit socioeconomical problem of the objector, the Dutch Fish Product Board. Thus, the problem of the objector is not addressed or eliminated by these technical investigations. Finally, as the diverging stakeholders' perceptions were not taken into account, we can conclude that the knowledge that is produced is an example of *superfluous knowledge*.

20.4.3 Use of knowledge

Regarding the use of knowledge, we found that there was no agreement on the argumentation for the final assessment. The project leader from the government stated that the results from the expert judgment (track 2) form a worst-case scenario and that the model results from track 1 are used to 'tune' these worst-case conclusions (Interview project leader RIKZ, 13 March 2006). However, the track 2 leader claimed that at the time the effects calculated in track 1 proved to be less strong than those assumed in the expert judgment, the whole track 2 argumentation should have been redone. Nevertheless, the government, to the contrary, claimed that '... the effects are less [than assumed in track 2], thus the effects are not significant' (Interview track 2 leader (consortium), 21 June 2006). Furthermore, the track 2 leader states that:

> ... too many other things [historical analyses; policy developments] were used as arguments to formulate the assessment. These arguments were not well [scientifically] underpinned and were not audited. This way the research results from track 1 and 2 are undermined. And it is a shame for the goodwill that was obtained from the scientists [participants in expert workshops, track 2], as they now have the feeling that the government does what it wants and does not really listen to them...

In short, the ecological experts (in track 2) claim that the final assessment should have been a scientific one. This is probably because the government asked the experts in track 2 to give a nonnormative/objective assessment of the effects on species and habitats (Minutes track 2 workshop, 19 May 2005). Thereby, the government made sure that the experts would handle the problem as a well-structured problem. The experts were assumed not to be stakeholders in the process. However, in the final assessment the government used the knowledge (from track 1 and 2) in a strategic way; thus, the problem was handled as a moderately structured (type 2) problem (see Figure 20.2). This caused that both parties look upon the use of knowledge in the final assessment differently. Also, the Netherlands Commission for Environmental Assessment concluded that from the reports it is not clear of the final assessment is agreed upon by all experts (Commissie voor de M.E.R., 2006).

As for stakeholders that were not directly involved in the investigations, the Wadden Sea Society (an environmental NGO) finds the conclusion of the Appropriate Assessment Wadden Sea too definite, as the models that are used are not validated and the effects are uncertain. The North Sea Foundation (also an environmental NGO) reacted that the sand extraction could also have a direct effect, which is not taken into account in the investigations. The Fish Product Board states that the conclusion that the effects on mud are not significant is subjective (Ministerie van Verkeer en Waterstaat, 2006).

20.5 Conclusions

This chapter is a case study of the decision-making process for the extension of Mainport Rotterdam. We focused on the studies for the impact of the port extension on a protected nature area (the Appropriate Assessment Wadden Sea). We found that this process can be characterized as a classical decision-making process as:

- The problem was perceived to be of a technical nature and was solved on the basis of rationality and objectivity.
- The two standard responses (information gathering and counter-expertise) to deal with substantive uncertainty in classical decision making were observed.
- The phase model and the criticism on it were recognized.

We analyzed how stakeholders' perceptions were dealt with and how knowledge was used.

The Dutch Fish Product Board objected to the extension of Mainport Rotterdam. This was their formal problem perception, which is based on ecological and procedural grounds. However, their implicit problem perception is a wider, socioeconomic problem. Nevertheless, the parties that executed the Appropriate Assessment Wadden Sea perceived the problem to be procedural and technical. Thus, the investigations focused only on the formal problem perceptions of the objector, not taking into account the implicit socioeconomical problem of the objector. Therefore, the socioeconomic problem of the Dutch Fish Product Board was not eliminated by these technical investigations. We conclude that diverging stakeholders' perceptions (multi-actor complexity) were not addressed, but that the investigations focused only on single-actor complexity. Thus, the knowledge that was produced is an example of *superfluous knowledge*.

Furthermore, with regard to the use of knowledge, we conclude that the ecological experts treated the problem as a well-structured problem, whereas the government treated it as a moderately structured problem. We also conclude that the government used the knowledge that was generated in the research tracks in a strategic way. This resulted in a lack of consensus on the use of knowledge in the final assessment.

We also observe the difficulties with stakeholders' perceptions and the use of knowledge that arise from classical decision making in the practice of Mainport Rotterdam. This constitutes empirical support for the literature that describes these aspects. Several authors claim that for these kind of complex problems, a process-oriented approach should be adopted, which pays attention to differences in stakeholders' perceptions and to interaction and communication (De Bruijn et al., 2002; Edelenbos et al., 2003; Koppenjan and Klijn, 2004). Our findings form an argument for practitioners in coastal management to choose a process-oriented approach for dealing with complex issues, instead of a classical decision-making approach.

Acknowledgments

This research is supported by the Technology Foundation STW, applied science division of NWO and the technology program of the Ministry of Economic Affairs of the Netherlands. We thank the National Institute for Coastal and Marine Management (RIKZ) for giving us the opportunity to observe the project 'from the inside'. Furthermore, we are grateful to all the participants for their cooperation and remarks and especially the participants with whom we did an in-depth interview. At last, we thank Hans Th. A. Bressers of the Centre of Clean Technology and Environmental Policy (University of Twente) for discussions and useful comments on the case study.

References

Arentsen, M.J., Bressers, H.T.A. and O'Toole, L.J. (2000) Institutional and policy responses to uncertainty in environmental policy: a comparison of Dutch and US styles. *Policy Studies Journal* **28** (3), 597–611.

Bennett, P., Cropper, S. and Huxham, C. (1989) Modelling interactive decisions: the hypergame focus. In: Rosenhead, J. (ed.). *Rational Analysis for a Problematic World: Problem Structuring Methods for Complexity, Uncertainty and Conflict*. John Wiley & Sons, West Sussex, pp. 283–314.

Boogerd, J.L.M. (2005) Van droge kennis naar natte natuur: de interactie tussen natuurwetenschap en beleid over verdroging, Ph.D. thesis. Eburon, Vrije Universiteit Amsterdam, Delft.

Commissie voor de M.E.R (2006) PKB Project Mainportontwikkeling Rotterdam (PMR) – Beoordeling van het milieurapport SMB en passende beoordelingen. rapportnummer 952–476 (in Dutch).

De Bruijn, H., Ten Heuvelhof, E.F. and In 't Veld, R.J. (2002) *Process Management: Why Project Management Fails in Complex Decision Making Processes*. Kluwer Academic Publishers, Dordrecht.

De Jong, M., Van Ledden, M., Van Zanten, M. and Heinis, F. (2005) Investigation into the effects of Maasvlakte 2 on the Wadden Sea for the purpose of the Appropriate Assessment and EIA – Summary Overall Approach Plan. Consortium 3|MV2, client: Port of Rotterdam and RWS/RIKZ.

De Jonge, V.N., Essink, K. and Boddeke, R. (1993) The Dutch Wadden Sea - a changed ecosystem. *Hydrobiologia* **265** (1–3), 45–71.

Deleon, P. (1999) The stages approach to the policy process – what has it done? Where is it going? In: Sabatier, P. (ed.). *Theories of the Policy Process*. Westview Press, Boulder, pp. 19–32.

Denters, B., Van Geffen, O., Huisman, J. and Klok, P.J. (2003) *The Rise of Interactive Governance and Quasi-Markets*. Library of public policy and public administration, Kluwer Academic Publishers, Dordrecht, 280 pp.

Dery, D. (1984) *Problem Definition in Policy Analysis, Sociology of the Sciences*, vol. 24. University Press of Kansas, Lawrence.

Douglas, M.T. and Wildavsky, A.B. (1982) *Risk and Culture: An Essay on the Selection of Technical and Environmental Dangers*. University of California Press, Berkeley.

Dunn, W.N. (1994) *Public Policy Analysis: An Introduction*, 2nd edn. Prentice-Hall, Englewood Cliffs.

Edelenbos, J. and Klijn, E.H. (2005) Managing stakeholder involvement in decision making: a comparative analysis of six interactive processes in the Netherlands. *Journal of Public Administration Research and Theory* **16** (3), 417–446.

Edelenbos, J., Monnikhof, R. and Van de Riet, O. (2003) A Double Helix approach: a proposal to forge a better integration of analysis and policy development. *International Journal of Technology Policy and Management* **3** (1), 1–21.

European Commission (2001) Assessment of plans and projects significantly affecting Natura 2000 sites – Methodological guidance on the provision of Article 6 (3) and (4) of the Habitats Directive 92/43/EEC. European Commission, Luxembourg.

Harte, M., Peters, B. and Van Zetten, R. (2005) The Wadden Sea Area in Perspective. RWS/RIKZ, The Hague.

Hawkesworth, M.E. (1988) *Theoretical Issues in Policy Analysis*. State University of New York Press, Albany.

Heinis, F., Van Der Vegte, J.W., De Vlas, J., Van Ledden, M. and Jager, Z. (2005) Impact of Maasvlakte 2 on the Wadden Sea and North Sea coastal zone – effects in the context of the Birds and Habitats Directives. Consortium 3|MV2.

Hisschemöller, M.J. (1993) De democratie van problemen: de relatie tussen de inhoud van beleidsproblemen en methoden van politieke besluitvorming, Ph.D. thesis (in Dutch). VU Uitgeverij, Amsterdam.

Hisschemöller, M., Groenewegen, P., Hoppe, R. and Midden, C.J.H. (1998) Kennisbenutting en politieke keuze: een dilemma voor milieubeleid? (in Dutch). Rathenau Instituut, Den Haag Werkdocument 65.

Hoppe, R. (1999) Argumentative turn – policy analysis, science and politics: from 'speaking truth to power' to 'making sense together'. *Science and Public Policy* **26** (3), 201–210.

Imeson, R.J. and Van Den Bergh, J. (2006) Policy failure and stakeholder dissatisfaction in complex ecosystem management: the case of the Dutch Wadden Sea shellfishery. *Ecological Economics* **56** (4), 488–507.

Koppenjan, J.F.M. (1990) Definiëring van complexe problemen door de overheid: balanceren tussen ruim en precies. *Beleidswetenschap* **4** (1), 21–45.

Koppenjan, J.F.M. and Klijn, E. (2004) *Managing Uncertainties in Networks: A Network Approach to Problem Solving and Decision Making*. Routledge, London.

Ministerie van V&W (2001) Milieu-effect rapport Project Mainport ontwikkeling Rotterdam (in Dutch). Available from http://www.verkeerenwaterstaat .nl/kennisplein/uploaded/ZH/2007-02/342252/Notitie_van_Bevindingen_bij_Deel_3_PKB_PMR_(2006).pdf.

Ministerie van Verkeer en Waterstaat (2006) Notitie van Bevindingen bij Deel 3 PKB PMR (2006) – Kabinetsstandpunt over inspraakreacties en het advies op de Strategische Milieubeoordeling en Passende Beoordeling Landaanwinning (in Dutch). Available from http://www.verkeerenwaterstaat.nl/kennisplein/uploaded/RIKZ/2006-02/144806/MER%20Hoofdrapport.pdf.

Miser, H.J. and Quade, E.S. (1985) *Handbook of System Analysis – Overview of Uses, Procedures, Applications, and Practice*. North-Holland, New York, Amsterdam, Oxford.

Parsons, W. (1995) *Public Policy: An Introduction to the Theory and Practice of Policy Analysis*. Edward Elgar, Cheltenham.

Project Mainport Rotterdam (2001) Birds and Habitats Directives – Request for advice from and exchange of information with the European Commission within the framework of the Birds and Habitats Directives.

Project Mainport Rotterdam (2006) Passende Beoordeling Landaanwinning - Deelrapport Speciale Beschermingszones Waddenzee en Noordzeekustzone (in Dutch).

Quade, E.S. (1989) *Analysis for Public Decisions*, 3rd edn. North-Holland, New York.

Raad van State (2005) Uitspraak zaaknummer 200307350/1. Afdeling Bestuursrechtspraak (in Dutch).

Rhodes, R.A.W. (1997) *Understanding Governance: Policy Networks, Governance, Reflexivity and Accountability*. Open University Press, Buckingham.

Rosenhead, J. (1989) *Rational Analysis for a Problematic World: Problem Structuring Methods for Complexity, Uncertainty and Conflict*. John Wiley & Sons, Chichester.

Sabatier, P.A. (1991) Toward better theories of the policy process. *Ps-Political Science & Politics* **24** (2), 147–156.

Sabatier, P.A. (1999) *Theories of the Policy Process*. Westview Press, Boulder.

Scharpf, F.W. (1997) *Games Real Actors Play: Actor-Centered Institutionalism in Policy Research*. Westview Press, Boulder.

Schön, D.A. (1983) *The Reflective Practitioner: How Professionals Think in action*. Basic Books, New York.

Teisman, G.R. (2000) Models for research into decision-making processes: on phases, streams and decision-making rounds. *Public Administration: Journal of the Royal Institute of Public Administration* **78** (4), 937–956.

Turnhout, E. (2003) Ecological indicators in Dutch nature conservation – science and policy intertwined in the classification and evaluation of nature, Ph.D. thesis. Aksant Publishers, Amsterdam.

Twaalfhoven, P.G.J. (1999) The success of policy analysis studies: an actor perspective: a search for success definitions based on cases in the field of transport and infrastructure in the Netherlands, Ph.D. thesis. Eburon, Technische Universiteit Delft, Delft.

Van Berkel, B.M. and Revier, J.M. (1991) Mussel fishery in the International Wadden Sea, consistent with wise use. *Landscape and Urban Planning* **20** (1–3), 27–32.

Van de Graaf, H. and Hoppe, R. (1996) Beleid en politiek: een inleiding tot de beleidswetenschap en de beleidskunde, 3e dr; 1e dr.: 1989 (in Dutch). Coutinho, Bussum.

Van de Riet, A.W.T. (2003) *Policy Analysis in Multi-Actor Policy Settings: Navigating between Negotiated Nonsense and Superfluous Knowledge*, Ph.D. thesis. Eburon, Technische Universiteit Delft, Delft.

Van Der Kleij, C.S., Hulscher, S.J.M.H. and Louters, T. (2003) Comparing uncertain alternatives for a possible airport island location in the North Sea. *Ocean and Coastal Management* **46**, 1031–1047.

Van Eeten, M.J.G. (1999) *Dialogues of the Deaf: Defining New Agendas for Environmental Deadlocks*, Ph.D. thesis. Eburon Publishing, Delft.

Van Heffen, O. (1993) Beleidontwerpen en omgevingsfactoren: vier alternatieve strategieën. In: Van Heffen, O. and Van Twist, M.J.W. (eds). *Beleid en wetenschap: hedendaagse bestuurskundige beschouwingen.* Samsom/Tjeenk Willink, Alphen a/d Rijn, pp. 67–81.

Verbeeten, T.C.M. (1999) Wijs met de Waddenzee? een onderzoek naar leerprocessen, Ph.D. thesis. Thela Thesis, Universiteit Utrecht, Amsterdam.

Chapter 21
Local Coastal Zone Planning and Stakeholder Participation in Norway

Knut Bjørn Stokke and Sissel Hovik

Abstract

Local planning is regarded as the most important tool for integrated coastal zone management in Norway, particularly due to the active participation of stakeholders. This chapter presents experiences with stakeholder participation in municipal planning in the coastal zone, focusing on different strategies for participation and the relationship between participation and competition for marine areas. In addition, we focus on the relation between stakeholder participation and the degree of influence. The findings are based on a study of ten 'best-practice' municipalities in terms of integrating the interests of coastal fisheries and aquaculture. In all ten municipalities, representatives from local fishermen and fish and shellfish farmers have participated quite actively in the planning processes. In the municipalities with low pressure on the marine area, we find a relatively closed participation strategy, dominated by informal and bilateral meetings with local fishermen and fish and shellfish farmers, respectively. The municipalities that have high degree of competition for the marine areas have all chosen a rather open planning strategy, involving a multitude of private actors in formal and public arenas. Furthermore, among the ten municipalities, we find a strong correlation between participation strategy and influence. The municipalities that chose a closed participation strategy gave priority to aquaculture, and to some degree to fishing. The municipalities that chose an open participatory strategy gave priority to multi-purpose interests, where also recreation, nature conservation and other coastal interests were integrated in the plan. At the end of the chapter we discuss the findings in light of democratic principles, such as equality and transparency.

21.1 Introduction

During the past decades, participation from societal stakeholders in public planning processes has been emphasised in the planning literature (Healey, 1997; Amdam, 2000). The same is true regarding nature resource management (Ostrom, 1990; Jentoft et al., 1998; Pinkerton, 2003). Citizen and stakeholder participation is regarded as important and desirable, because it tends to make the planning process and results more effective, democratic and legitimate (Pateman, 1970). In addition, to bring stakeholders representing different interests and values together and facilitate their interaction can contribute to conflict resolution and to the development of trust and reciprocity among diverse actors (Buanes et al.,

2005). Thus, participation is considered to contribute to the formulation and implementation of policies promoting integrated coastal zone management.

Participatory planning can be looked upon as an example of network governance (Healey, 1996). As for network governance in general, participatory planning does have some difficulties in fulfilling some crucial democratic principles, such as equality and accountability (Sørensen and Torfing, 2005). It is difficult, and often impossible, to secure that all affected actors have the same possibility to influence the decision, or the same chance to having their interests and values considered by the decision makers. This will weaken the legitimacy of the planning decisions (Scharpf, 1999). Furthermore, decisions made by partnerships or networks consisting of private and public actors make it hard both to know whom to hold accountable for the decision, and how (Sørensen and Torfing, 2005).

The fact that the coastal zone plans are finally decided by democratically elected local councils does reduce, but by no means eliminate, these problems (Peters and Pierre, 2000; Hovik and Vabo, 2005). It can be difficult for the council to make substantial changes in a plan formulated by the municipal administration in cooperation with powerful local stakeholders, and the ones who are negatively affected and have not been included in the process have no reason to accept the result. Transparent and open decision processes are therefore important to reduce these shortcomings (Kickert et al., 1997; Sørensen and Torfing, 2005). Buanes et al. (2005) identifies the importance of informal channels for participation in local coastal zone planning in Norway, such as direct contact with planning officers. The relative importance of this kind of informal channels contradicts the principles of transparency and public dialogue.

In Norway, integrated management of the coastal areas is primarily intended to be achieved through spatial planning carried out at local level in accordance with the Planning and Building Act (PBA). The responsibility for land use planning as well as the use and protection of marine areas is delegated to the municipalities. The municipalities have the authority to designate given marine areas for specific use or protection of specific interests, such as transportation, fisheries, recreation, nature conservation or aquaculture – combined or separate. The municipalities can alternatively leave some sea areas unplanned. Many coastal municipalities responded rather quickly to the revision of the PBA in 1989, which gave them the opportunity (but not an obligation) to produce spatial plans for the coastal areas and the sea out to the baseline (i.e. the straight line between the outer islets and reefs). By the end of 2005, 82% of Norway's 280 coastal municipalities had made such plans (The Directorate of Fisheries, 2005).

The PBA emphasises participation from private stakeholders and the general public. The municipal planning process is supposed to be a common arena for all relevant parties and interests. Affected individuals and groups shall be given an opportunity to participate actively in the planning process, and given the possibility of influencing the result. The municipalities are required to announce the start of the planning process for the general public, as well as to organise an open hearing on the plan proposal before the final decision in the local council. Within the framework of the law, the municipalities are expected to find participatory forms that suit their particular context and needs. What *active* participation (expansion of the formal minimum) actually is supposed to mean is not elaborated in the legislation, and this is in practice a question for the municipal authorities themselves to decide. In this chapter we concentrate on active forms of participation, expanding the minimum requirement of just responding to the planning proposals in the hearing phase.

Taking the above-mentioned theoretical considerations as a point of departure, we focus on the participation strategies within municipal coastal zone planning. Our first question is: What kind of strategies for private stakeholder participation have the 'best-practice' municipalities chosen? We are particularly interested in two dimensions: (1) Who is participating? (2) How do they participate? Emphasising the aspect of integration, we focus on the distinction between arenas for bilateral information and negotiations on the one hand, and multilateral (or public) deliberation on the other. The traditional consultation arena, where each stakeholder is giving a written comment on the proposed plan, is an example of the first. Meetings (formal or informal) between one stakeholder group (for example local fishermen) and the municipal planner are also examples of the first. Open public meetings and formal working groups with members from different stakeholders are examples of the second. In our best-practise cases, stakeholder participation always goes beyond the legal minimum of public consultation. We focus on whether this extended participation is 'private'/bilateral, or whether the municipalities (also) have arenas for public deliberative participation. According to Arnstein's (1969) ladder of participation, both strategies represent 'partnership', giving the private stakeholders potential power to influence the plan.

The increased expansion of aquaculture along the Norwegian coastal zone was an important reason why municipalities started to incorporate the marine areas in their planning. The need to balance this industry's growing demands for good locations with the fishermen's needs to maintain their access to important fishing grounds did bring forth the need for coastal zone planning at local level (Stokke et al., 2006). As the aquaculture industry is developing, through the cultivation of marine species and sea ranching, their need for locations is steadily expanding. Furthermore, other activities and interests demand for the marine area is also increasing, such as for recreation, tourism, sea transport and nature conservation. The competition for the marine area does, however, vary along the coast as well as within the different regions (Hovik and Stokke, 2007a, b). The trend is the more urbanised area, the more competition – especially related to outdoor recreation, sea transport, tourism, building of second homes and boathouses/marinas.

Related to the different contexts for local planning, our next question is: Does the municipal strategy for stakeholder participation vary, depending on the level of competition for marine area? Will the number of participating stakeholders increase as the number of interests (or competing activities) increase? And, will the arenas for participation be more public and open, as the number of interests increases?

Our third research question is: Does participation matter? Are the participating stakeholders given influence on the actual designated plan for the marine area? We operationalise the question of influence as to what degree the stakeholders are actually favoured with designated spaces in the coastal zone plan. Furthermore, we examine if there is a tendency that some stakeholders are more influential than others, and whether there is a link between the municipal strategy for participation and stakeholder influence.

21.2 Material and methods

The findings in this chapter are based on the study of the coastal zone plans and related documents in ten coastal municipalities; on interviews with representatives of the regional offices of the Directorate of Fisheries, the County Governor's environmental agencies, the county administration, the regional branches of the Norwegian Aquaculture Federation and the Fishermen's Federation: and on interviews with some of the municipal planners (Stokke et al., 2006). Moreover, representatives of local fishermen and fish farmers in a few municipalities were interviewed.

Data were gathered from a project in 15 municipalities covering all seven regions of the Norwegian Directorate of Fisheries. The choice of municipalities was made after recommendations from relevant regional actors. They were asked to name municipalities in their region, which in their opinion did meet some formulated criteria, such as involving the interests of fisheries and aquaculture in the planning process and developing and integrating biological knowledge and knowledge about the spatial interests of the fisheries and the aquaculture industry (Stokke et al., 2006). In this way, the chosen municipalities represented 'best-practice' cases, and not the coastal municipalities in Norway in general. The data collected from 5 of the municipalities were not sufficient to answer the questions of this chapter, and were therefore omitted from this chapter.

The ten municipalities are situated along the Norwegian coast from the western part (Sogn & Fjordane County) to the northern part (Finnmark County). Some of them are located in narrow fjords and others more exposed to the sea with open islands/archipelagos. As mentioned, the competition for marine areas varies, where pressure from a multitude of competing interests is highest in the (sub)urban areas. In addition, there are variations between the municipalities regarding population size, population growth, industrial base, amount of fishermen and planning experience (see Table 21.1). A common characteristic is the importance of the aquaculture industry, even though the relative dependency of the industry varies between the municipalities according to their general industrial base. Remote

Table 21.1 Co-variation between competition for the marine area and some other characteristics of the ten municipalities in the study.

	Degree of competition for marine area		
	High	Low	N
Population size 1.1.2008[a]			
Above 6000 inhabitants	3	1	4
Below 3000 inhabitants	1	5	6
Population change 1985–2007[a]			
Population growth	3	0	3
Population decline	1	6	7
Industrial base[b]			
Strong/complex	4	1	5
Weak	0	5	5
Importance of fisheries[c]			
Highly important	0	3	3
Less important	4	3	7
Planning experience			
Second or third generation	4	4	8
First-generation coastal zone plan	0	2	2

[a] *Source*: Statistics Norway 2008 (www.ssb.no/kommuner/).
[b] *Source*: Statistics Norway, municipality types 1994. Typology for industrial classification of the municipalities. Strong base: service production dominates, possibly in combination with manufacturing. (T alone or in combination with I or A). Weak: agriculture and/or fisheries play a dominant role. (L or F in combination with other types).
[c] Source as above: Highly important where fisheries play a dominant role. Less important: all other types.

municipalities with a low and declining population and a weak and unilateral industrial base are more dependent of aquaculture than more urbanised municipalities with a stronger and multilateral industrial base. Thus, there is a strong correlation between competition for the marine areas and the importance of industrial development based on utilisation of the local nature resources.

Due to the limited number of municipalities in the study, it was impossible to isolate the effects of competition for marine areas from effects of other characteristics. Thus, we have no ambitions to conclude on causal relations, but indicate some interesting connections we find in our data between competition for the marine area, participation strategies and degree of influence.

21.3 Results

21.3.1 *Stakeholder participation in the planning process*

In all the ten municipalities, representatives of local fishermen and fish and shellfish farmers have participated rather actively in the planning processes, which are not surprising, as the municipalities were selected in accordance with this criterion. Local fishermen and the aquaculture industry contribute with important information. The fishermen usually provide information about spots and areas important for the fisheries, such as fishing grounds and spawning areas and argue for the need to reserve these areas for fishing. The fish and shellfish farmers often identify areas suitable for aquaculture, and argue for the need to reserve these for possible future locations for the industry. Thus, both representatives of the coastal fisheries and aquaculture contribute with important and necessary input to the plans based on relevant knowledge. In addition, the regional offices of the Directorate of Fisheries, with a national responsibility for both coastal fisheries and aquaculture, have participated actively in the planning processes in all ten municipalities. They are responsible for the promotion of the interests of both these industries, for giving the municipalities' access to necessary spatial information, and, in addition, they coordinate the interests of the aquaculture and the coastal fisheries in cases where there are conflicts.

Representatives of other coastal interests, such as recreational interests, environmental groups or the tourist industry participate to a lesser degree than those that represent the interests of fishermen and fish and shellfish farmers. This result is confirmed by a study of municipal coastal zone planning in three Norwegian counties, particular regarding the contact with local environmental organisations (Hovik and Stokke, 2006). However, some municipalities do try to involve a range of other interests by arranging open meetings in the different communities within the municipal border. Furthermore, some municipalities involve a wider range of stakeholders in working groups that are responsible for the plan proposal or advisory groups giving input to this proposal. We find a clear tendency that the more private stakeholder groups involved, the more formalised, multilateral and public are the arenas for participation (see Table 21.2). Where the participation is limited to local fishermen and fish farmers, informal and bilateral arenas are dominating. These two particular stakeholder groups are, however, included in the more formal meetings or groups in municipalities with this kind of participation strategy.

We find two rather distinct participation strategies in these ten municipalities. Half of them have chosen an open strategy, where several different stakeholders are included in

Table 21.2 Participation strategy in coastal zone planning, according to degree of pressure on the marine area.

Pressure on the marine area	Who do participate?		How is the participation?		Participation strategy		N
	Limited to fishermen and fish farmers	Extended to other private actors	Multilateral	Bilateral	Open	Closed	
High	1	3	4	0	4	0	4
Low	5	1	1	5	1	5	6

formal and open arenas as working groups or advisory groups, in addition to the open meetings. The other half has chosen a closed participation strategy, where representatives of the fishing and fish-farming industries are invited to bilateral and often more informal meetings with the planning officer and/or the planning committee. Thus, our study confirms the findings in a recent study, concluding that informal participation, particularly direct contact with the municipal planner, is quite common in coastal zone planning in Norway (Buanes et al., 2005).

It was among the municipalities where there was relatively low pressure on the marine area that we found a closed participation strategy. The municipalities with stronger pressure and high competition for marine areas, characteristic for the urbanised municipalities in the study, had all chosen a rather open planning strategy, involving a multitude of private actors in formal and public arenas.

21.3.2 *Participation and degree of influence*

We examined to what extent the interests of aquaculture and fishing were particularly favoured in the municipal coastal zone plan, or whether the interests of nature and recreation were favoured. This was done by studying the distribution of sea areas designated to different purposes or combination of purposes in the legally binding planning map. Over time, the interests of both aquaculture and fisheries have been given stronger priority (Stokke et al., 2006). A majority of the ten municipalities designated larger areas for aquaculture, either as single purpose areas or as multi-purpose areas (combined with sea transport, recreation, nature conservation and fisheries) in the most recent revision of their coastal zone plan than in earlier versions. Furthermore, there was a trend in many municipalities of designating larger areas for the single purpose of fishing, meaning that these areas are protected from competing interests that may threaten commercial fishing inside the baseline. These two parallel trends might well be a result of an active participation from the fishermen and the aquaculture, with strong support from the Directorate of Fisheries.

In spite of these general trends, there was great variation in how the municipalities planned their sea areas. Some municipalities left most of their areas open for aquaculture, designated as multi-purpose areas including aquaculture. The plans identify where aquaculture should not be allowed due to conflicts with other coastal interests. This was characterised as flexible planning (Hovik and Stokke, 2007b). The category covering the largest areas is multi-purpose use, including aquaculture. Other municipalities had more detailed plans, designating specific areas where aquaculture is allowed. The category covering the largest

Table 21.3 Influence in coastal zone planning in ten municipalities, according to degree of pressure on the marine area and participation strategy.

Pressure	Participation strategy	Given interest particularly favoured[a]			N
		Aquaculture	Fisheries	Multi-purpose use	
High	Open	2	2	4	4
	Closed	0	0	0	0
Low	Open	0	0	1	1
	Closed	5	2	1	5

[a] Not exclusive categories, i.e. it is possible to favour one, two or all interests.

areas was multi-purpose use, excluding aquaculture. Some municipalities designated large areas for fishing. In some municipalities in northern Norway, single-purpose fishing areas dominated the marine part of the plan.

All these three strategies were represented among our ten municipalities (see Table 21.3). Half of them had the praxis of detailed planning, favouring multi-purpose use excluding aquaculture. Two of them, however, combined this with rather large areas designated for aquaculture, and two had rather large areas designated for fishing. The other five municipalities chose a strategy of flexible planning, opening the majority of their sea areas for aquaculture. Two of them had, in addition, rather large areas designated for fishing. The first praxis dominated among municipalities with strong pressure on their marine areas, the other among municipalities with low pressure.

The planning praxis did, however, even to a larger degree correspond with the participation strategy. The five municipalities with an open participation strategy had all chosen a praxis favouring multi-purpose use through a detailed planning strategy, and the five municipalities with a closed participation strategy had all chosen a praxis favouring aquaculture through a flexible planning strategy.

One interesting observation was the differences between small, remote municipalities with a lesser degree of competing interests in the coastal zone and more urbanised municipalities with a high level of competing interests. In the first group of municipalities, the informal ways of participation was dominant by separate and bilateral meetings between municipal planners and local fishermen and fish and shell farmers, respectively. The result of the planning processes in these municipalities was large areas designated to aquaculture and fisheries, either as single-purpose areas or as multi-purpose areas. In the second group of municipalities, a more formal participation, in the form of working groups and public meetings, including a wider range of stakeholders, dominated. The result of the planning processes in these municipalities was more complex and detailed plans regarding the specific use or protection of different interests in the coastal zone.

21.4 Discussion

Regarding our first question of strategies for stakeholder participation, we found two different strategies. These two strategies differ from each other regarding both dimensions we have examined; who are participating and how? The first was a strategy for open participation in both respects: Representatives of different stakeholders were invited to participate (i.e. fishermen, fish farmers, environmental and recreational interests, local community groups, local commercial groups, land owners, etc). They were invited to participate in formal and public arenas, such as public meetings or formal working groups (with open agenda and report). The second strategy was rather closed (compared with the first), as only representatives of fishermen and fish farmers were invited to participate in a more active manner, that goes beyond the legal minimum of public hearing. Furthermore, they participated mainly on informal arenas or in direct meetings between each stakeholder group and representatives of the municipality, with restricted public access to the agenda or the discussion of these meetings.

Regarding our second question, there was a clear relationship between the degree of competition for the marine areas in the municipality and the participation strategy. The municipalities with extensive competition had chosen an open participation strategy. This can be looked upon as a rational adaptation to a complex planning situation, which is a characteristic of the urban municipalities. In a situation with many actors and conflicting interests, deliberation on arenas where different actors have access can contribute to both conflict resolution and trust (Buanes et al., 2005). It can also contribute to solutions supported by all affected parties. The municipalities with less competition of the marine space were also often characterised by a weak industrial base. Therefore, there was a greater need to develop local industrial activities based on their particular nature and spatial resources. With this in mind, it was no surprise that this kind of municipalities gave high priority to the interests of fishermen and fish farmers. There are also often rather small municipalities with few administrative resources. Lack of administrative capacity to administer open and time-consuming planning processes can be one explanation why we found rather closed participation strategies in the smaller (and more remote) municipalities, and more open strategies in the larger (and more urbanised) municipalities (Buanes et al., 2005).

Regarding the third question, our data indicated that participation matters. In the ten municipalities, there was a strong correlation between participation strategy and how the specific plan actually favoured different interests. The municipalities that chose a closed participation strategy gave priority to aquaculture, and to some degree fishing. The municipalities that chose an open participatory strategy gave priority to multi-purpose interests, where also recreation, nature conservation and other coastal interests were integrated in the plan. As mentioned, this can be seen as a reasonable answer to the diverse challenges of the local communities. In remote municipalities, it is important to secure favourable conditions for the fishing and the aquaculture industry. In urbanised municipalities with high pressure on the marine area, it might be more important to strike a balance between a range of different, and often competing, interests.

21.5 Conclusion

We found a strong correlation between the active participation of local fishermen and fish and shellfish farmers, and their influence. These stakeholders contributed with important experience-based knowledge, which is considered essential to plan the marine areas. We also found that the municipalities' choice of participation strategy reflected different planning challenges the individual municipality is facing. In many municipalities, there was an urgent need to balance the interests of coastal fisheries and aquaculture, and active participation of these two interest groups in the planning processes was an important contribution to achieve conflict resolution and agreement. However, private stakeholders representing other interests than the fisheries and aquaculture do participate, to a lesser extent, in the planning processes. Therefore, there is a great risk that other interests, especially the interests of the environment and recreation, will be subordinating the interests of these two industries. A closed participation strategy resulting in a plan favouring the interests of the few stakeholders that were given access to the planning process will probably not contribute to the legitimacy needed for a sustainable and integrated solution in the long run.

In Norway, there is an increased pressure on the coastal zone from a multitude of different interests in almost all parts of the coastal Norway, not only in the (sub)urban parts, where the number of second homes and boathouses, etc., increases. In such a situation, it is important to involve all relevant private stakeholders, and emphasise open and deliberative planning processes in order to achieve integrated coastal zone management, as well as support from all affected interests. Participation through working groups indicates a higher degree of permanence and commitment through repeated interaction between the involved stakeholders. This kind of participation can be characterised as a communicative process, where different stakeholders can discuss and negotiate commons solutions (Healey, 1997). In addition, this is a more transparent and open planning process compared to the informal and bilateral contacts. Transparency can reduce the problems of inequality and lack of accountability which always will be a central issue in local planning processes and network governance (Kickert et al., 1997; Sørensen and Torfing, 2005), and thus increase the legitimacy and thereby promote a sustainable and integrated coastal zone management.

Acknowledgements

This chapter is based on findings from the research project 'evaluation of municipal coastal zone planning in Norway', which has been financed by the Norwegian Fishery and Aquaculture Industry Research Fund. An early draft of the chapter was presented at the International Symposium on Integrated Coastal Zone Management in Arendal, 11–14 June 2007. We thank our colleague Arne Tesli for useful comments to an earlier draft of this chapter.

References

Amdam, J. (2000) Confidence building in local planning and development. Some experiences from Norway. *European Planning Studies* **8** (5), 581–600.

Arnstein, S.R. (1969) A ladder of citizen participation. *American Institute of Planners Journal* **35** (2), 216–224.

Buanes, A., Jentoft, S., Maurstad, A., Søreng, S.U. and Karlsen, G.R. (2005) Stakeholder participation in Norwegian coastal zone planning. *Ocean & Coastal Management* **48** (9–10), 658–669.

Healey, P. (1996) The communicative turn in planning theory and its implications for spatial strategy formation. *Environment and Planning B: Planning and Design* **23** (2), 217–234.

Healey, P. (1997) *Collaborative Planning: Shaping Places in Fragmented Societies*. Macmillan Press, London.

Hovik, S. and Stokke, K.B. (2006) Regional kystsoneplanlegging. Et redskap for integrert kommunal kystsoneplanlegging? NIBR working paper 2006:119. The Norwegian Institute for Urban and Regional Research, Oslo.

Hovik, S. and Stokke, K.B. (2007a) Network governance and policy integration – the case of regional coastal zone planning in Norway. *European Planning Studies* **15** (7), 943–960.

Hovik, S. and Stokke, K.B. (2007b) Balancing aquaculture with other interests: a study of regional planning as a tool for ICZM in Norway. *Ocean and Coastal Management* **50** (11–12), 887–904.

Hovik, S. and Vabo, S.I. (2005) Norwegian local councils as democratic meta-governors? A study of networks established to manage cross-border natural resources. *Scandinavian Political Studies* **28** (3), 257–275.

Jentoft, S., Mc Cay, B. and Douglas, C.W. (1998) Social theory and fisheries co-management. *Marine Policy* **22** (4–5), 423–436.

Kickert, W.J.M., Klinj, E.H. and Koppenjan, J.F.M. (1997) *Managing Complex Networks: Strategies for the Public Sector*. Sage, London.

Ostrom, E. (1990) *Governing the Commons: The Evolution of Institutions for Collective Action*. Cambridge University Press, Cambridge, MA.

Pateman, C. (1970) *Participation and Democratic Theory*. Cambridge University Press, Cambridge, MA.

Peters, B.G. and Pierre, J. (2000) *Governance: Politics and the State*. St. Martin's Press, New York.

Pinkerton, E. (2003) Towards specificity in complexity: understanding co-management from a social science perspective. In: Wilson, D.C., Nielsen, J.R. and Degnbol, P. (eds). *The Fisheries Co-Management Experience: Accomplishments, Challenges and Prospects*. Kluwer Academic Publishers, Dordrecht, pp. 61–77.

Scharpf, F. (1999) *Governing in Europe: Effective and Democratic?* Oxford University Press, Oxford.

Sørensen, E. and Torfing, J. (2005) The democratic anchorage of governance networks. *Scandinavian Political Studies* **28** (3), 195–218.

Stokke, K.B, Hanssen, M. and Hovik, S. (2006) Kommunal kystsoneplanlegging. Et redskap for en balansert utvikling av havbruk og fiske. NIBR-report 2006:17. The Norwegian Institute for Urban and Regional Research, Oslo.

The Directorate of Fisheries (2005) Status for kystsoneplan og arealplan i sjø. 31.12.05. Available at www.fiskeridir.no/fiskeridir/content/download/7777/63516/file/kart_Status%20for%20kystsoneplan%202005.pdf.

Chapter 22
The Evolution of Governance Mechanisms for the Eastern Scotian Shelf Integrated Management Initiative

Joseph Arbour

Abstract

Located on the east coast of Canada off the province of Nova Scotia, the Eastern Scotian Shelf Integrated Management (ESSIM) initiative has established the governance structures and mechanisms to complete and implement an Integrated Oceans Management Plan for an area covering approximately 325 000 km^2. This is a diverse area covering a highly productive continental shelf and extending out to the margins of Canada's EEZ. The ESSIM project was initiated in 1998 as one of the first of the initiatives under Canada's Oceans Act. The plan addresses a wide range of issues including multiple ocean use conflicts, threats to the ecosystem, sensitive benthic habitats and endangered species. Ocean use pressures include the growth of the oil and gas sector, extensive and large commercial fishing interests, telecommunications, intense shipping activity and military operations. This plan is the product of 8 years of gathering scientific information, engaging stakeholders and all levels of government, development of methods, approaches and models and achieving broad consensus on the Plan. This chapter describes the evolution of the governance models that developed to support this initiative. These governance mechanisms now provide a solid foundation for steering the implementation of this first plan for Canada.

22.1 Introduction

In 1996, Canada put into place a legislative basis for management of its ocean domain through the Oceans Act, Statutes of Canada (1996) (http://laws.justice.gc.ca/en/showtdm/cs/O-2.4). This act enables the Minster of Fisheries and Oceans to provide leadership in developing a Government of Canada approach to our oceans and coasts. Canada is moving forward with implementation of the Oceans Act on a number of fronts. One of the most significant is the development of integrated ocean management plans in five priority areas across the country. The most advanced of these is the Eastern Scotian Shelf Integrated Management (ESSIM) initiative. The Plan is the product of work that began in 1998 and has involved extensive scientific review, stakeholder engagement and collaborative effort. It has been endorsed by stakeholders and governments alike and stands as a major achievement of Canada's Ocean Program. Essential to this success has been the development of a governance model to support this initiative. This chapter describes the evolution of that model.

The Eastern Scotian Shelf is a highly productive extension of the continental shelf extending outward from Nova Scotia some 200 km. One hundred and fifty million years ago it was a vast coastal plain; however, as continents realigned themselves the coastal plain submerged and the continental shelf took shape Breeze et al. (2002). Presently, the shelf represents an area of submerged land larger than Nova Scotia itself and rich in canyons, valleys, steep cliffs and gently sloping hills; notable are the submarine canyons along the shelf edge. Long known for its high fish productivity and therefore very active fishing industry, the area is now home to numerous ocean activities ranging from oil and gas exploration and development to undersea optic cables. The emergence of new and developing ocean industries also introduces conflicts and increases pressure on the ecosystem. These can prove to be detrimental to sustaining use of the resources and attracting new investment. The initiation of work on developing the integrated management framework for the Scotian Shelf began in 1998 shortly after the passage of the Oceans Act, Statutes of Canada (1996). In the time that has passed since its inception, the initiative has drawn together a tremendous amount of science, engaged a wide range of stakeholders and fostered a great deal of process development.

An essential element of the evolution of this initiative has been the establishment of a pragmatic and workable governance arrangement that can accommodate both the federal and provincial governments, be inclusive of all departments, draw in aboriginal communities and satisfy stakeholders. Setting this framework was not simple or quick; however, the result is now functioning quite well. The process and evolution of this governance framework is described here.

22.2 Legislative and policy framework

The development of Integrated Oceans Management Plans takes place under the authority given to the Minister of Fisheries and Oceans in Part 2, Section 31 of the Oceans Act. This section provides the minister with the authority to 'lead and facilitate the development of plans for the integrated management of all activities in or affecting estuaries, coastal waters and marine waters...' (Oceans Act, Statutes of Canada, 1996). The Act also enables the minister to establish some form of structure for implementation of these plans as defined in Part 2, Section 32: (a) establish advisory or management bodies and appoint or designate, as appropriate, members of those bodies, and recognize established advisory or management bodies; and (b) may, in consultation with other ministers, boards and agencies of the Government of Canada, with provincial and territorial governments and with affected aboriginal organizations, coastal communities and other persons and bodies, including those bodies established under land claims agreements, establish marine environmental quality guidelines, objectives and criteria respecting estuaries, coastal waters and marine waters.

The Act has been followed by the release of Canada's Oceans Strategy in 2002 that lays out the policy framework for the oceans programs. This document lays out the large ocean management framework that defines the geographical bounds of the oceans management initiatives in Canada. Currently five areas have been defined including the Pacific North Coast of British Columbia, the Beaufort Sea including the North West Territories and the Yukon, the Gulf of St. Lawrence which includes the five provinces of Quebec, New Brunswick, Prince Edward Island, Nova Scotia and Newfoundland, Placentia Bay/Grand Banks off Newfoundland and the Eastern Scotian Shelf off of Nova Scotia. The Oceans Strategy was accompanied by an operational policy framework for integrated management, which provided greater detail on what was intended by integrated management and how it should proceed. This document laid out the steps to integrated management (IM):

1. defining and assessing a management area,
2. engaging affected interests,
3. developing an integrated management plan,
4. endorsement of plan by decision-making authorities,
5. implementing the plan and monitoring and evaluating outcomes (DFO, 2002).

The Act and the Strategy establish the foundation upon which the initiative has been built. In 2005, Canada's Oceans Action Plan provided the impetus for putting the governance structures in place to move these initiatives forward.

22.3 The ESSIM project

The project was initiated in 1998 when the Minister of Fisheries and Oceans announced the planning exercise. As the first integrated management initiative with a primarily offshore focus ESSIM became a test bed for integrating large amounts of science and engaging very disparate interest groups from the oil and gas companies based in western Canada to the small community fishermen along the Nova Scotia Coast. The issues driving this project are diverse and significant. The Scotian Shelf has been home to a very dynamic and diverse fishing industry for hundreds of years. It has also seen in recent years a decline in the traditional ground fishery causing substantial upheaval in the industry. As new industries become active in the area, the pressures on the fishing industry have increased. These pressures come from industries such as oil and gas and telecommunications competing for ocean space. At the same time the Scotian Shelf has seen increasing shipping traffic including growth in the cruise industry. It is also an important operational area for Canada's Navy. One of the most significant concerns has been the impact of extensive seismic exploration activity along the full length of the shelf. It is an area that is home to numerous populations of marine mammals who are known to be sensitive to noise. Other conflicts arise around the concerns for ship strikes with whales as well as entanglement. During some months of the year, sea turtles visit the shelf edge and are threatened by long-line fishing activities.

Clearly, the traditional single industry use of the area (fishing) has changed dramatically in recent times and has resulted in a plethora of ocean use conflicts that either constrain the individual industries or threaten to damage the sensitive marine environment. The ESSIM project has been developed to implement a comprehensive planning regime that will help put in place the steps that can proactively help avoid these conflicts and/or provide a forum for resolution of issues after they have arisen.

The IM process as previously described presents the five stages of the process. These steps although appearing somewhat discrete and linear actually reflect an iterative process that moves generally from one to six with significant looping back around between each step. Defining and assessing the management area entails much collection and synthesis of science to set the foundation for establishing plan outcomes. The engagement step begins early and is continuous throughout the process. The development of the management plan is a clear step; however, implementing and monitoring is expected to lead to plan revisions in the future and represents iteration in the process based on the results of implementation and monitoring.

22.4 Governance

The term governance has extremely wide use and can carry a great many meanings. The following definition is drawn from the Governance Group of the World Bank: Governance can be broadly defined as the set of traditions and institutions by which authority in a country is exercised. This includes (1) the process by which governments are selected, monitored and replaced, (2) the capacity of the government to effectively formulate and implement sound policies and (3) the respect of citizens and the state for the institutions that govern economic and social interactions among them (http://go.worldbank.org/MKOGR258V0). The IM process described above and the generation of products and outcomes does not happen in a vacuum. It does require the establishment of a structure to provide guidance and approval. This is the governance structure. Governance in this context extends to both the structures that bring together not only the legal and institutional frameworks but also the multi-stakeholder dimensions of the projects. It is this duality that posed a distinct challenge with the ESSIM initiative. The structure that has evolved to support ESSIM has developed through an iterative and adaptive process that has responded to both stakeholder input as well as national policy development (Figure 22.1).

22.5 Steps in the evolution

A major challenge in the growth of the plan was bringing the very diverse stakeholders together. It is worth characterizing the wide range and scope of these stakeholders as it has a strong bearing on the structure that has evolved. Although the Scotian Shelf has traditionally been the domain of the large fishing fleets, it now includes a wide range of ocean-related industries. Most prominent among those are oil and gas exploration and development, telecommunications, shipping, tourism and conservation. As well, the very nature of the fishing industry has changed as traditional stocks have declined and the industry has diversified. The interest groups attached to these activities range from international, to North American to community and locally based. Adding to the complexity is the need to incorporate the two senior levels of government, i.e. federal and provincial and build a link for municipalities. Adding complexity is the fact that numerous federal and provincial departments and agencies play significant roles in the area of oceans and coastal management.

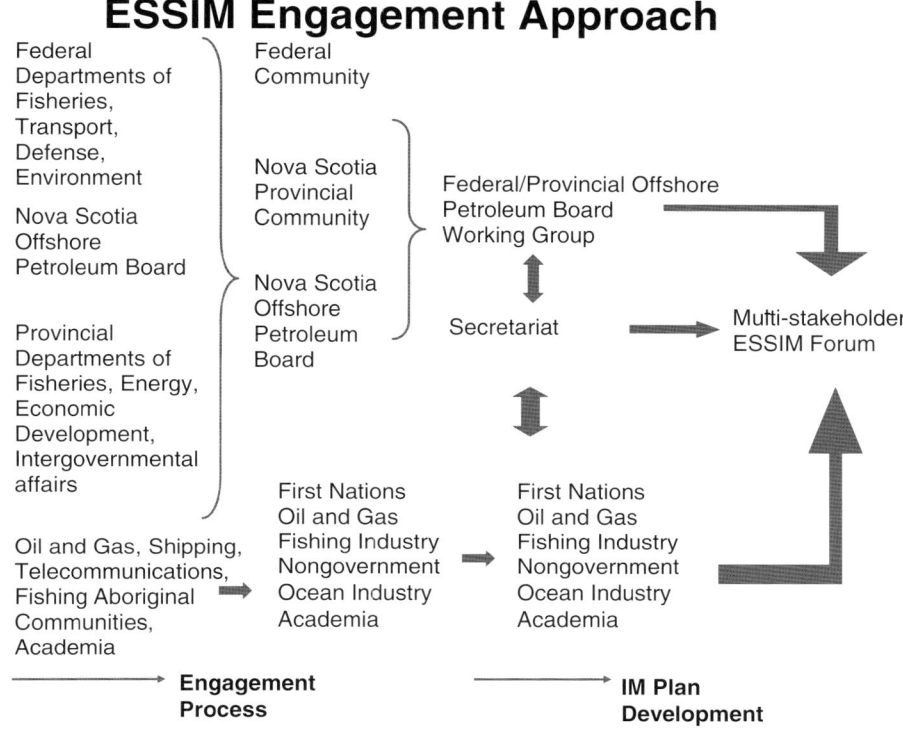

Figure 22.1 Engagement of stakeholders, DFO (2001) presentation to ESSIM Forum.

22.6 Government to government

As the coastal zone represents an area with the greatest mix of jurisdictional overlaps in the Canadian context, it was necessary as a first step to build the mechanisms that would establish the government-to-government link. The first step in this process was to draw other federal government agencies into the discussion and to secure their support. This was accomplished through a preexistent mechanism called the Regional Interdepartmental Committee on Oceans (RICO) that had representation at the program manager and director level from a wide range of federal agencies. The ESSIM initiative was taken to this group with the idea of creating a specific working group. During the first meeting of that group, the question was asked about the role of the provinces. The conclusion was that the working group should draw upon provincial interests as well, leading to an invitation to the province of Nova Scotia to participate. The next meeting, held in 2001, with provincial participation lead to an agreement to form the ESSIM Federal/Provincial Working Group. This group became a critical factor in advancing the initiative, as all development after that was discussed at this table with the objective of ensuring that it was workable at the broad federal and provincial level.

22.7 ESSIM Forum

As this initiative was in its very early stages, it was essential to get a feeling for the broad acceptance of stakeholders in the concept of integrated management. This need was driven by the increasing number of ocean users active in the Scotian Shelf area and the complexity of the policy/regulatory environment that they operated in. Getting all the players together led to this second element of the structure being initiated, the ESSIM Forum. This was effected through the gathering of all those with an active interest in the area through a series of large workshops, the first of which was held in February of 2002. Subsequent workshops were used to advance major milestones in the development of the ESSIM Plan. The first meeting of the ESSIM Forum provided an opportunity to gauge the interest in stakeholders in the development of an integrated management approach. The question back from Forum participants was: The concept sounds valid but what would this planning effort cover? Also, it was an opportunity to explore the concept of governance structure (Figure 22.2). The model presented was intended to provide recognition for the linkages to government, a secretariat and a larger planning group. The structure was also designed to allow for subgroups both within the planning group to address implementation and attached working groups to address specific issues. This model failed to meet a basic test, which is simplicity and transparency of functions. The Forum had significant difficulty understanding the relative roles identified in it. The next meeting of the Forum, held in 2003, involved presentation of a Strategic Framework (DFO, 2003), which laid out the concepts and subjects that would be covered in the Plan. This led to a healthy discussion on what a plan would include and a charge back from the Forum to develop a first draft of the plan before the next meeting.

The Strategic Framework identified the three main pillars of the plan: ecological, social and economic. It also laid out the basic principles upon which the plan would be produced DFO (2003). This document provided the basis upon which the ESSIM Forum recommended that the ESSIM Secretariat develop a first draft of a plan. During this session, first attempts at proposing a governance structure were presented for discussion. At this time,

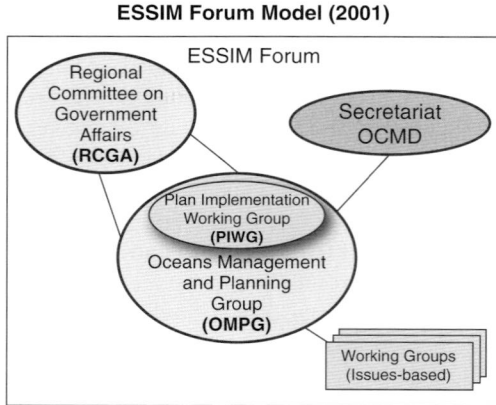

Figure 22.2 Initial ESSIM model DFO (2001) presentation to ESSIM Forum.

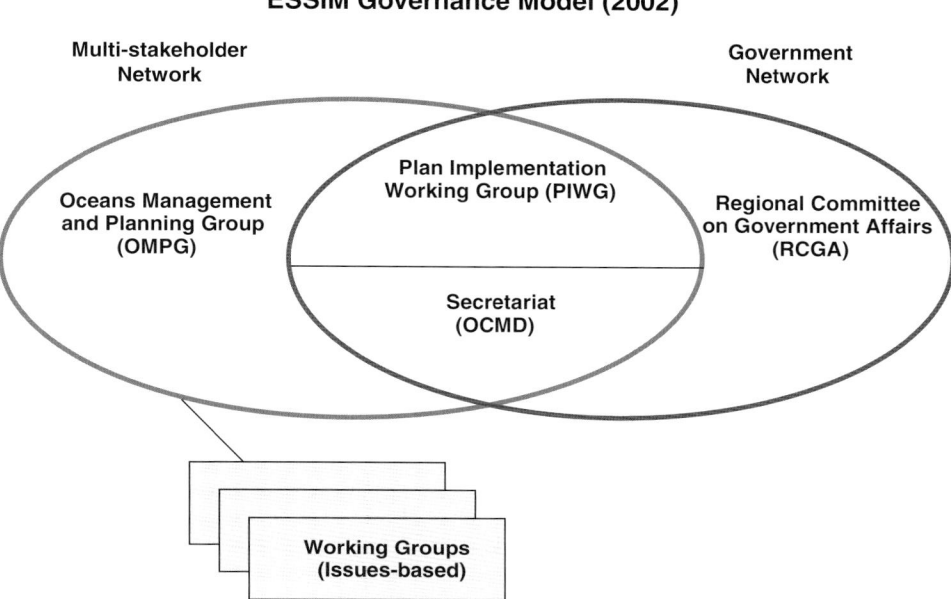

Figure 22.3 Governance model proposed in 2002, DFO (2002) presentation to ESSIM Forum.

a further version of governance was advanced and discussed at the Forum (Figure 22.3). Once again the intent was to represent the broader planning group in conjunction with the committee on government and the supporting groups. Once again the model failed to meet the basic criteria of clarity and simplicity to satisfy the many interests at the Forum. As the Strategic Framework and governance model were evolving, it became clear that there was a need to start to translate the more strategic considerations into more operational concepts around objectives. In order to begin this process, two multi-stakeholder working groups were established under ESSIM. The first to deal with ecological objectives and the second to deal with socioeconomic objectives. The concept of ecosystem objectives was injected into the process as a means of establishing clear and measurable direction on dealing with the environmental and ecological issues that were a product of the intensity of use on the Shelf. These objectives followed a structure and format developed nationally by the Department of Fisheries and Oceans (Jamieson et al., 2001). Similarly socioeconomic objectives were intended to provide the same type of information for the other two pillars of sustainable development.

The Ecosystem Objectives Working Group was drawn from the membership of the Forum and was very broad and multi-stakeholder based. The principle that was adopted was to pilot the concept of ecosystem objectives, assess geographic scope that could be addressed and to set out some foundations upon which the Integrated Oceans Management Plan could be based. The Socio-Economics Working Group was also drawn from the Forum membership; however, it was necessary to bring in additional expertise from the social sciences to develop an approach for this set of objectives. The objectives included developing a framework for objectives, looking at these from a geographic viewpoint and establishing a beginning

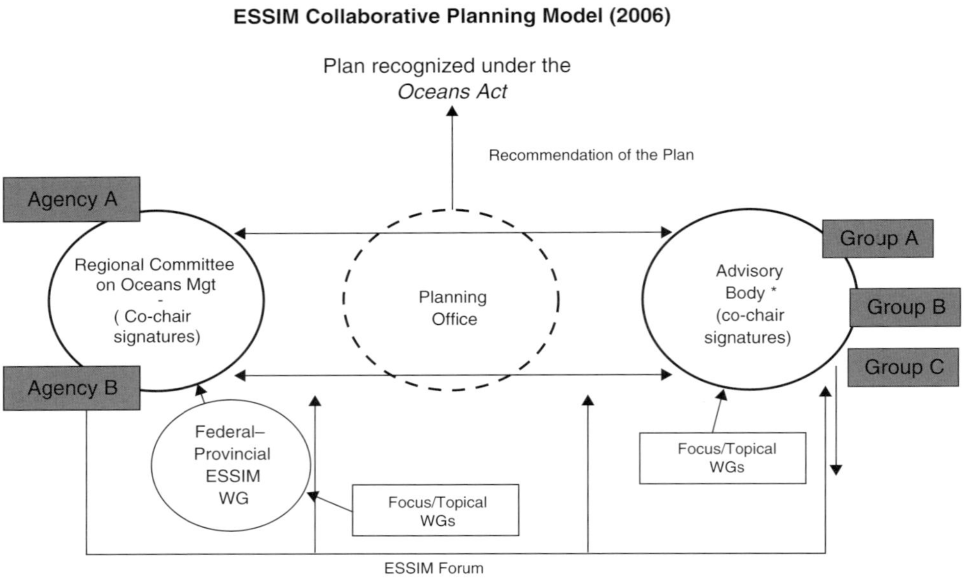

ESSIM Collaborative Planning Model (2006)

Figure 22.4 ESSIM collaborative planning model (DFO, 2005).

point. By the time of the third ESSIM Forum Workshop, several lessons became clear in the development of governance models. The most important development involved putting clarity behind the core elements of 'governance' in its most formal sense. This became important as the development of the plan involved several levels of government and numerous different regulatory jurisdictions. As such the evolution of the overall concept moved to one of a model for collaboration within which the governance component was clearly described. Governance in this case and for this project became refined as the legal and institutional frameworks and institutions that define the rules and polices under which activities are managed in the area. This is clear reference to the federal and provincial role as the two senior levels of government in Canada. This is a major component of the broader collaborative planning model (CPM) that has been developed for the ESSIM initiative as shown in Figure 22.4.

This model separates the structures for government from the broader multi-stakeholder mechanism – however, with clarity on the connection between the two and on the roles of both functions. This model was accepted at the third ESSIM Forum and became the basis for ongoing development of the Plan.

22.8 Regional Committee on Oceans Management

The need for the Regional Committee on Ocean Management (RCOM) was identified early in the process through discussions at Forum Workshops and with stakeholders. It was driven by the desire to have a clear conduit for the products of the planning process to be delivered to both levels of Government in Canada. As such the principle criteria were that

Table 22.1 Regional Committee on Ocean Management presentation to ESSIM SAC.

Canada	Nova Scotia
Environment Canada	Offshore Petroleum Board
Royal Canadian Mounted Police	Inter-governmental Affairs
Public Works and Government Services	Aquaculture and Fisheries
Fisheries and Oceans	Department of Energy
National Defence	Aboriginal Affairs
Indian and Northern Affairs Canada	Economic Development
Canadian Environmental Assessment Agency	Natural Resources
Atlantic Cooperation Opportunities Agency	Environment and Labour
Industry Canada	
Canadian Food Inspection Agency	
Transport Canada	
Justice Canada	
Parks Canada	
Natural Resources Canada	

it be at the most senior level and that it represents the broad spectrum of both federal and provincial interests. Extensive discussion went on through the Federal–Provincial Working Group on the approach to establish this next committee. The process was kick started by utilizing the existing federal structure known as the Nova Scotia Federal Council. This includes representatives of all federal agencies with a regional presence in the province of Nova Scotia. Subsequent to approaching Federal Council and in order to address the provincial interests, the concept was taken to a senior committee of heads of federal agencies and deputy ministers from the provincial government. Through these discussions, it was determined that the best approach was to create a stand alone Regional Committee on Oceans Management that would include federal and provincial departments with an interest in the oceans. Representation would be at the highest level, i.e. regional director generals for the federal government and deputy ministers for the provincial government. The membership is listed in Table 22.1.

22.9 Stakeholder Advisory Council

The Stakeholder Advisory Council (SAC) is a critical element of the collaborative planning approach developed for ESSIM. It was formulated through discussions at the Forum Workshops as well as in direct bilateral discussions with stakeholders. The principle behind the SAC is that it is representative of all the stakeholder groups. Representation at the SAC is determined by the different groups, including the federal and provincial governments. In some cases, existing mechanisms were used to designate representation such as an existing fishing industry roundtable. In other cases, groups were brought together with the ESSIM planning office acting as facilitator. The overall numbers for representation were proposed initially through the planning process and then modified after the first meetings to reflect requests from both the fishing industry and the nongovernmental organization sector. The makeup of the SAC is shown in Table 22.2.

Table 22.2 Stakeholder Advisory Council presentation to ESSIM SAC (established in 2005).

Federal Government	3	Oil and Gas Industry	2
Province of Nova Scotia	3	Conservation Groups	3
Province of Newfoundland and Labrador	1	Community Groups	2
Offshore Petroleum Board	1–2	Academia	2
Municipal Government	1–2	Transportation	1
Aboriginal Groups	1–2	Telecommunication	1
Fishing Industry	4	Tourism	1
		Citizen at Large	1

This organizational structure is now active and has resulted in the completion of Canada's first integrated ocean management plan. The success of the model thus far is in the collaboration that has gone into this plan. The next steps, development of specific action plans, will provide further test of the integrity of the model.

22.10 Conclusions

It is unlikely that there is any one size fits all governance models that work in all cases. A number of key principles can be drawn from the experience gained with the ESSIM initiative. These are:

- Understand the governance elements versus collaboration elements.
- Ensure all parties are involved in the design and implementation.
- Utilize existing mechanisms as much as possible.
- Ensure adequate support to sustain the development of mechanisms.
- Allow sufficient time for the model to mature.
- Be adaptable.

The experience with ESSIM has demonstrated that senior levels of government, the private sector and nongovernmental organizations can coalesce around shared objectives, through a collaborative planning model that clearly sets out the governance elements of the process.

References

Breeze, H., Fenton, D.G., Rutherford, R.J. and Silva, M.A. (2002) The Scotian Shelf: an ecological overview for ocean planning. *Canadian Technical Report of Fisheries and Aquatic Sciences* **2393**, 259.

DFO (2001) *Presentation to 1st ESSIM Forum, Engagement of Stakeholders*. Halifax, Nova Scotia.

DFO (2002) *Policy and Operational Framework for Integrated Management of Estuarine, Coastal and Marine Environments in Canada*. Oceans Directorate, Ottawa.

DFO (2003) The Eastern Scotian Shelf Integrated Management Initiative, A Strategic Planning Framework for the Eastern Scotian Shelf Ocean Management Plan. A Discussion Paper prepared for the 2nd ESSIM Forum Workshop ESSIM Secretariat.

DFO (2005) *Presentation to ESSIM Forum, ESSIM Collaborative Planning Model*. Halifax, Nova Scotia.

Jamieson, G., Oboyle, R., Arbour, J., Cobb, D., Courtney, S., Gregory, R., Levens, C., Munro, J., Perry, I. and Vandermuelen, H. (2001) *Proceedings of the National Workshop on Objectives and Indicators for Ecosystem Based Management.* Canadian Science Advisory Secretariat Proceedings Series 2001/09, Department of Fisheries and Oceans, 142 pp.

Oceans Act, Statutes of Canada (1996) An Act respecting the oceans of Canada (assented to 18 December 1996).

Chapter 23
Climate Change, Coastal Communities and Governance: Developing Solutions for Change in Australia

Melissa Nursey-Bray

Abstract

Climate change is a global problem that affects everyone. Climate change science has established that changes in snow melt regimes, hydrological cycles, fish stocks, nutrient and heat flows, coral bleaching and sea level rise are all effects that will impact coastal communities. The costs of addressing these changes, however, are likely to fall disproportionately on local governments, industries, communities and workers. This chapter outlines the ways in which local governments in Tasmania, Australia, are responding to these changes. It is argued that the implementation of solutions to climate change will require not only good science but also the development of institutional strategies and political solutions that will build the socio-ecological capacity of local government.

23.1 Introduction

Climate change is an example of a global systemic problem as its causes are initiated anywhere on earth, and the effects felt worldwide (Frederick and Gleick, 1999). Indeed, today, 23% of the world's population lives within 100 km of the coast, a figure that is projected to increase to 50% by 2030. Estimates from the Intergovernmental Panel for Climate Change (IPCC) highlight that there will be heightened risks of and exposure to extreme weather events, related health issues and increasing social disruption as a result of climate change. The costs of addressing and resolving the effects of climate change are likely, moreover, to fall disproportionately on local governments, industries, communities and workers (Adger et al., 2005). Specifically, Australia is predicted to sustain huge coastal impacts as a result of climate change (Institution of Engineers, 1991; CSIRO, 2002; Walsh et al., 2002; Allen Consulting, 2005; Bird and Dominey-Howes, 2006; Stern, 2006). Projections for Australia are for a hotter climate with more frequent extreme weather events. The temperatures are predicted to warm between 0.4–2°C by 2030 and 1–6°C by 2070 compared to 1990. Australians will experience more days over 35°C, and perhaps up to three times as many by 2070. The frequency and duration of events such as heavy rains, cyclones, floods and droughts will intensify, and it is expected that there will be a rise in sea level of 9–88 cm by 2100 compared to 1990 (Allens Consulting, 2005). While these effects will not be uniform across the nation, all places will experience some change to conventional weather patterns. There is a clear need to explore ways to minimise these adverse impacts (Greenwald et al., 2001).

This chapter presents the results of a risk perception study that documented how local governments in Tasmania, Australia, perceived and then dealt with the problem of climate change at local scales. The chapter reports on whether or not local governments have the institutional and social flexibility to respond to the effects of climate change predicted by the science. Thematic analysis of the data aimed to explain what views engineers and planners held in relation to the state of scientific knowledge about climate change, how they understood or experienced risk and uncertainty and how they constructed the role of local government in this context. The study was also designed to find out what implications climate change would have on policy making. The chapter concludes with some general reflection on whether research findings can help us ascertain and understand the level of social–ecological resilience within local governments in Australia, indeed the world. In this context, societies that respond best to change are understood as having social-ecological resilience (Holling, 1996; Holling et al., 1998).[1] While it is understood that resilience studies construct resilience in different ways, for example engineering versus ecological resilience (Holling, 1996), this chapter reflects on resilience as defined by Folke (2006):

> (i) the amount of disturbance a system can absorb and still remain within the same state or domain of attraction, (ii) the degree to which the system is capable of self organisation (versus lack of organisation, or organisation forced by external factors), and (iii) the degree to which the system can build and increase the capacity for learning and adaptation.

Factoring in resilience as part of coastal zone management planning enables policy makers to incorporate the robustness of natural, social and economic systems to change and can help to provide a platform for suggested solutions. Folke (2006) notes that the advantage of the resilience perspective is that it shifts policies from those that aspire to control change in systems assumed to be stable, to managing the capacity of social–ecological systems to cope with, and adapt to shape change.

This research also builds on other case studies of local government response to climate change (Lorenzoni et al., 2000; Crabbe and Robin, 2006; Lazarow et al., 2006; Local Government Association of Queensland, 2007; Gurran et al., 2008). For example, Crabbe and Robin (2006) undertook institutional research within local governments in eastern Ontario, and showed that institutional adaptation of water-related infrastructure was necessary to effectively respond to climate change impacts in the region. It takes as its starting point the position that local government plays a fundamental role in the resolution of the issue. They show that local government's adaptive capacity is determined by a composite of factors including (i) the range of available technological options, (ii) the available resources and their distribution across the municipal population, (iii) the structure of critical institutions and the

---

[1] For further information on resilience *per se* and literature, the web site for the Resilience Alliance and the online journal Conservation Ecology are a very useful starting point.

criteria for decision-making, (iv) the human and social infrastructures, (v) the access to risk-spreading mechanisms, (vi) the ability of decision-makers to manage credible information and their own credibility and (vii) the public's perception of the source of the impact.

23.2 Method

This research took a qualitative case study approach. The case study approach is one commonly used by social scientists as it provides an empirical basis for investigating current issues in real life contexts (Guba and Lincoln, 1981). It facilitates understandings between the boundaries of the object of inquiry and its context, and enables the use of multiple sources of evidence, thus building research credibility and viability. In 2005–2006, research was conducted to assess and analyse the flexibility within local government regions in Tasmania, in order to react to climate change impacts (Nursey-Bray, 2006). Current predictions show that Tasmania, as with the rest of Australia, will feel the effects of climate change, particularly sea level rise (DPIWE, 2004, 2006). This research was based on the following three objectives: (a) the determination of what institutional arrangements and legislative frameworks exist within Australian local governments to provide for management of climate change, (b) the determination of local government perceptions about the impact of climate change on local communities and (c) the determination of whether the relationship between existing institutional arrangements and legislative frameworks in relation to climate change and other natural resource management arrangements provided any opportunity for local government to build capacity to provide for management of climate change.

Each local government in Tasmania received an introductory letter inviting them to participate in the research program. Of these, 22 of the 29 councils agreed to participate. In-depth qualitative semi-structured interviews were then conducted using both snowball and purposive sampling techniques (Denzin and Lincoln, 1998; Wodak and Meyer, 2001). The research was primarily conducted as a risk perception exercise, designed to understand how the notion of risk constructed or framed decision-making. Thematic and conversation analysis were applied to synthesise results (Wodak and Meyer, 2001). Evaluation of this project was by the use of triangulation (Denzin and Lincoln, 1998) and face validity, thus ensuring the validity and reliability of the information (Guba and Lincoln, 1981). The planners and engineers for each council were interviewed in each case, as it was considered that they have the most influence and day-to-day decision-making capacity basis in relation to the issue of climate change. Specifically, this chapter presents the results from the interviews conducted within the coastal local governments (14 out of a possible 20) within Tasmania.

23.3 Results

Overall the research indicated that with a few exceptions, local governments in Tasmania felt that (i) climate change is and will remain an issue and (ii) by and large they would like to do something about it. Indeed, institutional analysis (which was confirmed by respondents) indicated that within the policy and jurisdictional framework within Tasmania there is the potential to negotiate and implement climate change management regimes. For example, there are many triggers that could be amended or used to implement statutory change.

The State Policies and Projects Act 1993, the State Coastal Policy 1996 (currently in revision), the Land Use Planning Act 1982 and each individual local government planning scheme provide room to plan for and manage climate change. Moreover, the Australian Local Government Association (ALGA), many energy and building companies and the Australian government have put together many toolkits and codes to support locally based climate change management. Nonetheless, these opportunities are not being followed up.

Many planners thought this was because there was not the political or socio-economic will to do so, or as noted by a planner from the north-west of Tasmania: 'We need to make concrete decisions, not ones that are based on nebulous advice and that also fly in the face of politics'. This is consistent with many other studies worldwide that have identified a number of institutional roadblocks to coastal local government planning (Middle, 2002; Nursey-Bray, 2006). Moreover, these options while offering opportunity do not provide statutory direction. The State Coastal Policy 1996, for example, does not currently offer strong direction in relation to climate change, so local governments have little statutory mandate that forces them to act in this regard. While the new State Coastal Policy will include climate change management provisions, the reality is that this policy will not be applied for at least another 10 years (C. Rees, 2007, personal communication). Another problem is that while the local scientific reports put councils 'on alert', as research reports they are not enforceable, hence not defendable to ratepayers in the event of development decisions. The best local planners can do is to point out vulnerable areas to prospective developers and ratepayers. There is also significant confusion over who is responsible for (and who pays for) what. Many engineers pointed out that they did not receive development applications at the point of purchase so were not in a position to advise the purchaser in relation to climate change scenarios. Many articulated a fundamental confusion over local, state and federal responsibility in relation to climate change jurisdictional management as shown by the following comment: 'We are the 7th biggest council in Tassie. We might survive when government legislates reform regarding sea level rise, but what about the others? They are too small to survive and they are not going to be able to pay in proportion to the problem. We can only pass community by laws with the assistance of the state. So all we got right now are codes of practice'.

Ultimately, as the following quote shows, in relation to climate change management at a local government level, there *is* institutional flexibility but overall, 'We only have ad hoc tools – bits and pieces, there actually is no integrated and structured approach to climate change or anything else'. This begs the question: why not? What is inhibiting the effective uptake of the climate change management tools currently available and preventing the development of locally specific responses? This research found that there were three core reasons for this.

23.3.1 Key finding one: uncertainty about the scale of climate change impacts acted as inhibitor to institutional reform

A core finding from the research was that uncertainty about the scale of climate change impacts at local levels acted to inhibit management action on the ground. Uncertainty about climate change was the key driver for local government approaches and perceptions about climate change. This finding is consistent with studies that show uncertainty is the key issue in climate change planning (Yohe et al., 2004). Uncertainty arises from insufficient, inaccurate or unavailable data, external developments and cross-boundary issues, and the unpredictability of human behaviour (Carter et al., 1999; Jones, 2001; Westmacott, 2001; Ha-Duong et al., 2007). In this context, this uncertainty also acted as an inhibitor to implementation of institutional reform. This was manifest in many ways. Almost all councils pointed out the need for, or conversely the lack of, data at local scales about the impacts of climate change. This lack of information in turn was seen to hamper decision-making – without that baseline, planners in particular were uncomfortable about making decisions based on 'predictions' rather than 'fact'. Some engineers considered climate change management a waste of time until they had better certainty about its effects. One engineer added: 'There is still scepticism of likelihood of effects of climate change occurring at local levels ... due to uncertainty out there we don't know what to cater for – adaptation or mitigation?' Interestingly, many also saw climate change as a potential opportunity for Tasmania, hence derided the negativity associated with climate change rhetoric, indicating climate change impacts might be positive for Tasmania: 'Climate change will increase Tasmania's market share for goods like our wine, soft fruits ... worldwide so there is a good opportunity there'. As such they were again unprepared to make planning or development decisions, based on the precautionary principle, when perhaps it would emerge there was no need to do so. As one planner reiterated: 'At a global perspective, most hope that no, climate change is not a problem but they will believe in it if it affects them locally. Until they are certain though, any planning decisions that we make relating to climate change will be unpopular if they hamper people getting what they want'.

Another overriding concern was in relation to the source of climate change – was it natural or human induced? As one respondent noted: 'Well, there's no doubt global warming is happening but the key is – is it human or natural and are we aggravating a natural process?' This lack of clarity was again employed to reinforce lack of action on the issue. If it was naturally induced, many engineers in particular preferred to exercise the option of 'doing nothing' until structural physical changes needed to be enacted. Until then, trying to change behaviour was not encouraged.

Other key aspects of uncertainty fashioned local government discourse about action on climate change, such as the issue of how to manage for what one respondent called the 'time lag effect'. Many planners pointed out the disjuncture between the need to make immediate decisions in the context of uncertain but potential future climate effects/impacts (and their scale). In the words of one planner: 'The time lag thing is very difficult – it is really hard to make decisions in terms of effects that might take 10 s of 100 s of years perhaps to mature or manifest. So a move in the right direction is going to be very slow.'

Finally, many council respondents felt the core issue was about the communication of the issue of climate change to different sectors, from real estate agencies, to developers, and even their own elected councillors. An engineer went so far as to wryly comment that deciding to communicate the risk of climate change was a gamble even for a nation of gamblers: 'How do you convey risk? It's the hardest thing for us – as well as the fact that in the national psyche we are a nation of gamblers, so do we as punters take risks or gambles in relation to climate change?' It was clear that until action on climate change was acceptable to the rate-paying public, decision-making would be hampered by public perception and aggression towards having to 'pay' for climate change impacts at the point of development or planning approval.

23.3.2 Key finding two: responding to climate change is a risk to day-to-day management

The second key finding of this research is that climate change management per se was often constructed as a risk where it impacted on day-to-day business. The reality was that while most staff in local government wanted in principle to do something about climate change issues, they were hampered by the real politick of day-to-day management of other core management issues. If climate change could be effectively incorporated into existing business regimes, they appeared open to that possibility. However, where climate change management emerged as a threat to existing business, it was seen as a significant risk to day-to-day management. In this context, discussion focussed on the range of issues currently facing local government in Tasmania, issues echoed across Australia in other studies.

The most significant issue was the struggle many councils were having in accommodating the residential boom, a result of 'sea changers' coming from mainland Australia to settle in coastal Tasmania, and the requisite problems, i.e. water supply, infrastructure and overall development control issues attached to this phenomenon. The notion of a sea change expresses a decision by many to have a change of life: to move away from busier activities and to slow down and live life, with a focus on family, recreation and friends (Burnley and Murphy, 2004; National Sea Change Task Force, 2007). There was not always time to consider climate change impacts, such as flooding, erosion, subsidence and sea level rise, causing damage to assets (public and private) and infrastructure unless they were included as part of potential day-to-day management. This reticence is partly explained by the fact that sea changers have power due to their capacity to invest heavily in local areas, and thus a disproportionate capacity to influence decision-making, even if they are applying to build in an area subject to the impacts of climate change: 'Most local government is affected by the sea change mentality. These are peoples who have freed up their superannuation, they have punch power, they descend on the coast, and they bring expectations with them. They are used to having what they want, can pay for it and have a high-level expectation of local government and our services . . . it creates an intense political climate for us'. This is a new challenge, and as the following quote shows, not one all local governments are ready for: 'For many years, we had a stable if declining community, last 10 years actually, now we have had a mini boom – our population has grown by 1% a year in the last 3 years. Many people here now are from elsewhere and we now need new settlement strategies'.

Given the political environment, respondents were clear that managing climate change was a risky political enterprise because it raised a number of related sociocultural problems which they would then have to solve. For example, many planners were worried about the relationship between the issues of litigation, compensation and equity. Engineers and planners both felt they were placed in an invidious position when they were being asked to make decisions that incorporated climate change predictions. If they did take account of climate change predictions, they would be charged with being inequitable: 'There are major issues with cumulative development. Can you knock back the seventh house when you have already permitted the previous six? We've got to be seen to be equitable across the board'. On the other hand, if they failed to take account of climate change, they were worried they would face litigation for compensation in the future: 'We have, however, insurance issues, premiums, civil liabilities and legislation. What legal basis has local government to refuse applications on basis of climate change? We have no legal standing but if we don't make the decision in light of climate change it looks like we are being negligent ... All councils are facing the same issue on the east coast ... people will say to us later – we weren't told, we want compensation, we're baby boomers and we get what we want'.

23.3.3 Key finding three: a number of local governments are already active in establishing climate change initiatives

Finally, this research found that a few councils were already active in both acknowledging and taking action on climate change in Tasmania. Of these, three examples are provided below as they encapsulate the diversity of responses possible to this issue.

23.3.3.1 Clarence Council

Clarence Council, located near Hobart, has commissioned a socio-economic assessment and response project for climate change impacts on their foreshore. Based on a risk assessment methodology, the outputs from this project will inform the development of a coastal management strategy. The strategy will address climate change impacts such as storm surges, coastal erosion and sea level rise that may occur over the next 20–100 years. The planners and engineers for this region have already experienced the effects of sea level rise on their foreshores and efforts to develop strategies to respond is supported by the local community. The support is forthcoming because of the damage already sustained by sea level rise, which has included erosion and flooding. The social and institutional assessment project will complement and integrate with a scientific consultancy that involves identifying the hazards, analysing the risks, evaluating the risks and identifying and assessing possible responses including their costs.[2]

23.3.3.2 Kingborough Council

Development in Kingborough Council is administered via the Kingborough Planning Scheme. This is the primary regulatory mechanism by which Council can take account of climate change. The Kingborough Planning Scheme addresses climate change and sea level rise in Schedule 1 of the Environmental Management Schedule, which says:

> To avoid or mitigate the impacts of any potential rise in the level of the sea or ocean along the coast and inshore, particularly with respect to existing and future physical and social infrastructure.

The Council has adopted a precautionary approach in order to prevent development occurring in high-risk locations. As such, despite the absence of any mapping that specifically identifies areas at risk, the scheme defines high-risk areas as being any land 30 m from the high water mark or less than 3 m above Australian Height Datum. Moderate risk areas are between 30 and 100 m from high water mark or less than 3 m above Australian Height Datum. All development in both high and moderate risk areas is discretionary and must meet some fairly stringent standards. The Planning Scheme also identifies that any permitted development must occur 30 m back from the rear of any sand dune. Kingborough has also embarked on a risk assessment process to enable it to assess risk in relation to

---

[2] This project is now completed and the results can be viewed in full at the web site for Clarence Council: http://www.ccc.tas.gov.au/site/page.cfm?u=807.

climate change impacts. The risk assessment issue mainly relates to a mapping exercise to identify areas at risk across the municipality from flooding, storm surge, sea level rise and bushfire. This work has included estimations of the potential economic loss to Council owned assets that might be caused by predicted sea level rise. This is based on an 800-mm rise over 100 years. Council estimates indicate that overall sea level rise has the potential to cost 50 million US$, and as such there is an incentive to implement management measures sooner rather than later. Kingborough Council has now developed a decision making tool designed to assist local government to make decisions about climate change in their coastal areas (Nursey-Bray and Kingborough, 2009). At the time of writing, all these projects were at various stages of development and completion.

23.3.3.3 West Tamar Council

West Tamar Council is another local government active in addressing climate change issues through amendments to its planning scheme. In this case, an application for a retirement village along the banks of the West Tamar River was rejected on the basis that it was within an area subject to the possibility of a 1-in-10-year flood and was also likely to be subject to climate change impacts. This decision was unanimously supported by the West Tamar Council in 2006. The West Tamar Planning Scheme (2007) now has formal sections addressing climate change. For example, in the strategic directions sections that look at the protection of unique and finite resource, key action and indicator A5 is to 'investigate the potential implications of sea level rise caused by climate change on the foreshore of the estuary and coast and ensure that development takes sea level rise into account'. Similarly, Section S1.4.1 on flooding and storm surge notes:

> Building location and design alleviates potential damage to or loss of buildings and other works to human safety and (a) does not cumulatively increase the risk of flood to other land and (b) takes into account potential sea level rise due to global warming.

23.4 Discussion

While inherent capacity exists to enable local governments to facilitate change, this chapter has shown that there are still institutional, political and structural blocks to instigating change when adapting to climate change impacts. These blocks occur primarily because local governments in Tasmania construct climate change management itself as a risk in its uncertainty, to day-to-day management, and to the communication of the issue of climate change to their respective communities. Local governments in Tasmania then, while evincing many of the characteristics located in Folke's definition of resilience quoted earlier, need a policy framework to assist them to transform these characteristics into creative solutions to the management blocks identified by this research and build societal resilience overall. The adaptive capacity, therefore, of local governments to respond to the climate challenge is an important determinant of whether or not they will be able to respond to climate change along their coasts (Smit and Wandel, 2006). Adaptive capacity will be determined by an array of factors including (i) the range of available technological options, (ii) the available resources and their distribution across the municipal population, (iii) the structure of critical institutions and the criteria for decision-making, (iv) the human and social infrastructures, (v) the access to risk-spreading mechanisms, (vi) the ability of decision-makers to manage credible information and their own credibility and (vii) the public's perception of the source of the impact (Crabbe and Robin, 2006).

In the context of effective and integrated coastal management planning in Tasmania, building the resilience of local governments to respond to climate change impacts in their region can be achieved through the implementation of adaptive management. The concept of resilience has implications for policy because it requires a shift in thinking towards an acknowledgement that humans are a part of the environment. This encourages the development of policy that provides the conditions that will enable effective adaptive governance structures to evolve and be implemented. Adaptive management is based on the assumption that circumstances change and thus so must management (Leach, 2006). It is a technique that provides a framework to continually improve management practice and deliver environmental outcomes within socio-economic contexts. Adaptive management builds on environmental assessment techniques to deal with uncertainty, and access information sets on partially known processes, making it ideal for climate change management. The fluid nature of adaptive management also suits the dynamics of working with the changing quality of coastal areas. Adaptive management techniques also enable planners to communicate the results of planning to ensure a shared understanding, not only of the successes but the challenges of management trials. In so doing, they are able to better work together to achieve ways forward that build on previous experience. Adaptive management will also address the current tendency of local governments for 'management-by-crisis'. At present, most local governments will respond on an 'as needs' basis to the incursions caused by sea level rise, flooding and storms (McInnes et al., 2003; Paton et al., 2005; Bird and Dominey-Howes, 2006).

In this context, an adaptive management framework may be appropriate – one that can guide planners at the evaluation stage of coastal management programs and assist them to incorporate climate change management within coastal planning schemes. This framework should have four components: (i) environmental and social scoping, (ii) planning, (iii) implementation and (iv) evaluation. For example, in the environmental scoping stage, local governments could invest in local and regional studies of climate change impacts so as to reduce the uncertainty that bedevils current planning. Within the planning stage, local government could decide to consolidate disaster management strategies and aim to invest in adaptation and mitigation methodologies. Implementation of effective communications about climate change would build both community resilience and social capacity to respond to climate change impacts while accepting local government management initiatives. An effective management review and evaluation of how climate change and coastal zone management strategies are working together will ensure that local government exposure to hazards and future risks will be reduced.

23.5 Conclusion

It is indisputable that local governments have a major role to play in resolving climate change impacts along their coast. Institutional barriers to integrated coastal management are perhaps the most intractable. Even for the most well-intentioned local government, investing in change, particularly change that may be confronting, there are many roadblocks to breakthrough.

There is a high likelihood that the costs of addressing climate change will fall disproportionately on local governments. Responding to climate change in coastal contexts requires not only good science (Patrinos and Bamzai, 2005), but also the development of institutional strategies and political solutions that address the social, cultural and economic factors that profoundly influence how a problem of this magnitude can be resolved at local levels. In Australia, as elsewhere, the resolution of climate change is beset by many institutional obstacles, and as this research has shown, many conceptual blocks as a result of risk perceptions about climate change impede or influence action on the ground. The research also confirms other studies which highlight the need to consider that risk perception as well as risk per se is a crucial part of climate change management, for (as in this case) the perception of risk is what guides decision-making, whether that perception is accurate or not (Lowe, 2006). The adoption of an adaptive management planning framework will enable local governments to plan in accordance with the principles of continuous improvement and feedback and as such help to minimise their exposure to the risks and challenges faced by climate change. In this context, the insights from this research and the suggested adaptive management framework could be applied in many coastal local government areas worldwide. Effective and integrated coastal zone management that is responsive to this latest global challenge to our coasts, that of climate change, must occur and build on the results of this and other research to do so.

Acknowledgements

This research was made possible through a grant from the Australian Maritime College, Tasmania. I thank Rob Palmer for editing advice and ideas on the text. I also thank all the individuals and organisations who agreed to be part of this project, especially Jenni Rigby from the Tasmanian Local Government Association, Alasdair Wells from the Climate Change Unit within the Tasmanian Government, Phil Watson from the Clarence Council and Tony Ferrier from Kingborough Council.

References

Adger, W.N., Hughes, T., Folke, C., Carpenter, S. and Rockstrom, J. (2005) Social-ecological resilience to coastal disasters. *Science* **309**, 1036–1039.

Allen Consulting Group (2005) Climate change risk and vulnerability promoting an efficient adaptation response in Australia. Final Report to the Australian Greenhouse Office, Department of the Environment and Heritage, 159 pp.

Bird, D. and Dominey-Howes, D. (2006) Tsunami risk mitigation and the issue of public awareness. *The Australian Journal of Emergency Management* **21** (4), 29–36.

Burnley, I. and Murphy, P. (2004) *Sea Change: Movement from Metropolitan to Arcadian Australia*. University of New South Wales Press, Sydney, 308 pp.

Carter, T., Hulme, M. and Viner, D. (eds) (1999) Representing uncertainty in climate change scenarios and impact studies. *Proceedings of the ECLAT-2*, Helsinki Workshop, 14–16 April 1999, Climatic Research Unit, Norwich, 128 pp.

Crabbe, P. and Robin, M. (2006) Institutional adaptation of water resource infrastructures to climate change in eastern Ontario. *Climatic Change* **78**, 103–133.

CSIRO (2002) *Climate Change and Coastal Communities*. CSIRO, Victoria, Australia, 8 pp.
Denzin, N.K. and Lincoln, Y.S. (eds) (1998) *Handbook of Qualitative Research*. Sage, London, 188 pp.
DPIWE (2004) *Sea level Change Around Tasmania*. DPIWE, Hobart, 4 pp.
DPIWE (2006) *Draft Climate Change Strategy for Tasmania*. Tasmanian Government, 4 pp.
Folke, C. (2006) Resilience: the emergence of a perspective for social-ecological systems analyses. *Global Environmental Change* **16**, 253–267.
Frederick, K.D. and Gleick, P.H. (1999) *Water and Global Climate Change: Potential Impacts on U.S. Water Resources*. The Pew Center on Global Climate Change, Arlington, 55 pp.
Greenwald, J., Roberts, B. and Reamer, A. (2001) *Community Adjustment to Climate Change Policy*. The Pew Center on Global Climate Change, Arlington, 52 pp.
Guba, E. and Lincoln, Y.S. (1981) *Effective Evaluation*. Jossey-Bass Publishers, San Francisco, 320 pp.
Gurran, N., Hamin, E. and Norman, B. (2008) *Planning for Climate Change: Leading Practice Principles and Models for Sea Change Communities in Coastal Australia*. Prepared for the National Sea Change Taskforce, University of Tasmania.
Ha-Duong, M., Swart, R., Bernstein, L. and Petersen, A. (2007) Uncertainty management in the IPCC: agreeing to disagree. *Global Environmental Change* **17** (1), 8–11.
Holling, C.S. (1996) Engineering resilience versus ecological resilience. In: Schulze, P. (ed.). *Engineering Within Ecological Constraints*. National Academy, Washington, DC, pp. 41–44.
Holling, C.S., Berkes, F. and Folke, C. (1998) Science, sustainability and resource management. In: Berkes, F. and Folke, C. (eds). *Linking Social and Ecological Systems: Management Practices and Social Mechanisms for Building Resilience*. Cambridge University Press, Cambridge, MA, pp. 342–362.
Jones, R. (2001) An environmental risk assessment/management framework for climate change impacts assessments. *Natural Hazards* **23**, 197–230.
Lazarow, N., Fearon, R., Souter, R. and Dovers, S. (2006) Coastal management in Australia: key institutional and governance issues for coastal natural resource management and planning. A collection of essays on the key institutional and governance challenges and issues for integrated coastal planning and management in Australia. CRC for Coastal Zone, Estuary and Waterway Management, Indooroopilly, NSW, Australia, 132 pp.
Leach, G. (2006) *Enabling Adaptive Management for Regional Natural Resource Management*. CRC Coastal Zone Estuary and Waterway Management, Indooroopilly, NSW, Australia, 78 pp.
Local Government Association of Queensland (2007) *Adapting to Climate Change: A Queensland Local Government Guide*. Local Government Association of Queensland Inc., Australia, 56 pp.
Lorenzoni, I., Jordan, A., Hulme, M., Turner, K. and O'Riordan, T. (2000) A co-evolutionary approach to climate change impact assessment. *Global Environmental Change* **10**, 57–68.
Lowe, T. (2006) *Vicarious Experience vs. Scientific Information in Climate Change Risk Perception and Behaviour: A Case Study of Undergraduate Students in Norwich, UK*. Tyndall Centre for Climate Change, 143 pp.
McInnes, K.L., Walsh, K.J.E., Hubbert, G.D. and Beer, T. (2003) Impact of sea-level rise and storm surges on a coastal community. *Natural Hazards* **30**, 187–207.
Middle, G. (2002) *Institutional Arrangements, Incentives and Governance – Unlocking the Barriers to Successful Coastal Policy Making*. Cooperative Research Centre for Coastal Zone Estuary & Waterway Management, Indooroopilly, NSW, Australia, 91 pp.
National Committee on Coastal and Ocean Engineering (1991) *Guidelines for Responding to the Effects of Climatic Change in Coastal Engineering Design*. Institution of Engineers, Australia, 47 pp.
National Sea Change Task Force (2007) A policy framework for coastal Australia. A Discussion Paper. National Sea Change Task Force, Canberra, 15 pp.

Nursey-Bray, M. (2006) Institutional capacity and climate change: local government in Tasmania, Research Report No. 1, 2006. Australian Maritime College, Tasmania, 35 pp.

Nursey-Bray, M. and Kingborough Council (2009) *Climate Change Risk Assessment: A Practical Tool to Assist Local Government Priority Setting.* Australian Maritime College, Australia.

Paton, D., Smith, L. and Johnson, D. (2005) When good intentions turn bad: promoting natural hazard preparedness. *The Australian Journal of Emergency Management* **20**, 25–30.

Patrinos, A. and Bamzai, A. (2005) Policy needs robust climate science. *Nature* **438**, 285.

Smit, B. and Wandel, J. (2006) Adaptation, adaptive capacity and vulnerability. *Global Environmental Change* **16**, 282–292.

Stern, N. (ed.) (2006) *Stern Review on the Economics of Climate Change.* HM Treasury, London, 170 pp.

Walsh, K.J.E., McInnes, K.L. and Abbs, D.J. (2002) Sea level rise projections and planning in Australia. *Proceedings of Coast to Coast 2002,* Tweed Heads, Australia, pp. 1–4.

Westmacott, S. (2001) Developing decision support systems for integrated coastal management in the tropics: is the ICM decision-making environment too complex for the development of a useable and useful DSS? *Journal of Environmental Management* **62** (1), 55–74.

Wodak, R. and Meyer, M. (eds) (2001) *Methods of Critical Discourse Analysis.* Sage, London.

Yohe, G., Andronova, N. and Schlesinger, M. (2004) To hedge or not against an uncertain climate. *Science* **306**, 416–417.

Chapter 24
Map of Coastal Zone Vulnerabilities to Wave Actions Application to Aveiro District (Portugal)

Carlos Coelho, Maria-José Granjo and Cristina Segurado-Silva

Abstract

Coastal erosion is a problem of many coastal zones on the globe. In Portugal, the coastal municipalities of the Aveiro district, situated at the northwest of the Atlantic Coast, are critical places. The present chapter produces a digital map of vulnerable territories. This map is based on different parameters related to the distance to the shoreline, topographic data, geology, geomorphology, ground cover, wave height, tide levels, history rates of erosion/accretion and human intervention effects on the coast. The isolated analysis of those parameters, which are assumed to be the main agents in the vulnerability evaluation, allows us to formulate different classes and an objective classification. Based on this classification and on the weighting of the importance of each parameter, a global map of vulnerabilities is presented. The goal of this representation of coastal vulnerabilities is to produce an auxiliary tool for the management and planning of coastal zones, thereby supporting political priorities of an intervention.

24.1 Introduction

Energetic sea action (waves, tides and currents) plays a significant role in the vulnerability analysis of coastal zones. In these zones, there is a natural dynamic, which is often amplified or anticipated by human interventions and has the reverse effect of conflicting with human activities and occupation. As a result of these dynamic processes, there is a generalised coastal erosion situation, with land loss, that is very critical in some sectors and is of great concern, especially in a medium to long term (Coelho et al., 2006b). In fact, the natural transport and nourishment of sediments to the coast is impaired by sand extractions of rivers and beaches, fluvial regularisation and barrier effect of dams. The continuous destruction of the dune system and its vegetation impede the accumulation of sands, representing one more contribution to the instability of those natural defences. Adding to these, the generalised sea level rise and the intensification or coastal areas urbanisation represent negative factors.

Climate change and anthropogenic factors have been the primary causes for the increasing degradation of coastal ecosystems. In order to address this degradation, it is important to understand the status and distribution of coastal populations, land cover, fragile ecosystems and the causes of their vulnerability. A reliable assessment of the current status of the global coastal environment is long overdue. The major constraint has been a widespread lack of accurate and timely data at the global level. However, recent advances in spatial data gathering and processing techniques, including satellite remote sensing and geographic information systems, have started to assist the scientific community in overcoming these constraints and allow to develop a preliminary coastal vulnerability index (UNEP, 2005).

The costs of coastal defences are increasing worldwide and it is necessary to understand the involved phenomena, assessing the behaviour of those interventions in a longer time perspective (Coelho, 2005). It is necessary and urgent to establish a coastal width that considers the dynamics of the coast, the general erosions and the respective vulnerability. In this chapter, the vulnerability is understood as the sensibility of the coastal areas to waves, tides and sea currents energy. The analysis of this vulnerability and its evolution with time is essential in management and planning terms, especially in high risk and vulnerable regions. As a first approach, vulnerability maps built on the basis of historical and recent data may be used as coastal planning and managing tools (Coelho et al., 2006b). Other important factors which are not considered in this analysis are oil spills and pollutant discharges (related with navigation), diffuse pollution, seismic actions (including tsunamis), the exploitation of living resources and tourism.

The study of vulnerabilities demands a great amount of information and a multidisciplinary approach that should include engineering, environment, socioeconomics and political sciences (Nicholls, 1998). The first basic step towards a better management of anthropogenic impacts is to identify and map the vulnerabilities. A map of vulnerabilities can help reach certain goals, such as the reduction of human lives losses, reduction of damage in structures and coastal buildings, preservation of natural environments, increase in evacuation capacities, location of new structures and buildings out of danger areas and relocation of existing structures and buildings.

This chapter presents one methodology to classify the vulnerability of coastal areas to sea actions energy, including a description of all parameters that are considered in the vulnerability analyses, its classification and weight importance in the global results.

24.2 Methods

To make the map of vulnerability, some parameters need to be identified and described. Table 24.1 shows the classification of each parameter based on individually defined criteria. Vulnerability classification ranges from 1 (very low) to 5 (very high). For accurate classification of coastal vulnerability, those several factors are then weighted in a unique global result.

24.2.1 *Vulnerability parameters description*

Certain parameters need to be evaluated separately before integration in the vulnerability map. Those parameters are distance to shoreline, topographic elevation, geology, geomorphology, ground cover, maximum significant wave height, maximum tidal range, average rates of erosion/accretion and anthropogenic actions.

Table 24.1 Classification of the vulnerability parameters.

Vulnerability	Very low 1	Low 2	Moderated 3	High 4	Very high 5
Distance to shoreline (m)	>1000	>200 ≤1000	>50 ≤200	>20 ≤50	≤20
Topography elevation (m)	>30	>20 ≤30	>10 ≤20	>5 ≤10	≤5
Geology	Magmatic rocks	Metamorphic rocks	Sedimentary rocks	Non-consolidated coarse sediments	Non-consolidated fine sediments
Geomorphology	Mountains	Rocky cliffs	Erosive cliffs, sheltered beaches	Exposed beaches, plain	Dunes, river mouths, estuaries
Ground cover	Forest	Vegetation, cultivated	Non-covered	Rural urbanised	Urbanised or industrial
Anthropogenic actions	Shoreline stabilisation intervention	Intervention without sediment sources reduction	Intervention with sediment sources reduction	Without interventions or sediment sources reduction	Without interventions but with sediment sources reduction
Maximum significant wave height (m)	<3.0	≥3.0 <5.0	≥5.0 <6.0	≥6.0 <6.9	≥6.9
Maximum tidal range (m)	<1.0	≥1.0 <2.0	≥2.0 <4.0	≥4.0 ≤6.0	>6.0
Average rates of erosion/accretion (m/yr)	>0 Accretion	>−1 ≤0	>−3 ≤−1	>−5 ≤−3	≤−5 Erosion

24.2.1.1 Distance to shoreline

The variability of vulnerabilities is naturally conditioned by the distance to the shoreline. A coastal waterfront is highly exposed to waves, tides and currents energy, whereas that effect weakens with distance to the shore increases (Coelho, 2005).

24.2.1.2 Topography elevation

The topography defines the geometry of the study area, referring quotas relative to the mean sea level. Lower-lying areas will be more vulnerable to sea actions than areas with a high shore face (Coelho, 2005). In Gornitz et al. (1997), a classification of altitudes was adopted providing five categories. These categories correspond, respectively, to the classification of frequent floods, probable floods, occasional swash, remote swash and finally very unlikely swash, as the vulnerability decreases.

24.2.1.3 Geology

The geology evaluates the category of rocks and sediments, allowing a classification based on the behaviour and scale of hardness of the minerals that compose them. The adopted classification was adapted from Gornitz et al. (1997). It is natural that the areas where the sediments are smaller and not consolidated represent higher vulnerabilities than rocky areas.

24.2.1.4 Geomorphology

The geomorphology is the study of landforms, including their origin and evolution, and the processes that shape them. Coelho (2005) adapted the work of Andrade and Freitas (2002) and incorporated some of the classifications from Gornitz et al. (1997) to provide a new classification that allows the attribution of classification indexes as a function of the geomorphology distinguishing sandbanks, dunes and lagoons as high vulnerability areas and cliffs and mountains as low vulnerability areas.

24.2.1.5 Ground cover

The ground cover represents possible changes to the natural behaviour of the soil, depending on the land use or covering. According to Coelho (2005), the higher the level of change of the natural conditions of the ground, the higher its vulnerability. In its natural and permeable condition, a ground covered with more vegetation can be less vulnerable (the vegetation represents a natural defence to the soil, unlike a naked soil and consequently without capacity of natural retention of sediments).

24.2.1.6 Anthropogenic actions

The anthropogenic actions that influence the natural behaviour of the coastal dynamics should include not only human direct actions, with the goal of defence and reduction of vulnerabilities, but also negative effect of interventions, often with indirect relationships within the coastal area. According to Veloso-Gomes and Taveira-Pinto (1997), the classification of this type of vulnerabilities should depend on the reduction of the sediment sources (due to extraction of sands, dredging, hydroelectric dams, breakwaters and groins), the degree of defence interventions already carried out and on the effectiveness of these interventions. In this case, smaller vulnerabilities are found in areas where defences exist, such as groins, longitudinal revetments, artificial nourishments, artificial dunes and fences for sand retention.

24.2.1.7   Maximum significant wave height

Wave characteristics are extremely variable in space and time. However, the range of significant wave heights that take place on a coastal area represents an indicator of the energy of the waves and it links directly with the potential of sediments transport. It is for that reason that a larger movement of sediments happens in storm situations (Coelho, 2005). To classify that energy capacity, the maximum wave height observed at breaking was chosen. This way, more sheltered areas will be less vulnerable. For classification of this parameter, the proposal of Gornitz et al. (1997) was adopted.

24.2.1.8   Maximum tidal range

Tides are periodic vertical movements of the sea surface, of astronomic origin, that will depend on the configuration of the oceans, bays, sea entrances in land, etc. Its impact as coastal sediments transport agent and as cause of erosion should be considered. The sea level surface changes during the day, defining the position of the shoreline. Gornitz et al. (1997) propose a division of vulnerabilities depending on the observed spring tides amplitude in the coast.

24.2.1.9   Average rates of erosion/accretion

Movement rates of the shoreline are the type of measures mostly used by scientists and engineers, appearing in the planning illustrations to indicate the dynamics and the dangers in the coast (Dolan et al., 1991). The rates of shoreline movement represent the history of the last evolution and they can be an indicator of the future tendencies. Gornitz et al. (1997) propose a division of vulnerabilities depending on the erosion and accretion rates for the U.S. West Coast. Conjugating proposals of other authors, Coelho (2005) suggests a classification that is considered appropriate to the generality of the situations, mainly based on different degrees of erosion.

24.2.2   Vulnerability parameters classification

The methodology presented intends to be a simple tool for easy classification of the nine described parameters. Unfortunately, the lack of data or the subjectivity of the evaluation can represent different classes for the results. Table 24.1 defines the categories for each vulnerability parameter and tries to summarise the possible classifications. Any one that knows the coastal area in study should be able to use the table, even those with difficulties to evaluate some of the parameters. In fact, the importance of some small errors on the classification of one or two parameters is negligible in the global result (Coelho et al., 2006a).

24.2.3 Global vulnerability evaluation

The integration of all parameters represents the global vulnerability to waves, tides and currents energy, and cannot be objective. Naturally, several parameters considered relevant in the evaluation of vulnerabilities of the coastal areas to sea energy have different importance and they should represent variable weights in the final classification of vulnerability. However, the weight that should be attributed to each parameter is not necessarily equal in all the places of the coastal areas (Coelho et al., 2006a). Studies accomplished by Coelho et al. (2006a), Segurado-Silva and Granjo (2006) and opinion surveys reveal different perspectives in the form of attribution of the relative importance of each parameter. However, it is generally accepted that the distance of 5 km represents a reasonable limit for a very low vulnerability. A width of 100 m is recognised as the distance that should be considered in a degree of similarity with the other parameters. In this closer stretch of 100 m, the distance to shore, the topography and the geology are the parameters with highest weight, representing a factor of 6/28 each. According to Coelho et al. (2006a), the tidal range and the anthropogenic actions are the parameters considered with the smallest importance, corresponding to a factor of 1/28. All other parameters have a factor of 2/28. Between the two referred limits, a linear gradual increase of the importance is admitted for the distance to the shoreline, considering also a decreasing of the importance of all the other parameters. Table 24.2 summarises the weights attributed to each one of the parameters. In the equations presented in Table 24.2, d represents the distance to the shoreline, expressed in meters. For each coastal site, the nine parameters should be classified with 1 to 5 levels of vulnerability and weighted accordingly (Table 24.2). For inland points, the parameters

Table 24.2 Weight of each of the parameters in the global classification of vulnerability, depending on the distance to the shoreline.

Vulnerability parameters	Distance to shoreline, $d/(m)$		
	<100	100–5000	>5000
DS – distance to shoreline	0.214	$\frac{6}{28} + \frac{0.665(d-100)}{4900}$	0.879
TE – topography elevation	0.214	$\frac{6}{28} - \frac{0.665(d-100)}{4900} \frac{6}{22}$	0.033
GL – geology	0.214		0.033
GM – geomorphology	0.071		0.011
GC – ground cover	0.071	$\frac{2}{28} - \frac{0.665(d-100)}{4900} \frac{2}{22}$	0.011
WH – maximum significant wave height	0.071		0.011
EA – average rate of erosion/accretion	0.071		0.011
TR – maximum tidal range	0.036	$\frac{1}{28} - \frac{0.665(d-100)}{4900} \frac{1}{22}$	0.005
AA – anthropogenic actions	0.036		0.005

Table 24.3 Limits of global vulnerability, in function of the distance to the shoreline.

DS	5	4		3		2		1
Distance (m)	<20	20–50	50	200	200	1000	1000	5000
Maximum	5.00	4.79	4.57	4.54	4.32	3.99	3.65	1.48
Minimum	1.86	1.64	1.43	1.46	1.23	1.34	1.00	1.00

that are dependent only on coastal characteristics, such as waves, tides, erosion/accretion, rates or anthropogenic actions should be classified with the same level of the coast.

Table 24.3 represents the maximum and minimum limits that can be obtained at different distances of the shoreline. To produce this table, all of the parameters are classified with the value of 5 (maximum) or 1 (minimum), respectively (with the exception of the classification of the parameter DS, that naturally will be dependent on the distance to the shoreline).

It is verified that, for general situations, the global classification of vulnerability is necessarily very low (maximum <1.5) starting from a superior distance to 5000 m to the shoreline. Starting from 200 m, it will be impossible to obtain very high values of global vulnerability (maximum <4.5). Up to 50 m, a classification of very low vulnerability can never be obtained (minimum >1.5).

24.3 Results

The exposed methodology was applied in the coastal municipalities of the Aveiro district (Espinho, Ovar, Murtosa, Aveiro, Ílhavo and Vagos). The municipalities of Santa Maria da Feira and of Estarreja were also included, as well as of two smaller places (Angeja and Frossos), as a way to homogenise the width of a strip of about 20 km (see Figure 24.1).

24.3.1 *Characterisation of the coast of the Aveiro district*

In general, the whole studied coastal stretch corresponds to a plain surface of low altitude, consisting of exposed beaches and dunes. The lagoon of Aveiro is dominant in the analysed coastal area and therefore extensive sediment deposits are present, forming islands, with high number of channels of small depth and of irregular outlines.

Between Espinho and Vagos, the coast is essentially of the sandy type, where a negative balance of sediments is verified, reflecting the serious problems of coastal erosion. In fact, a great amount of all of the problems of this coastal area is related with the coastal erosion with origin in the negative budget of sediments. The littoral located south of the mouth of the Douro River observed a reduction of the sediments, which is verified since the construction of the dams along the river, and is worsened by the dredging, which removes volumes of sand to the coastal system (EUrosion, 2003).

The whole coast of the district is extraordinarily dynamic and the human occupation in spite of being very asymmetrical has been contributing strongly to the destruction of the natural protections against the erosion phenomenon. To combat the regression of the shoreline, multiple defence structures, were built since the 1970s, making the coast artificial in some more critical sectors, namely in Espinho, Esmoriz, Cortegaça and Furadouro and

Figure 24.1 Study area, corresponding to the coastal municipalities of the district of Aveiro.

in the area between Costa Nova do Prado and Vagueira. These interventions led to a relative stability, due to the dissipation of the energy of the waves in the longitudinal revetments and to the accumulation of sand up drift of the groins.

24.3.2 Individual vulnerability maps representation

Based on the parameters and its classification classes presented in Table 24.1, nine maps were elaborated to represent each one of the referred parameters.

For analysis of the parameter related to the distance to the shoreline, military maps, to the scale 1:25 000, of the Geographical Institute of the Army (IGE) were used, corresponding to the Aveiro district. Based on the shoreline of the military maps, lines were marked at the distances of 20, 50, 200 and 1000 m, defining the limits of the different vulnerability classifications. To represent the analysis of the topographic parameter, the same maps were used, representing the lines at levels 0, 5, 10, 20 and 30 m, referring to the topographical zero (mean sea water level). The study of the geology was supported in geological maps, to the scale 1:50 000, of the district of Aveiro of IGE. For the definition of the areas of different vulnerability related to the geomorphology, the military maps were again used in substitution of the geomorphologic maps (1:50 0000). The representation of the vulnerability associated with the ground cover was based on the division presented in the maps of soil occupation of the Portuguese Geographical Institute (IGP), as referred to the information of August 1990. The vulnerability grades were diluted by approach to the higher vulnerability degree once the used maps presented excessive detail for the intended analysis type. The representation of the remaining parameters is of the linear type, since they only depend on evaluations of the behaviour close to the shoreline. The identification of the defence interventions, necessary in the classification of the anthropogenic actions, was done based on military maps, aerial photos and field observations of the places. Concerning to the analysis of the wave height parameter, the whole coast is classified with maximum vulnerability, as some data of the Hydrograph Institute (IH), for a located buoy seaward of Leixões, represent maximum values superior to 6.9 m. Based again on IH data, values of tide range are assumed in the western Portuguese coast between 3 and 4 m.

The analysis of the parameter related to rates of erosion and/or accretion was accomplished with values proposed by several authors and presented in EUrosion (2003). Classification was adopted considering average values due to the existence of several studies, with different time analysis intervals and different evaluation of erosion and/or accretion rates. To mitigate that difficulty and the gap of values in some coastal stretches, aerial photos and military maps were also used.

Based on the concepts and methodology exposed, the maps of Figure 24.2 represent the several vulnerability degrees corresponding to the individual analysis of each parameter.

24.3.3 *Global vulnerability map*

Figure 24.3 presents the map of global vulnerability corresponding to the coastal municipalities of the Aveiro district. For the representation of the map, a grid of points was created, allowing the classification of all the parameters in an isolated way, in agreement with the maps of Figure 24.2. For the linear representation parameters, the points located landward were attributed the same value as those points close to shore. For that reason, the limits to north and south of the represented coastal stretch were not classified and they are not represented in Figure 24.3. By application of the weights presented in Table 24.2, the global vulnerability was obtained in each point of the grid. By interpolation of the values obtained in the points of the grid, curves of equal global vulnerability were defined for the values of 1.5, 2.5, 3.5 and 4.5. Those curves represent the limits between vulnerability classifications, allowing the map representation illustrated in Figure 24.3. In the analysis of Figure 24.3, the stretches corresponding to several vulnerability degrees present a width relatively uniform, and a gradual decrease as the distance to the shoreline increases. The results show that the maximum vulnerability includes coastal areas corresponding to beach areas. Only the municipalities of Espinho, Aveiro and part of Murtosa do not present very high vulnerabilities, because of low classifications in the parameters of erosion rates and of anthropogenic actions. The degree of high vulnerability (3.5–4.5) represents extensive areas, presenting widths between 200 and 1300 m. The medium vulnerabilities (2.5–3.5) and low vulnerabilities (1.5–2.5) are about 2500 m of width each. In the area of Espinho, the very low vulnerability (<1.5) is reached about 2500 m from the shoreline.

24.4 Discussion

The vulnerability analysis methodology presented was applied to the coastal sector of Aveiro district. Nine vulnerability parameters were considered and classified, based on individually

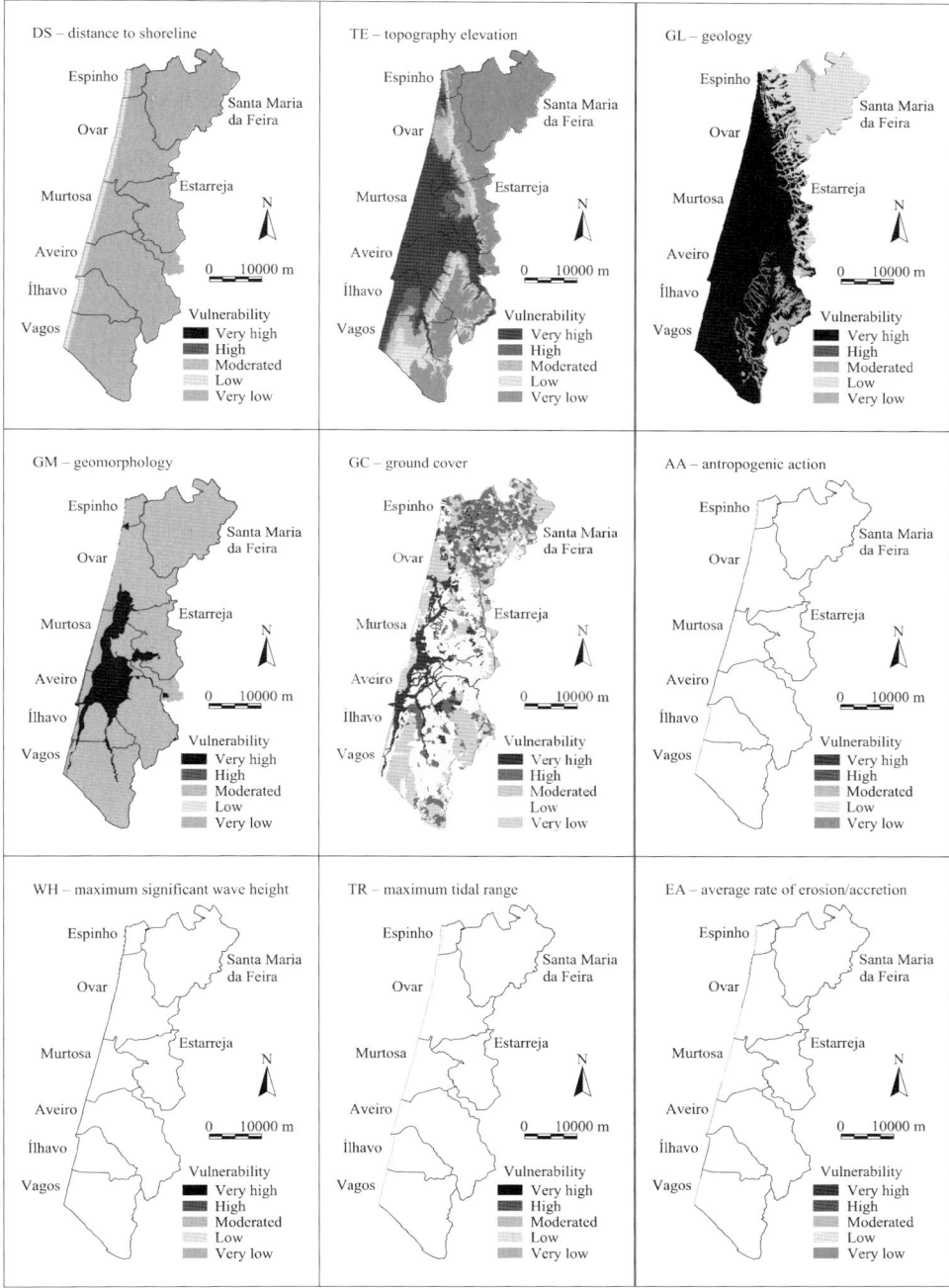

Figure 24.2 Representation of the vulnerability parameters for the district of Aveiro (based on Segurado-Silva and Granjo, 2006). (To see this figure in colour, please see colour plate 24)

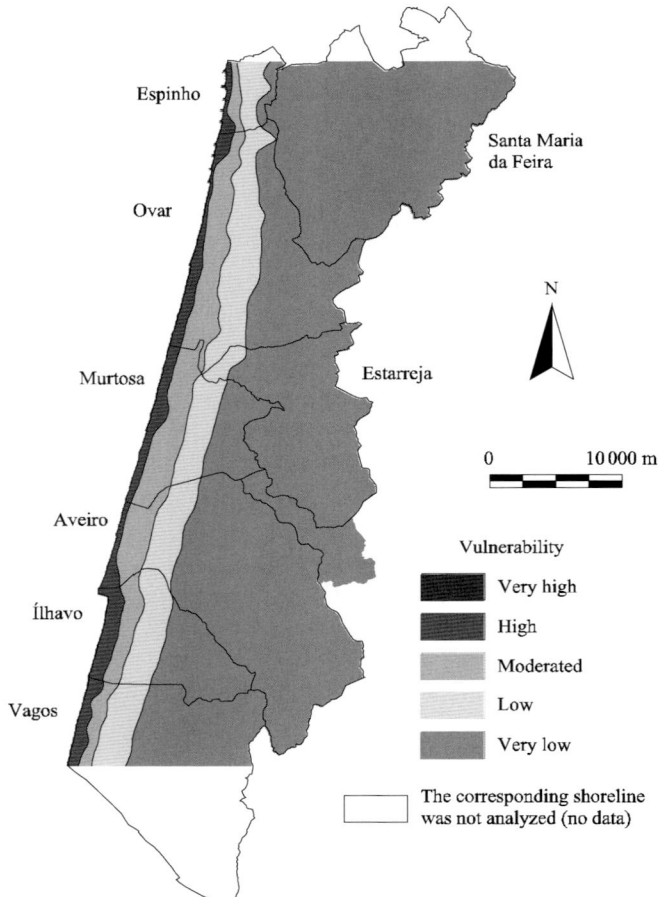

Figure 24.3 Representation of the map of global vulnerability for the Aveiro district.

defined criteria. Aiming the development of a tool with quick and intuitive vulnerability visualisation of coastal categories, vulnerability maps were built. The accomplished map for the district of Aveiro allows the establishment of zones that indicate several degrees of vulnerability relative to sea energy and is useful for management purposes as it allows the definition of priorities and establishment of intervention criteria for the coast. The district of Aveiro is characterised essentially by low quotas, with predominance of beach areas, sand dune and dunes, where considerable erosion rates are observed and high landscape alteration is observed due to intensive land use. This way, the vulnerability classification is high for long stretches of coast.

Management measures and specific planning for high vulnerability areas can be defined as the relocation of existent constructions, the prohibition of construction of new buildings, the promotion of preservation campaigns of the natural defences and the periodic monitoring of the defence works already existent (maintaining a low cost for the respective maintenance). It is also necessary a periodic revaluation of the vulnerability map relative to some parameters, since they depend on the shoreline evolution, on the changes on land use and on the human activities in the coastal area, and because these can have consequences in the classification of global vulnerability. Finally, the application of this methodology to other areas of heterogenic characteristics is considered fundamental, for its total validation. In fact, the use of this methodology to a small and homogeneous stretch as the Aveiro district does not represent big differences in the global classification. However, it is important to realise that even for this small area, the limit to very low vulnerabilities varies from 2500 m far from the shoreline in Espinho, up to almost 5000 m in Murtosa.

The application of the global vulnerability classification methodology to other places, which can be rockier, elevated or less energetic, will result in areas with smaller stretches of high vulnerability. The evaluation of the differences in the results is important and can only be evaluated on a larger scale. Thus, the methodology will be applied to other areas. The vulnerability analysis does not represent the potential negative consequences that can happen in a coastal area. It is also possible to identify coastal stretches which are highly vulnerable to sea energy but without risk of exposure due to the absence of human occupation, or without economic and environmental assets. At the moment, work on identification of a global map of consequences is being developed, which will include cross effects of vulnerabilities and consequences, in a final risk classification.

24.5 Conclusion

In general, the entire coast represents areas of difficult administration, where complex phenomena happen and an intense dynamics model the coast. The constant instability of these areas associated with the pressures due to human, social and economical interests leads to the need of possessing tools for management, planning and administration. Careful application of vulnerability classification criteria together with improved shoreline evolution predictions can help decision-makers in the planning and management of coastal zones, contributing to a better quality of life in those regions (Veloso-Gomes and Taveira-Pinto, 1997).

References

Andrade, C.F. and Freitas, M.C. (2002) Coastal zones. In: Santos, F.D., Forbes, K. and Moita, D. (eds). *Climate Change in Portugal – Scenarios, Impacts and Adaptation Measures – SIAM Project*. Gradiva, Lisbon, 456 pp.

Coelho, C. (2005) *Riscos de Exposição de Frentes Urbanas Para Diferentes Intervenções de Defesa Costeira*. Ph.D. Thesis, University of Aveiro, 404 pp.

Coelho, C., Cabarrão-D'Albuquerque, M. and Veloso-Gomes, F. (2006a) Aplicação de uma Classificação de Vulnerabilidades às Zonas Costeiras do Noroeste Português. 8 Congresso Nacional da Água, Figueira da Foz, 12 pp.

Coelho, C., Silva, R., Veloso-Gomes, F. and Taveira-Pinto, F. (2006b) Risk analysis approach for the Portuguese West Coast. *Risk Analysis 2006 – Fifth International Conference on Computer Simulation in Risk Analysis and Hazard Mitigation*, Malta, pp. 251–262.

Dolan, R., Fenster, M.S. and Holme, S.J. (1991) Temporal analysis of shoreline recession and accretion. *Journal of Coastal Research* **7**, 723–744.

EUrosion (2003) *A European Initiative for Sustainable Coastal Erosion Management – Guidelines for Developing Local Information Systems – Study Cases of River Douro – Mondego Cape*, Vol. 1. EUrosion Project, Universitat Auntònoma de Barcelona, Instituto de Hidráulica e Recursos Hídricos, 182 pp.

Gornitz, V.M., Beaty, T.W. and Daniels, R.C. (1997) *A Coastal Hazards Data Base for the U.S. West Coast*. Environmental Sciences Division Publication No. 4590, ORNL/CDIAC-81, NDP-043C, 78 pp.

Nicholls, R. (1998) Coastal Vulnerability Assessment for Sea Level Rise: Evaluation and Selection of Methodologies for Implementation. Technical Report TR 98002, Publication of Caribbean Planning for Adaptation to Global Climate Change (CPACC) Project, 39 pp.

Segurado-Silva, C. and Granjo, M.J. (2006) Vulnerabilidades das zonas costeiras às acções energéticas do mar. Final degree work (not edited), University of Aveiro. Available at http://vulnerabilidades.no.sapo.pt/index.html.

UNEP (2005) *Assessing Coastal Vulnerability: Developing a Global Index for Measuring Risk*. United Nations Environment Programme, 54 pp.

Veloso-Gomes, F. and Taveira-Pinto, F. (1997) A Opção 'Protecção' para a Costa Oeste Portuguesa. Colectânea de Ideias Sobre a Zona Costeira de Portugal, Associação Eurocoast, Portugal, pp. 163–190.

Chapter 25
Managing Coastal Vulnerability: New Solutions for Local Government

Timothy F. Smith, Benjamin Preston, Cassandra Brooke, Russell Gorddard, Deborah Abbs, Kathleen McInnes, Geoff Withycombe, Craig Morrison, Beth Beveridge and Tom G. Measham

Abstract

Coastal research and management often has an issue-specific focus, with little attention paid to the interdependencies among the various issues. Climate variability is one such issue that is impacted by, and impacts on, several other areas (e.g. coastal processes, infrastructure, health and regional economies). These interdependencies create challenges for local governments to scale up management responses to tackle issues at a regional scale. Critical to this process of scaling up is the concept of adaptive capacity, which is affected by a range of contextual, institutional and procedural factors. Research is being undertaken to support local governments in the Sydney region to improve their adaptive capacity to deal with the impacts of climate variability and change; hence, the chapter is more about testing 'new solutions' than about reporting on them. The goal of this research is to work with local governments to determine key vulnerabilities and their capacity to adapt in order to manage these risks at a regional scale.

25.1 Introduction

Coastal areas throughout the world are exposed to the impacts of climate change. The authors argue that managing response to climate change in these areas is as much a social issue as one of understanding the biophysical processes. In order to discuss the management of climate variability and change in coastal areas, the chapter has been broken into five sections: firstly, brief discussion of some of the recognized problems of integrated coastal zone management (ICZM); secondly, discussion of new research approaches for ICZM; thirdly, background on the Australian Greenhouse Office National Climate Change Adaptation Program; fourthly, discussion of the application of a new research approach to the 15 local governments that make up the Sydney coastal zone; and lastly, articulation of the benefits of such an approach.

While this chapter deals with climate change adaptation, or adaptive responses to climate change, the authors recognize the importance of needing to tackle climate change through a combination of adaptation (adapting to the drivers), mitigation (slowing the drivers) and prevention (reducing the drivers) strategies. Although the focus of the chapter is on adaptation, by using the systems approach to understanding climate change interdependencies critical issues related to mitigative and preventative strategies can also be identified. To facilitate sustainable management as a self-regulating form of governance, it is critical that fatalism is not encouraged through a focus only on adaptation, rather than a combination of adaptation, mitigation and prevention.

25.2 The problems

ICZM problems have been documented by many authors (e.g. Cicin-Sain and Knecht, 1998; Kay and Alder, 1999) and often relate to social and institutional issues. These issues have been further emphasized in a recent monograph on coastal management in Australia (Lazarow et al., 2006). From the literature and the authors' own experience, the following list highlights some of the key ICZM impediments:

- Linear decision making – rather than systems thinking
- Adversarial modes of decision making – rather than consensus building
- Mismatches of jurisdictions, benefits, costs and implementation
- Subordination of public interest to a special interest
- Lack of coordination and trust – intra- and interinstitutional 'silos'
- Institutional inertia – due to sector-specific mandates
- Piecemeal or symptoms approaches to problems
- Fragmented management responses
- Ineffective use of science

These impediments compound the inability to deal with the complexity inherent in coastal systems.

It is this complexity that raises the following management and research objectives:

- Managing for multiple uses and benefits
- Working with communities using a participatory approach and with a view to capacity building
- Dealing with complex interactions and feedbacks across atmospheric, terrestrial and marine, as well as biophysical, economic and social components, which requires a systems approach
- Dealing with uncertainty, which requires adaptive management approaches and explicit treatment of uncertainty and risk
- Needing to combine all these considerations

So, the question becomes: 'how do we achieve all these objectives simultaneously?' Certainly, there is a place for traditional disciplinary-specific approaches to research; however, alternate methodologies are also required to work within these uncertain and complex systems.

25.3 A research approach for ICZM

Addressing climate change in coastal zones is an uncertain and complex challenge, which in part relates to community expectations and preferences, and creates a need for a participatory approach to adaptation. As briefly discussed in the previous section, the tendency for sectoral approaches to decision making (e.g. within local government directorates) has led to a narrow conception of issues and consequent ineffective solutions to problems. This is demonstrated by the limited use of systems approaches to address climate vulnerability and adaptation. Coupled with this have been difficulties in implementing regional management approaches in large urban areas – creating situations where regional-scale issues (e.g. infrastructure, biodiversity and health) may not be adequately and effectively dealt with, science is poorly linked to management, and key interventions may be overlooked. The innovation proposed by this project was to develop and trial a transferable method for a systems approach to regional climate change impact assessment and adaptation in metropolises.

ICZM is complex, and issues such as climate change exacerbate this complexity and uncertainty through increased system interactions. In order to study these systems, the authors proposed a research approach with a focus on (i) the drivers of change, (ii) impacts on sustainability, (iii) identification of high-leverage interventions and (iv) analysis of adaptive capacity. While the authors focus on adaptive capacity, we also recognize the importance of mitigative capacity (Yohe, 2001). It is through this approach that science can maximize impact (Funtowicz and Ravetz, 1991; Ravetz, 1999), and through a process of social learning (Milbrath, 1989; Lee, 1993) it can transform data to information to knowledge to impact (Davenport, 1998; Smith and Smith, 2006). The research approach is currently being applied to better understand and manage the impacts of climate change in the coastal zone of Sydney, Australia.

25.4 Department of Climate Change National Climate Change Adaptation Program

The Australian Government began a four-year program (2004–2008) (the National Climate Change Adaptation Program) to begin addressing climate change adaptation. The program has three aims:

- To help Australians understand the likely impacts of climate change
- To develop practical tools to support decision making on adaptation
- To assist in planning ahead to reduce the risks and capture opportunities

While the initial investment in the program is modest (Australian $14.2 million), the Australian Government has recently committed an additional Australian $164 million dollars to climate change adaptation research and implementation. The research described by the authors represents one of five programs funded under the original program, and compliments other research being conducted in Clarence (Tasmania), the Gold Coast (Queensland), Western Port (Victoria) and a joint project in Victoria and the Australian Capital Territory.

25.5 Application of the method in the Sydney coastal zone

The focus of the Sydney project is to build adaptive capacity within local governments through a systems approach to understanding climate change issues. The study utilizes a

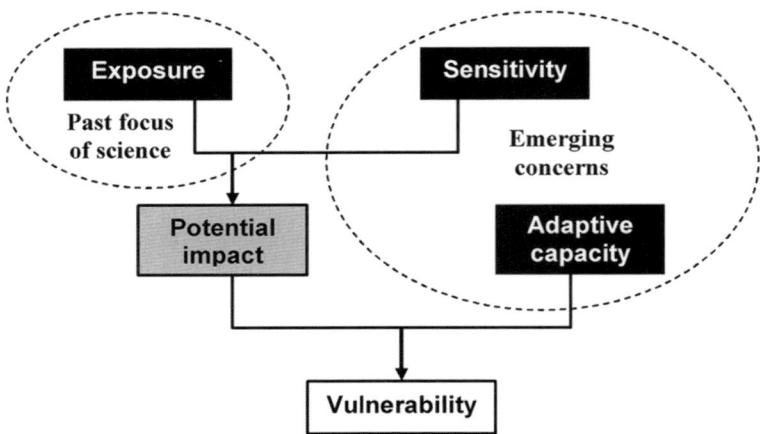

Figure 25.1 Model of vulnerability (Allen Consulting, 2005).

relative vulnerability-based assessment of climate change to approach climate and coastal vulnerability largely from a social perspective, with the aim of reducing vulnerability through building adaptive capacity (Figure 25.1), whereby adaptation is used to expand the coping range of an activity, sector or species and therefore reduce vulnerability to climate change (Figure 25.2). Vulnerability for the purposes of this project is defined as '... the degree to which a system is susceptible to, or unable to cope with, adverse effects of climate change, including climate vulnerability and extremes' (IPCC, 2001).

The project has five stages through which adaptive capacity is assessed. These stages consist of (i) mapping relative vulnerabilities, (ii) systems conceptualization, (iii) identification of key issues, (iv) identification of perceptions of barriers and opportunities to manage the key issues and (v) an institutional analysis of adaptive capacity in relation to crosscutting barriers and opportunities. Another key consideration of the project is the potential transferability to other large urban regions, which is facilitated through the involvement of a project reference group, including key policy and decision makers in other jurisdictions, at all stages of the research process.

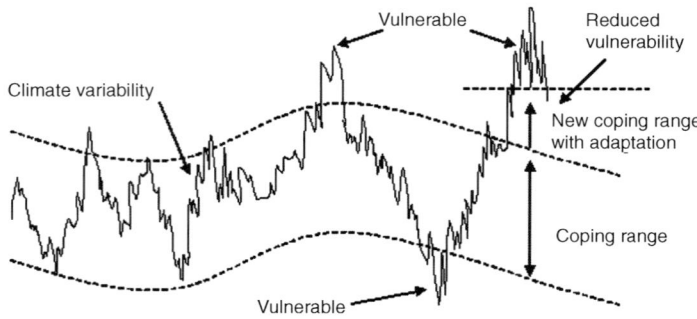

Figure 25.2 Example of extending a coping range through adaptation (Jones and Mearns, 2005).

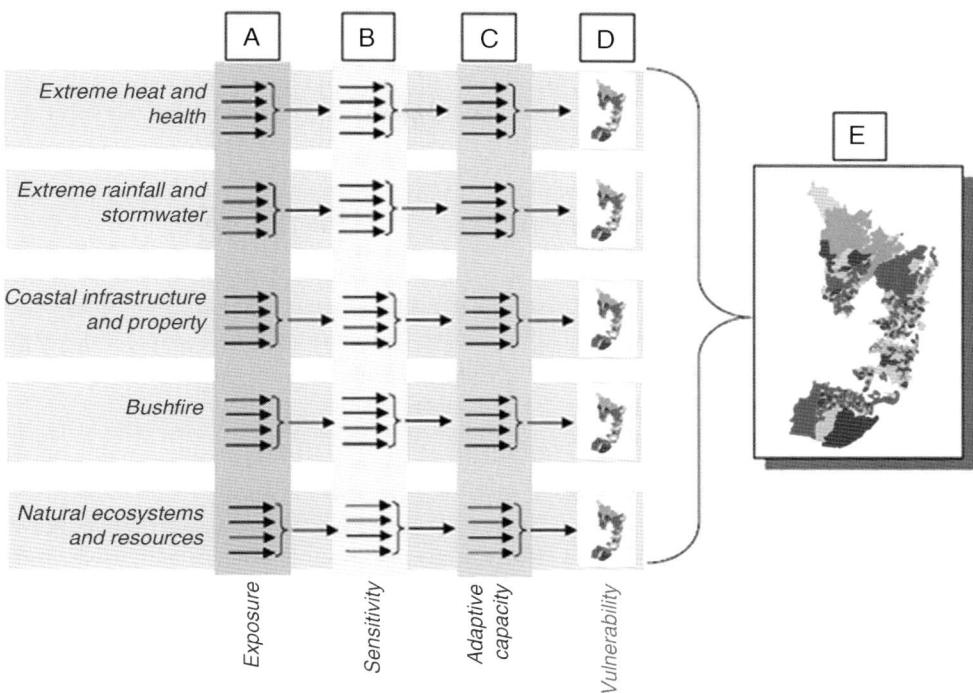

Figure 25.3 Illustration of the process of determining relative vulnerability for the 15 member local governments of the Sydney Coastal Councils Group (Preston et al., 2007). (To see this figure in color, please see color plate 23.)

25.5.1 *Mapping relative vulnerabilities*

In order to stimulate discussion among the 15 member local governments of the Sydney Coastal Councils Group and to raise their general awareness of climate change issues, a template for mapping relative vulnerability among the 15 local governments was created. The relative vulnerability assessment integrated existing indicators representing system exposure, sensitivity and adaptive capacity (Figure 25.1). Selected indicators represented a range of biophysical and socioeconomic data sets across a range of spatial scales. Simplistic examples of indicators used include temperature increases (exposure), proportion of population who are elderly (sensitivity) and relative wealth (adaptive capacity) (Figure 25.3). The relative vulnerability maps were also compared with surveys from each council on their perceptions of their individual vulnerability in order to ground-truth the modeling outputs and to also provide stimulus for workshop discussion.

25.5.2 *System conceptualization*

The second phase of the project consisted of workshops with each of the 15 local governments that comprise the Sydney Coastal Councils Group. The workshops consisted of a presentation on the relative vulnerability mapping outputs to stimulate thinking on exposure, sensitivity and adaptive capacity in relation to climate change. The workshops then focused on interactive input from the council participants (including staff and elected members). Participants identified system drivers, impacts and management responses and the linkages between them in order to develop a systems diagram or mental model of the climate change considerations for each local government area (Figure 25.4).

25.5.3 *Identification of key issues*

Using the system conceptualization process, participants then explored perceived key issues in terms of direct and indirect drivers of those issues (Figure 25.5), and the direct and indirect consequences (i.e. flow-on effects) arising from those issues (Figure 25.6). Using this approach, workshop participants were able to identify their priority issues.

25.5.4 *Perceptions of barriers and opportunities*

Once the priority issues affecting local governments were identified, each council then worked through their perceptions of the barriers and opportunities affecting their ability to manage those key issues (Table 25.1). The perceived barriers and opportunities were then used to provide an indication of adaptive capacity.

25.5.5 *Institutional analysis of adaptive capacity*

To date, 13 of the 15 workshops with the local governments have been completed. After the crosscutting barriers and opportunities have been analyzed across the 15 local councils, institutional analysis will be used to determine actual adaptive capacity to the issues in terms of contextual, structural and procedural criteria (Bellamy et al., 2005). A proposal will also be submitted to monitor and evaluate climate change adaptation interventions arising from the earlier stages of the research. The monitoring and evaluation component would allow for detailed assessment of the challenges, successes, failures and innovations arising from the local government interventions in order to facilitate the implementation of climate change initiatives and build adaptive capacity in other regions.

25.6 Conclusion

Climate change exacerbates the challenges of ICZM. The authors propose a systems-based method to work with local governments in the Sydney region to build their adaptive capacity to understand and manage these challenges. The method consists of mapping relative vulnerability, conceptualizing the system, identifying priority issues, identifying barriers and opportunities to managing those priority issues and analyzing adaptive capacity to manage those issues.

The benefits of the method are that (i) research and management is treated in an integrated way, (ii) environmental problems are analyzed within social systems, (iii) it identifies potential unintended consequences of short-term actions that may exacerbate problems in the long term and (iv) it identifies high-leverage interventions – getting more impact for

Managing Coastal Vulnerability: New Solutions for Local Government

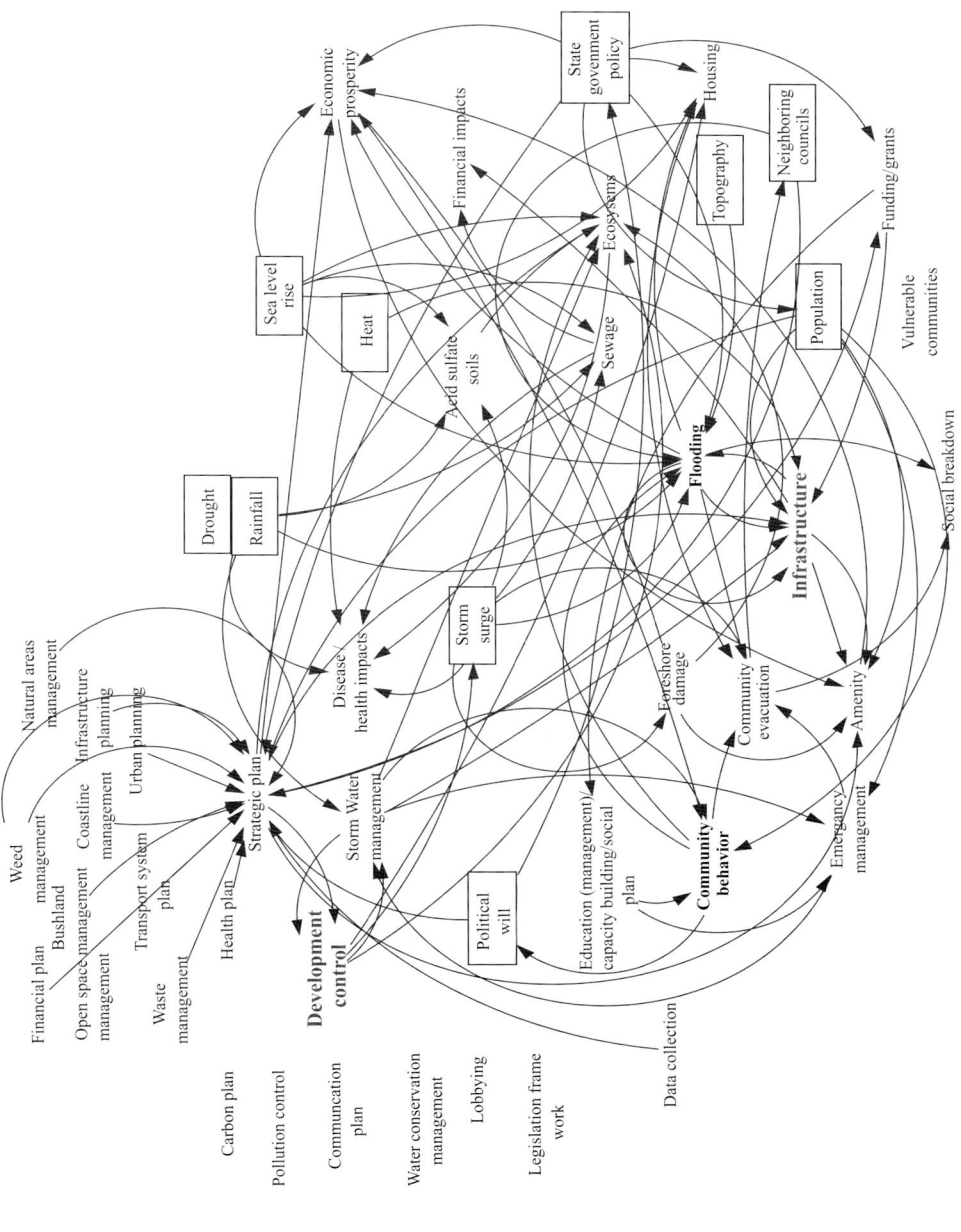

Figure 25.4 Example of a system conceptualization output showing climate change drivers, impacts and management responses.

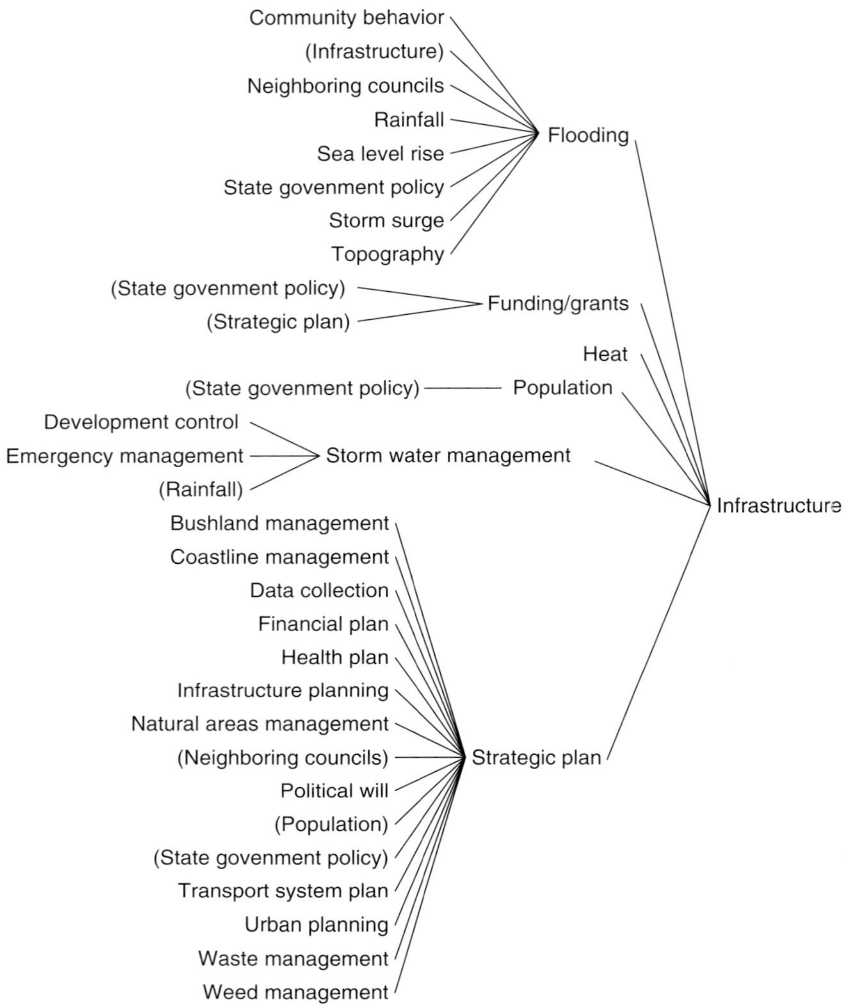

338 Integrated Coastal Zone Management

```
                    Community behavior
                       (Infrastructure)
                    Neighboring councils
                               Rainfall
                                           Flooding
                          Sea level rise
                   State govenment policy
                             Storm surge
                              Topography
         (State govenment policy)
                                          Funding/grants
                      (Strategic plan)
                                               Heat
             (State govenment policy) ——— Population
      Development control
    Emergency management ——→ Storm water management
                (Rainfall)
                                                              Infrastructure
                  Bushland management
                  Coastline management
                        Data collection
                         Financial plan
                           Health plan
                Infrastructure planning
               Natural areas management
                (Neighboring councils) ——— Strategic plan
                         Political will
                           (Population)
              (State govenment policy)
                Transport system plan
                        Urban planning
                      Waste management
                      Weed management
```

Figure 25.5 Illustration of direct and indirect drivers of infrastructure impacts.

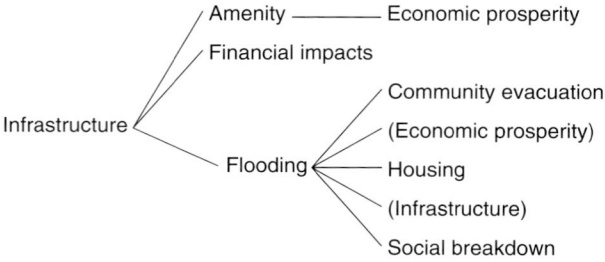

Figure 25.6 Illustration of direct and indirect consequences arising from impacts on infrastructure.

Table 25.1 Illustration of barriers and opportunities to infrastructure impacts.

Barriers	Opportunities
• Aging infrastructure • High cost of maintenance and low capacity to fund new infrastructure • No standards recapacity required • Restricted space and scope to increase capacity/Brownfield • Political cutting back S94 funds • Uncertainty of science – planning for future needs • Topography • Some infrastructure would have adverse impacts on the city • Forced reliance on other infrastructure provides for institutional barriers, competing interests, legislative inconsistency	• Good cross-unit linkages internally • New development brings funding opportunities and opportunities to ensure climate change needs are met (S94) • Improved technology, design higher standards • Good frameworks for funding and gaining new infrastructure (legislation) • Share information with other organization and councils • Innovative thinking • New technology to improve environmental outcomes, e.g. storm water reuse • Proximity to public transport thoroughfare to city • Special levy (to ensure climate change needs are met) and storm water levy

the time and money invested. The method essentially promotes adaptive management of complex and uncertain systems through a participatory and dynamic learning approach.

Acknowledgments

The authors thank the local government workshop participants. The authors also thank the anonymous reviewer whose comments improved the manuscript. The authors acknowledge the National Climate Change Adaptation Program of the Department of Climate Change (Australian Government) which provided funding for the research.

References

Allen Consulting (2005) Climate change risk and vulnerability. Report to the Australian Greenhouse Office, Department of Environment and Water Resources, Canberra.

Bellamy, J., Smith, T., Taylor, B., McDonald, G., Walker, M., Jones, J. and Pero, L. (2005) *Criteria and Methods for Monitoring and Evaluating Healthy Regional Planning Arrangements*. Tropical Savannas Management CRC, Darwin.

Cicin-Sain, B. and Knecht, R.W. (1998) *Integrated Coastal and Ocean Management: Concepts and Practices*. Island Press, Washington, DC.

Davenport, T.H., De Long, D.W. and Beers, M.C. (1998) Successful knowledge management projects. *Sloan Management Review* **39** (2), 43–57.

Funtowicz, S.O. and Ravetz, J.R. (1991) A new scientific methodology for global environmental issues. In: Costanza, R. (ed.). *Ecological Economics: The Science and Management of Sustainability*. Columbia University Press, New York, pp. 137–152.

IPCC (Intergovernmental Panel on Climate Change) (2001) Climate change 2001: impacts and adaptation. Contribution of Working Group II to the Third Assessment Report of the Intergovernmental Panel on Climate Change. Cambridge University Press, Cambridge, MA.

Jones, R. and Mearns, L. (2005) Assessing future climate risks. In: Lim, B., Spanger-Siegfried, E., Burton, I., Malone, E.L. and Huq, S. (eds). *Adaptation Policy Frameworks for Climate Change: Developing Strategies, Policies and Measures, United Nations Development Programme.* Cambridge University Press, Cambridge, MA.

Kay, R. and Alder, J. (1999) *Coastal Planning and Management.* Routledge, New York.

Lazarow, N., Souter, R., Fearon, R. and Dovers, S. (eds) (2006) *Coastal Management in Australia: Key Institutional and Governance Issues for Coastal Natural Resource Management and Planning.* CRC for Coastal Zone, Estuary and Waterway Management, Brisbane.

Lee, K.N. (1993) *Compass and Gyroscope: Integrating Science and Politics for the Environment.* Island Press, Washington, DC.

Milbrath, L.W. (1989) *Sustainable Society: Learning Our Way Out.* SUNY Press, New York.

Preston, B.L., Smith, T., Brooke, C., Gorddard, R., Measham, T., Withycombe, G., McInnes, K., Abbs, D., Beveridge, B. and Morrison, C. (2007) *Mapping Climate Change Vulnerability in the Sydney Coastal Councils Group.* Prepared for the Sydney Coastal Councils Group and the Australian Greenhouse Office, Melbourne.

Ravetz, J.R. (1999) What is post-normal science? *Futures* **31**, 647–653.

Smith, T.F. and Smith, D.C. (2006) Institutionalising adaptive learning for coastal management. In: Lazarow, N., Souter, R., Fearon, R. and Dovers, S. (eds). *Coastal Management in Australia: Key Institutional and Governance Issues for Coastal Natural Resource Management and Planning.* CRC for Coastal Zone, Estuary and Waterway Management, Brisbane, pp. 115–120.

Yohe, G.W. (2001) Mitigative capacity – the mirror image of adaptive capacity on the emission side. *Climatic Change* **49**, 247–262.

Index

Adaptive capacity, 9, 158, 162–8, 307, 313, 331, 333–6
Advancing Coastal Ecosystem Governance, 261
Adversarial modes of decision making, 332
Agenda, 21, 3, 20, 170, 173, 175, 254
Akaike information criterion (AIC), 57, 58
Anthropogenic impacts, 145, 146, 319
Appropriate Assessment procedure, 266, 273, 274

Before-After-Control-Impact-Pairs (BACIP), 180, 184
Benthic foraminifera, 97–9, 102, 106, 107, 110
Biodiversity Convention, 165, 166
Bioshields, 131–4, 140, 143

Canada's Oceans Strategy, 14, 15, 297
Classical decision-making, 267, 268, 270, 280
Climate change, 1–10, 17–21, 158, 160, 162, 170, 173, 174, 214, 215, 231, 306–15, 318, 331–7, 339
Coastal degradation, 116
Coastal Ecosystem Governance, 253, 261, 263
Coastal erosion, 10, 27, 152, 256, 312, 318, 324
Coastal shrimp aquaculture, 116
Coastal vulnerability, 231, 319, 331, 334
Coastal Zone Canada Association (CZCA), 19
Code of Conduct for Responsible Fisheries, 165
Commercial fishing, 12, 163, 181, 184, 290, 295
Coral reefs, 7–10, 27, 173, 202, 261
Cross-shore dynamics, 39, 45

Decision-Making Process, 69, 71, 133, 173, 266–8, 271, 277, 279, 280
Dredged material, 237–42, 244–9

Ecological restoration, 131
Economic sustainability, 124, 125, 190
Environmental Policy, 157, 158, 161, 166, 168, 281
Environmental Protection Agency (EPA), 100, 219, 220
Environmental restoration, 237, 238, 241, 248
Environmental stress, 97, 98, 112
Essential fish habitats (EFH), 28

EU indicator, 69
Eutrophication, 27, 31, 79, 80, 87, 90, 93, 94, 98, 121, 214, 220, 256, 262, 276
Evolution of Contemporary Coastal Management, 253, 254

Food and Agriculture Organisation (FAO), 11, 125, 165, 166, 221, 224, 226

Generalised Additive Models (GAM), 57, 59
Geomorphology, 47, 241, 318, 319, 320, 321, 323, 325
GIS modelling, 58, 60
Global Forum on Oceans, Coasts and Islands, 16, 256
Global Programme of Action for the Protection of the Marine Environment, 11
Global Vulnerability Evaluation, 323
Global Vulnerability Map, 326
Global warming, 1, 2, 4, 5, 6, 7, 11, 14, 18, 27, 158, 159, 160, 161, 162, 310, 313
Governance, 5, 21, 72, 78, 122, 128, 157, 159, 160, 162–7, 171, 176, 212, 216, 253, 255, 257–9, 261–3, 286, 293, 295–8, 300–302, 304, 306, 314, 332
Governance Mechanisms, 295
Governing system, 163, 164, 165, 166, 167

Habitat construction, 248
Heavy metals, 97, 98
Hyperbenthos, 79, 80, 83, 84, 86–94

Infauna, 79, 80, 83, 84, 86–94
Institutional sustainability, 123, 124
Intergovernmental Panel on Climate Change (IPCC), 4, 5, 7, 9, 21, 158, 161, 162, 306
International Convention on Biological Diversity, 11, 21, 172, 173, 235
International Council for the Exploration of the Seas (ICES), 3, 19, 29, 31
International Legal Framework, 170, 172

Large Marine Ecosystems (LMEs), 21, 258, 259
Large Ocean Management Areas (LOMAs), 13, 16
Legislative and Policy Framework, 296

Index

Linear decision making 332
LITDRIFT, 39
LITPACK, 39
Lobster reserves, 178, 179, 180, 181, 182, 183, 184, 185, 186
London Dumping Convention, 237
Longshore dynamics, 39, 43

Macro-algae habitat, 50
Management tools, 159, 164, 309
Mangrove Action Plan (MAP), 223, 231
Mangrove degradation, 222, 223
Marine Protected Areas (MPAs), 11, 13, 16, 21, 30, 178, 179, 189, 191, 192, 193, 263, 272, 273, 279
Marine Spatial Planning, 76
Migratory Birds Convention Act, 17
Modelling of marine habitats, 57, 58

Non-government organizations (NGOs), 10, 31, 131, 132, 133, 135, 136, 137, 140, 141, 142, 143, 181, 184, 219, 233, 234, 280, 299, 303, 304
Northwest Atlantic Fisheries Organization's (NAFO), 16

Oceans Act, 14, 15, 16, 17, 18, 19, 20, 295, 296, 297, 302
Offshore wind farms, 207, 208, 209, 211, 212, 213

Polychlorinated Biphenyls (PCBs), 97, 100, 104, 111
Polycyclic Aromatic Hydrocarbons (PAHs), 97, 98, 100, 104, 105, 108, 109, 110, 111, 112, 248
Precautionary principle, 57, 63, 165, 166, 174, 310
Predictive modelling, 60, 62

Sediment dynamics, 35, 37, 39, 41, 43, 47
Small-Scale Trawlers, 189, 195
Social–ecological–economic interactions, 120, 126
Spatial planning, 27, 30, 31, 72, 76, 211, 213, 214, 216, 217, 286
Stakeholder participation, 161, 179, 285, 287, 289, 291
Stakeholders' perceptions, 266, 267, 269, 270, 278, 279, 280, 281
SWAN, 37, 39

Total Organic Carbon (TOC), 97, 100, 105, 111

UN Fish Stock Agreement, 165
United Nations Convention on the Law of the Sea (UNCLOS), 14, 15, 16, 226
United Nations Educational, Scientific and Cultural Organisation (UNESCO), 10
United Nations Environment Program (UNEP), 10, 173
United Nations Law of the Sea Convention (UNCLOS III), 16
Urbanization, 145, 146, 147, 148, 150, 151, 153
US Coastal Zone Management (CZM) Act, 254

Vulnerability Parameters Description, 319

Water Framework Directive, 76, 94
World Commission on Environment and Development, 3
World Summit on Sustainable Development (WSSD), 11, 256

X-ray Sedigraph, 100